FRAMINGHAM STATE COLLEGE
3 3014 00396 9484

D0886950

Ecological Studies, Vol. 163

Analysis and Synthesis

Ecological Studies

Volumes published since 1997 are listed at the end of this book.

Springer
Berlin
Heidelberg
New York
Hong Kong
London
Milan
Paris
Tokyo

R. Valentini (Ed.)

Fluxes of Carbon, Water and Energy of European Forests

With 92 Figures, 3 in Color, and 46 Tables

Springer

Prof. Dr. Riccardo Valentini
University of Tuscia
Department of Forest Environment and Resources
Via S. Camillo de Lellis
01100 Viterbo
Italy

Cover illustration by courtesy of Dennis and Nicole Baldocchi

ISSN 0070-8356
ISBN 3-540-43791-6 Springer-Verlag Berlin Heidelberg New York

Library of Congress Cataloging-in-Publication Data

Fluxes of carbon, water, and energy of European forests / R. Valentini (ed.)
p. cm. -- (Ecological studies, ; v. 163)
Includes bibliographical references (p.)..
ISBN 3540437916 (alk. paper)
1. Carbon cycle (Biogeochemistry)--Europe. 2. Forest ecology--Europe. I. Valentini, R. (Riccardo), 1959.

QH344 .F58 2002
577.3'44'094--dc21 2002030469

Springer-Verlag Berlin Heidelberg New York
a part of Springer Science+Business Media GmbH

http://www.springer.de

Printed in Germany

Production: Friedmut Kröner, 69115 Heidelberg, Germany
Cover design: *design & production* GmbH, Heidelberg
Typesetting: Kröner, 69115 Heidelberg, Germany

31/3111 – 5 4 3 2 1 – Printed on acid free paper SPIN 11370604

Preface

After years of technological development and its important achievements to make our life easier and more comfortable, human society is going to face one of the most difficult challenges of the last century: to stabilize the concentration levels of greenhouse gases in the atmosphere to prevent harmful effects on the climate system.

Through a delicate balance between photosynthesis and respiration, terrestrial ecosystems, and in particular forests, are today thought to take up a significant part of the carbon dioxide emissions in the atmosphere, sometimes called the "terrestrial carbon sink". However, the location, magnitude, and vulnerability of the carbon dioxide sink of the terrestrial biota are still uncertain.

The suite of traditional tools in an ecologist's toolbox for studying ecosystem productivity and carbon balance include leaf cuvettes, whole-plant and soil chambers for gas exchange, and biomass and soil carbon inventories. While each of the cited methods has distinct advantages, they are limited with regards to their ability to measure net carbon dioxide exchange of the whole ecosystem across a variety of time scales.

This book present a compendium of results of a European project (EUROFLUX), funded by the European Commission through its fourth framework program, aiming to elucidate the role of forests in continental carbon balance. For the first time, a novel technology, eddy covariance, has been used in a network of investigation sites ranging from Scandinavia to the Mediterranean region to provide direct measurements of carbon, water, and energy fluxes between forests and the atmosphere.

Use of the eddy covariance method by specialists and non-specialists has increased in recent years. A citation search of published papers that index the term 'eddy covariance' produced over 200 records and over 500 papers referring to the analogous term 'eddy correlation'. We hope that this book will provide a useful compendium of both methodological advancements as well as critical analyses of results on atmosphere–biosphere interactions in European forests, which could stimulate new scientific questions and provide substantial information to policy makers engaged today in negotiations of the Kyoto protocol and related international conventions for climate and biodiversity protection.

October 2002

Viterbo, Italy *Riccardo Valentini*
Bruxelles, Belgium *Anver Ghazi, Claus Brüning, Panagiotis Balabanis*

Contents

1 **EUROFLUX: An Integrated Network for Studying the Long-Term Responses of Biospheric Exchanges of Carbon, Water and Energy of European Forests** 1
R. VALENTINI

1.1 The Background 1
1.2 The Network of Sites 2
1.3 The Methodology 5
1.4 EUROFLUX-Specific Objectives 6
Further Reading 8

2 **Methodology for Data Acquisition, Storage, and Treatment** 9
M. AUBINET, R. CLEMENT, J.A. ELBERS, T. FOKEN, A. GRELLE, A. IBROM, J. MONCRIEFF, K. PILEGAARD, Ü. RANNIK, C. REBMANN

2.1 Introduction 9
2.2 Theory 10
2.3 Material 11
2.4 Data Acquisition, Computation, and Corrections 17
2.4.1 Procedure Outline 17
2.4.2 Computation of the Uncorrected Fluxes 18
2.4.3 Correction for Frequency Response Losses 20
2.4.4 Effect of Low Frequency Cut on Fluxes 21
2.4.5 Effect of High Frequency Cut on Fluxes 23
2.5 Quality Control 26
2.5.1 Raw Data Analysis 26
2.5.2 Stationarity Test 26
2.5.3 Integral Turbulence Characteristics Test 27
2.5.4 Energy Balance Closure 27
2.6 Spatial Representativeness of the Fluxes 29
2.7 Night Flux Corrections 30

Annex: Technical Details About the Software Validation Set 32
References 33

3 Measurement of Soil Respiration 37
H. Lankreijer, I.A. Janssens, N. Buchmann, B. Longdoz, D. Epron, S. Dore

3.1 Introduction 37
3.1.1 Measuring Soil Respiration 38
3.1.2 Modeling 40
3.2 Soil CO_2 Efflux Measurements in the EUROFLUX Project . . 41
3.2.1 Chamber Systems 41
3.2.2 Meteorological and Profile Gradient Systems 44
3.2.2.1 Eddy Covariance Below the Canopy (EC) 44
3.2.2.2 Flux-Profile Measurements (P) 44
3.2.2.3 CO_2 Profile in the Soil (SP) 45
3.2.2.4 Additional Soil Measurements 45
3.3 Comparison of Systems 45
3.4 Discussion 47
3.5 Conclusions 49
References 49

4 Deciduous Forests (Beech): Carbon and Water Fluxes, Balances, Ecological and Ecophysiological Determinants . 55
A. Granier, M. Aubinet, D. Epron, E. Falge, J. Umundsson, N.O. Jensen, B. Köstner, G. Matteucci, K. Pilegaard, M. Schmidt, J. Tenhunen

4.1 Introduction 55
4.2 Water Transfer and Water Balance 56
4.2.1 Sap Flow in Beech, Tree Transpiration 56
4.2.2 Sap Flow Profile in Trunks 57
4.2.3 Range of Sap Flow and Among Tree Variation 58
4.2.4 Canopy Conductance for Water Vapor: Effect of Radiation, Vapor Pressure Deficit, Temperature, LAI, and Drought . . . 58
4.2.5 Annual Water Balance 60
4.3 Carbon Fluxes and Carbon Balance 61
4.3.1 Net Ecosystem Exchange 61
4.3.2 Ecosystem Respiration 62
4.3.3 Soil Respiration 64
4.3.4 Daily Net Ecosystem Exchange 64

4.3.5 Annual Carbon Balance, Intersite and Interannual Comparison . . . 68
4.4 Conclusions . . . 68
References . . . 69

5 Coniferous Forests (Scots and Maritime Pine): Carbon and Water Fluxes, Balances, Ecological and Ecophysiological Determinants . . . 71
R. Ceulemans, A.S. Kowalski, P. Berbigier, H. Dolman, A. Grelle, I.A. Janssens, A. Lindroth, E. Moors, U. Rannik, T. Vesala

5.1 Introduction . . . 71
5.1.1 Aims and Objectives . . . 71
5.1.2 Scots Pine and Maritime Pine as Representatives of the Genus *Pinus* in the EUROFLUX Network . . . 71
5.2 Description of Experimental Sites and Data Collection . . . 75
5.3 Results and Observations . . . 77
5.3.1 Climate . . . 77
5.3.2 Radiation . . . 77
5.3.3 Energy Budget . . . 81
5.3.4 Carbon Dioxide Exchanges . . . 85
5.4 Conclusion . . . 95
References . . . 96

6 Spruce Forests (Norway and Sitka Spruce, Including Douglas Fir): Carbon and Water Fluxes, Balances, Ecological and Ecophysiological Determinants . 99
C. Bernhofer, M. Aubinet, R. Clement, A. Grelle, T. Grünwald, A. Ibrom, P. Jarvis, C. Rebmann, E.-D. Schulze, J.D. Tenhunen

6.1 Introduction . . . 99
6.2 Material and Methods . . . 100
6.3 Results . . . 102
6.3.1 Radiation . . . 102
6.3.2 Momentum . . . 103
6.3.2.1 Heat Storage . . . 107
6.3.2.2 Soil Heat Flux . . . 108
6.3.2.3 Canopy Heat Flux . . . 108
6.3.3 Energy Fluxes . . . 110

6.3.4 Carbon Fluxes 114
6.4 Summary and Conclusion 120
References 122

7 **Evergreen Mediterranean Forests: Carbon and Water Fluxes, Balances, Ecological and Ecophysiological Determinants** . 125
G. Tirone, S. Dore, G. Matteucci, S. Greco, R. Valentini

7.1 Introduction 125
7.2 Materials and Methods 127
7.2.1 Site Description 127
7.2.2 Dendrometric Survey 128
7.2.3 Eddy Covariance Measurements 129
7.2.4 Meteorological and Micrometeorological Measurements . . 129
7.2.5 Ancillary Measurements 130
7.3 Biomass and Productivity Dynamics of the *Qercus ilex* Forest Stand 130
7.4 Soil and Canopy Carbon Fluxes 134
7.4.1 Daily Carbon Fluxes 134
7.4.2 Seasonal Carbon Fluxes 135
7.5 Energy Partition at Canopy Level 138
7.6 Canopy Ecophysiology 140
7.6.1 Light Response of Canopy Photosynthesis 142
7.6.2 Temperature and Soil Water Response of Ecosystem Respiration 142
7.6.3 Water Exchanges 144
7.7 Conclusions 147
References 147

8 **A Model-Based Study of Carbon Fluxes at Ten European Forest Sites** 151
E. Falge, J. Tenhunen, M. Aubinet, C. Bernhofer, R. Clement, A. Granier, A. Kowalski, E. Moors, K. Pilegaard, Ü. Rannik, C. Reb

8.1 Introduction 151
8.2 Methods 152
8.2.1 Observations of NEE at the EUROFLUX Sites 152
8.2.2 Flux Model Description and Parameterization 154
8.2.3 Foliage Gas Exchange 155
8.2.4 Ecosystem Respiration 159

8.3 Results 161
8.3.1 Model Comparison 161
8.3.2 Performance of Nighttime Flux Extrapolation Method at Monthly and Daily Time Scales 165
8.4 Discussion and Future Directions 172
8.5 Conclusions 174
References 175

9 A Model-Based Approach for the Estimation of Carbon Sinks in European Forests 179
D. Mollicone, G. Matteucci, R. Koble, A. Masci, M. Chiesi, P.C. Smits

9.1 Introduction 179
9.2 Biome-BGC Model 181
9.3 Validation of Biome-BGC 181
9.3.1 Data for Model Validation 182
9.3.2 Validation and Calibration 183
9.3.3 Discussion 185
9.4 Carbon Budget Information System: Methods and Data . . . 186
9.4.1 Biome-BGC Model and CBIS 187
9.4.1.1 Input Data 189
9.4.1.2 Soil Database 189
9.4.1.3 Meteorological Database 190
9.4.2 "Clustered" Forest Tree Species Maps 190
9.4.3 Ecophysiological Constants of Forest Clusters 193
9.5 Results and Discussion 196
9.6 Conclusions 201
References 202

10 Factors Controlling Forest Atmosphere Exchange of Water, Energy and Carbon 207
A.J. Dolman, E.J. Moors, T. Grunwald, P. Berbigier, C. Bernhofer

10.1 Introduction 207
10.2 Radiation Balance 208
10.3 Turbulent Exchange 211
10.4 Water Use of Forests 213
10.4.1 Wet Canopy Evaporation 213
10.4.2 Dry Canopy Evaporation 215

10.5 Surface Conductance Photosynthesis Relations 219
10.6 Conclusions 221
References 223

11 The Carbon Sink Strength of Forests in Europe: Results of the EUROFLUX Network 225
R. Valentini, G. Matteucci, A.J. Dolman, E.-D. Schulze, P.G. Jarvis

11.1 Introduction 225
11.2 Spatial Distribution of Carbon Fluxes Across Europe 226
11.3 Interpreting the Spatial Distribution of Carbon Sinks/Sources 230
References 232

12 Climatic Influences on Seasonal and Spatial Differences in Soil CO_2 Efflux 235
I.A. Janssens, S. Dore, D. Epron, H. Lankreijer, N. Buchmann, B. Longdoz, J. Brossaud, L. Montagnani

12.1 Modeling the Temporal Variability of soil CO_2 Efflux 235
12.1.1 Temperature Responses 236
12.1.2 Sensitivity of Empirical Models to the Type of Temperature Regression 237
12.1.3 Moisture Responses 238
12.1.4 Sensitivity of Empirical Models to the Type of Moisture Regression 240
12.1.5 Additional Comments on Empirical Models 242
12.2 Spatial Variability Among the EUROFLUX Forests 243
12.2.1 Effect of Precipitation 245
12.2.2 Effect of Soil Temperature 245
12.2.3 Effect of Site Productivity 249
12.3 Importance of Roots in Soil CO_2 Efflux 250
12.4 Conclusions 252
References 252

13 Conclusions: The Role of Canopy Flux Measurements in Global C-Cycle Research 257
R. Valentini, G. Matteucci, A.J. Dolman, E.-D. Schulze

13.1 Introduction 257
13.2 The Role of Canopy Flux Measurements in Global C-Cycle Research 261
References 266

Subject Index 269

Contributors

M. Aubinet

Unit of Physics, Faculty of Agricultural Sciences, 8 Av. de la Faculté,
5030 Gembloux, Belgium

P. Balabanis

European Commission – DG Research, 200 Rue de la Loi,
1049 Bruxelles, Belgium

P. Berbigier

INRA-Bordeaux, Centre de Recherches Forestières,
Unité de Bioclimatologie, B.P. 81, 33883 Villenave d'Ornon, France

C. Bernhofer

Dresden University of Technology, Institute of Hydrology and Meteorology,
Pienner Str. 9, 01737 Tharandt, Germany

J. Brossaud

ILE AS CR, Prici 3b, 60300 Brno, Czech Republic

C. Brüning

European Commission – DG Research, 200 Rue de la Loi,
1049 Bruxelles, Belgium

N. Buchmann

Max Planck Institute for Biogeochemistry, Carl-Zeiss-Promenade 10,
07745 Jena, Germany.

R. Ceulemans

University of Antwerpen, UIA, Department of Biology, Research Group of Plant and Vegetation Ecology, Universiteitsplein 1, 2610 Wilrijk, Belgium

M. Chiesi

CNR– IATA, Piazzale delle Cascine 18, 50145 Firenze (FI), Italy

R. Clement

Edinburgh University, Darwin Building, Mayfield Road,
Edinburgh, EH9 3JU, UK

A.J. Dolman

Free University of Amsterdam, Faculty of Earth and Life Sciences,
Department of Geo-Environmental Sciences, De Boelelaan 1085,
1081 HV Amsterdam, The Netherlands

S. Dore

University of Tuscia, Department of Forest Environment and Resources,
Via S.Camillo de Lellis, 01100 Viterbo (VT), Italy

J.A. Elbers

Alterra, Winand Staring Centre, Department of Agricultural Research,
SC-DLO, 6700 AC Wageningen, The Netherlands

D. Epron

Université de Franche-Comté, Institut des Sciences et des Techniques
de l'Environnement, Laboratoire de Biologie et Ecophysiologie,
B.P. 71427, 25211 Montbéliard Cédex, France

E. Falge

University of Bayreuth, Department of Micrometeorology and Department
of Plant Ecology, Universitätsstr. 30, 95440 Bayreuth, Germany

T. Foken

University of Bayreuth, Department of Micrometeorology,
Universitätsstr. 30, 95440 Bayreuth, Germany

A. Ghazi

European Commission – DG Research, 200 Rue de la Loi,
1049 Bruxelles, Belgium

A. Granier

INRA, Unité d'Ecophysiologie Forestière, 54280 Champenoux, France

S. Greco

University of Tuscia, Department of Forest Environment and Resources,
Via S. Camillo de Lellis, 01100 Viterbo (VT), Italy

A. Grelle

Swedish University of Agricultural Sciences, Department for Production
Ecology, Box 7042, 5007 Uppsala, Sweden

T. Grünwald

Institute of Hydrology and Meteorology, Dresden University of Technology,
Tharandt, Germany

J. Gudmundson

Environmental Department, Agricultural Research Institute,
Keldnaholti, 112 Reykjavík, Iceland

A. Ibrom

University of Göttingen, Institute of Bioclimatology,
37077 Göttingen, Germany

P. Jarvis

University of Edinburgh, Institute of Ecology and Environmental Research,
Darwin Building, Mayfield Road, EH9 3JU, UK

I. A. Janssens

University of Antwerpen, UIA, Department of Biology, Research Group of
Plant and Vegetation Ecology, Universiteitsplein 1, 2610 Wilrijk, Belgium

N.O. Jensen

Plant Biology and Biogeochemistry Department, Building 309,
Risoe National Laboratory, P.O. Box 49, 4000 Roskilde, Denmark

R. Köble

Institute for Environment and Sustainability, European Commission - Joint Research Centre, 21020 Ispra (VA), Italy

B. Köstner

Institute of Hydrology and Meteorology, Technische Universität Dresden, Plenner Str. 9, 01737 Tharandt, Germany

A. S. Kowalski

University of Antwerpen, UIA, Department of Biology, Research Group of Plant and Vegetation Ecology, Universiteitsplein 1, 2610 Wilrijk, Belgium

H. Lankreijer

Lund University, Department of Physical Geography and Ecosystem Analysis, Box 118 HS3, 221 00 Lund, Sweden

A. Lindroth

University of Lund, Department of Physical Geography, Lund University, Box 118,22100 Lund, Sweden

B. Longdoz

Unit of Physics, Faculty of Agricultural Sciences, 8 Av. de la Faculté, 5030 Gembloux, Belgium

A. Masci

University of Tuscia, Department of Forest Environment and Resources, Via S. Camillo de Lellis, 01100 Viterbo (VT), Italy

G. Matteucci

Institute for Environment and Sustainability, European Commission - Joint Research Centre, 21020 Ispra (VA) Italy

D. Mollicone

Institute for Environment and Sustainability, European Commission - Joint Research Centre, 21020 Ispra (VA), Italy

J. Moncrieff

Edinburgh University, Darwin Building, Mayfield Road,
Edinburgh, EH9 3JU, UK

L. Montagnani

Forest Services, Autonomous Province of Bolzano, 39100 Bolzano, Italy

E.J. Moors

Alternna, Winand Staring Centre, Department of Agricultural Research,
SC-DLO, 6700 AC Wageningen, The Netherlands

K. Pilegaard

Plant Biology and Biogeochemistry Department, Risø National Laboratory,
P.O. Box 49, 4000 Roskilde, Denmark

Ü. Rannik

University of Helsiniki, Department of Physics, P.O.Box 64,
00014 Helsinki, Finland

C. Rebmann

Max-Planck-Institut für Biogeochemie, Carl-Zeiss-Promenade 10,
07745 Jena, Germany

M. Schmidt

Universität Bayreuth, Lehrstuhl für Pflanzenökologie,
95440 Bayreuth, Germany

E.-D. Schulze

Max-Planck-Institute for Biogeochemistry, Carl-Zeiss-Promenade 10,
07745 Jena, Germany

P.C. Smits

Institute for Environment and Sustainability, European Commission –
Joint Research Centre, 21020 Ispra (VA), Italy

J.D. Tenhunen

University of Bayreuth, Department of Plant Ecology,
95440 Bayreuth, Germany

G. Tirone

University of Tuscia, Department of Forest Environment and Resources,
01100 Viterbo (VT), Italy

R. Valentini

University of Tuscia, Department of Forest Environment and Resources,
Via S. Camillo de Lellis, 01100 Viterbo (VT), Italy

T. Vesala

University of Helsinki, Department of Physics, P.O.Box 64,
00014 Helsinki, Finland

1 EUROFLUX: An Integrated Network for Studying the Long-Term Responses of Biospheric Exchanges of Carbon, Water, and Energy of European Forests

R. VALENTINI

1.1 The Background

Much of what we know about the contemporary global carbon budget has been learned from careful observations of the atmospheric CO_2 mixing ratio and $^{13}C/^{12}C$ isotope ratio ($\delta^{13}C$), interpreted with global circulation models. From these studies we have learned important facts such as that about one third of the annual input of CO_2 to the atmosphere from fossil fuel combustion and deforestation is taken up by the terrestrial biosphere (Keeling et al. 1996), that a significant portion of the net uptake of CO_2 occurs at mid-latitudes of the Northern Hemisphere, and, in particular, north temperate terrestrial ecosystems are implicated as a large sink (Tans et al. 1990). The methods used have provided the necessary global and continental scale perspective for carbon balance calculations, but their value in addressing small temporal and spatial changes in the carbon balance is rather limited.

The net carbon exchange of terrestrial ecosystems is the result of a delicate balance between uptake (photosynthesis) and losses (respiration), and shows a strong diurnal, seasonal, and interannual variability. Under favorable conditions, during daytime the net ecosystem flux is dominated by photosynthesis, while at night, and for deciduous ecosystems in leafless periods, the system loses carbon by respiration. The influence of climate and phenology can in some cases shift a terrestrial ecosystem from a sink to a source of carbon.

Furthermore, the global and continental scale techniques are of limited use in addressing one of the key questions raised by the Kyoto protocol, namely, how to calculate the changes in "carbon stocks" associated with land-use changes and forestry activities during the commitment period. Indeed, one of the major effects of land-use changes, including the afforestation, reforestation, and deforestation of land, is changes in soil organic matter, both as

Ecological Studies, Vol. 163
R. Valentini (Ed.) Fluxes of Carbon,
Water and Energy of European Forests

buildup and decomposition. The changes in stocks of soil carbon in a 4–5 year period are unfortunately within the errors of the survey techniques used for most ecosystems. Remote sensing techniques also appear inadequate for such purposes, since they have limited capability of estimating below-canopy processes such as soil respiration.

In this context, the direct, long-term measurement of carbon fluxes by the eddy covariance technique, particularly if applied in conjunction with other ecosystem level studies, offers a distinct possibility of assessing the carbon sequestration rates of forests and of land-use changes activities at the local scale (Valentini et al. 2000). The technique can also provide a better understanding of the vulnerability of the carbon balance of ecosystems to climate variability and at the same time can be used to validate ecosystem models and provide parameterization data for land surface exchange schemes in global models. The correct application of the technique, however, depends on particular requirements which will be discussed later in Chapter 2.

Under the IVth Framework program Environment and Climate, the European Commission funded a consortium of European institutions (EUROFLUX, ENV4-CT95-0078) to investigate the long-term biospheric exchanges of carbon, water, and energy of European forests. Since 1996 automated eddy covariance measurements of CO_2 fluxes have now been made routinely over 15 forests in Europe.

1.2 The Network of Sites

The sites of investigation represent a range of forest ecosystems in Europe, which differ for species composition, type of management, climate, age, and geographical location. The map of the sites is presented in Fig. 1.1, while their main features are listed in Table 1.1.

The sites are distributed along a north–south transect, going from 38° to 60°N latitude and from about 8°W to 27°E longitude. The selected sites fall into four main climate classes: Mediterranean, Temperate, Temperate-Oceanic, Temperate-Continental and Boreal. The major forest biomes are constituted by deciduous (beech), coniferous (pine, spruce), and broad-leaved evergreen (oak) forests. In particular, *Fagus sylvatica* L. and *Pinus* spp. extend from the Mediterranean up to the Nordic region. *Spruce* spp. sites are also distributed along an ecoclimatic gradient, in this case lying mostly in an east–west direction, from the United Kingdom to North Sweden, crossing Germany. In the Mediterranean region the site of evergreen oak is located in Italy (*Quercus ilex* L.).

In Fig. 1.2, the representativeness of the EUROFLUX sites is displayed in a climate diagram showing how the sites cover a wide range of temperature and precipitation.

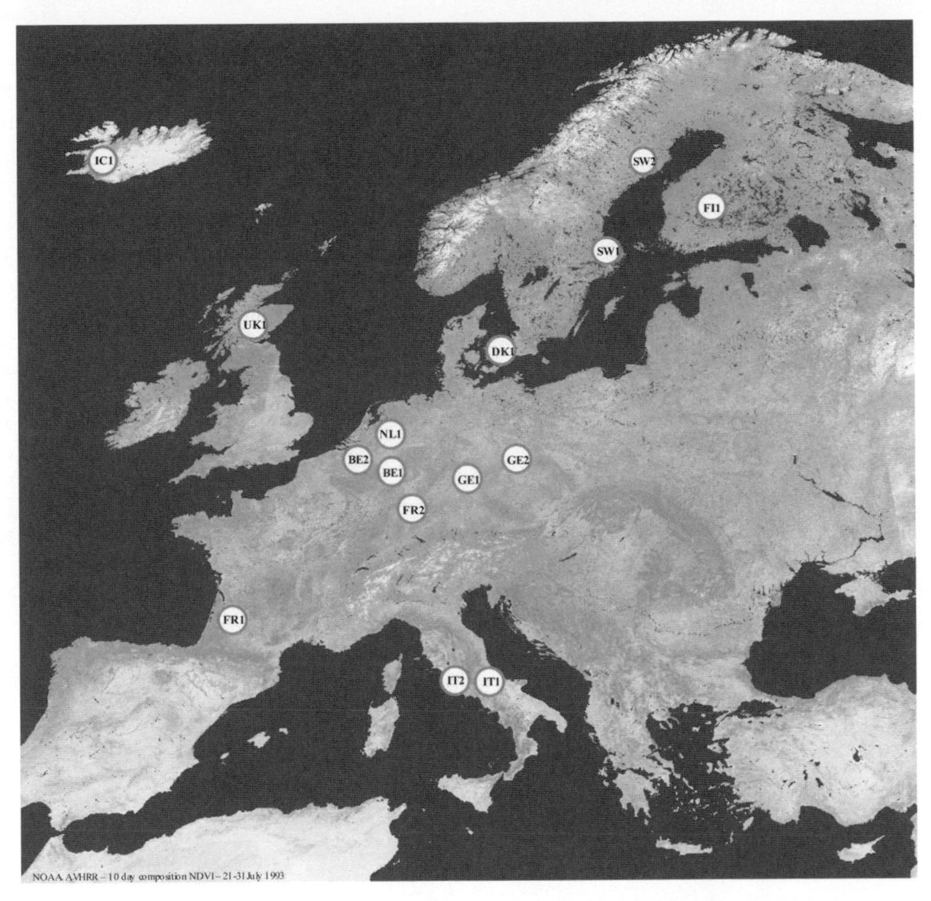

Fig. 1.1. Euroflux network sites distribution in Europe (see Table 1 for details)

After the example of EUROFLUX, the first network of this kind to be initiated, and the development of methodological standards, new networks appeared in various regions of the world. In 1998, the network approach was expanded in the US (AMERIFLUX) and plans are being developed to implement similar networks in Brazil (the Large-Scale Biosphere Atmosphere Experiment in Amazonia), Southeast Asia (the GEWEX Asian Monsoon Experiment), Japan, and Siberia. These major sites are now forming a global network, FLUXNET, with standard measurement protocols and data quality and storage systems.

Table 1.1. Main characteristics of Euroflux sites

Country	Site Name	Site ID	Latitude	Longitude	Elevation (m asl)
Italy	Collelongo	IT1	41° 50' 57.7" N	13° 35' 17.3" E	1550
Italy	Castelporziano	IT2	41° 42' 20.9" N	12° 22' 38.3" E	3
France	Le Bray, Bordeaux	FR1	44° 43' 01.59" N	00° 46' 09.49" W	60
France	Hesse	FR2	48° 40' 27.18" N	07° 03' 52.62" E	300
Denmark	Soroe	DK1	55° 29' 13" N	11° 38' 45" E	40
Sweden	Norunda	SW1	60° 05' 10" N	16° 13' 13" E	45
Sweden	Flakaliden	SW2	64° 06' 46" N	19° 27' 25" E	225
Germany	Hainich	GE1	51° 04' 45.14" N	10° 27' 07.83" E	445
Netherlands	Loobos	NL1	52° 10' 04.28" N	05° 44' 38.25" E	52
UK	Griffin	UK1	56° 36' 23.59" N	03° 47' 48.55" E	340
Germany	Tharandt, Anchor Station	GE2	50° 57' 49" N	13° 34' 01" E	380
Belgium	Vielsalm	BE1	50° 18' 32" N	05° 59' 55" E	450
Belgium	Brasschaat, 'De Inslag' Forest	BE2	51° 18' 33" N	04° 31' 14" E	16
Iceland	Gunnarsholt	IC1	63° 50' N	20° 13' W	78
Finland	Hyytiälä	FI1	61° 50' 50.7" N	24° 17' 41.14" E	170
Italy	Renon/Ritten (Bolzano)	IT4	46° 35' 16.16" N	11° 26' 04.9" E	1730
Germany	Leinefelde	GE3	51° 19' 41.74" N	10° 22' 04.2" E	440

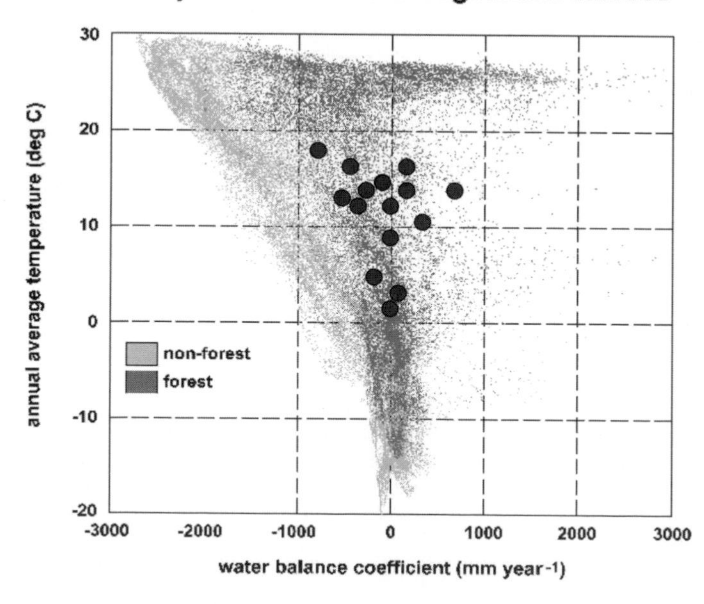

Fig. 1.2. Distribution of Euroflux sites in a climate-space diagram

Dominant species	Annual precipi-tation (mm)	Annual mean air tempera-ture (°C)	Stand age in 2000 (years)	Canopy height in 2000 (m)	Stem density in 2000 (./ha)	LAI in 2000 (m^2/m^2)
Fagus sylvatica	1180	6.3	105	25	900	5
Quercus ilex	550	15.3	30	10-15	1500	3.4
Pinus pinaster	950	13.5	30	20	500 (before storm 1999)	2.6
Fagus sylvatica	885	9.2	35	13	3800	5.5 to 7
Fagus sylvatica	510	8.1				4.75
Picea abies/Pinus sylvestris	527	5.5	100	25	600	5
Picea abies	587	1.9	31	8	2100	2
Fagus sylvatica	750	7	0-250	33	334	-
Pinus sylvestris	786	9.8	80	14	620	3
Picea sitchensis	1200	8	20	9	2150	8
Picea abies	820	7.5	108	26	480	7.6
Fagus sylvatica/ Pseudotsuga menziensii	1000	7.5	*Fagus* 90; *Pseudo-tsuga* 60	*Fagus* 27; *Pseudo-tsuga* 35	*Fagus* 145; *Pseudo-tsuga* 49	5.1
Pinus sylvestris/Quercus robur	750	10	71	21	556	
Populus trichocarpa	1120	3.6	7	1	10000	2.5
Pinus sylvestris	640	3.5	39	14	2500	2-3
Picea abies	1008	3.8	up to 181	up to 29	280 (DBH >12 cm)	4
Fagus sylvatica	750	7	122	34	224	

1.3 The Methodology

The flux stations measure the net flux of carbon entering or leaving the ecosystem. This is the flux which, if summed annually, provides the estimate of Net Ecosystem Exchange (NEE), and thus provides a direct measurement of the annual ecosystem carbon balance, excluding disturbances by harvest and fire (Net Biome Productivity).

The innovation in the technique is that:

- it is possible to measure carbon and energy fluxes directly, without destructive sampling, using a highly standardized technology, allowing comparison across ecosystems on the same basis;
- the measurements integrate over an area of approximately 1 km^2 which is the typical scale of vegetation stands, and where land management also usually occurs;

- it is possible by means of turbulence modeling to define the extension and shape of the source area of integration of fluxes;
- fluxes are measured continuously every 30 min on a daily basis extending to years, providing insight into the seasonal and annual variability;
- it is possible to validate the ecosystem model of carbon and energy exchanges on a range of ecosystems, climate, and European regions;
- in fulfilling fetch requirements, it is possible to compare stock change methods with flux measurements for carbon accounting as well as to separate carbon exchanges by different ecosystem components.

1.4 EUROFLUX Specific Objectives

The EUROFLUX program aimed to investigate the relevant biosphere–atmosphere gas exchanges of European forests in response to a wide recognition in the climate science community of the need for long-term studies of atmospheric exchanges of water and carbon of terrestrial ecosystems. Furthermore, EUROFLUX represented a unique framework of datasets for analyzing the response of the ecosystems to climatic and biotic perturbations under a variety of conditions, including extreme events, for biogeochemical model evaluation and improvement.

Additionally, the impact of climate on forests and the effects of forest management on the atmospheric exchanges was considered with particular attention. Forest management in Europe is highly dependent on local and regional needs, and rather scarce are the studies concerning coordinated policies and recommendations about such management that are able to interface with ongoing requests for a "sustainable forestry" – which desires today are expanded to include the efficiency of biogeochemical cycling and the mitigation of greenhouse gas emissions.

In synopsis the general objectives of the present volume can be summarized as follows:

1. *characterize long-term fluxes and energy exchange of representative European forests* in order to provide useful parameters for global and regional climate modelers and to analyze the variables that determine energy partitioning by forests in different climatic conditions, including extreme events and stress conditions (such as water limitation, nutrient deficiency, etc.);
2. *determine the sink strength of European forests for carbon* and analyze the variables that determine the gains and the losses of carbon from forests of differing vegetation composition and in different climate regions;
3. *analyze the response of European forest water and carbon fluxes to climatic factors* in order to aid regional scale modeling designed to predict impacts of global environmental change on forest ecosystem function;

4. *provide objective data for the validation of forest models* related to growth, partitioning of primary production, water cycling, and hydrology;
5. *recommend management strategies for the conservation of carbon stores in forests.*

The project was carried out by 11 participants who represent the main expertise in the field of biospheric fluxes research in Europe, covering ten countries (Italy, France, Belgium, Holland, Denmark, Germany, Sweden, Finland, Iceland, and the JRC research center).

The material collected in this book represents the synthesis of 3 years of research and is organized into several sections. The first section describes the methodological advancements of the project and is divided into two chapters dealing with the eddy covariance method and soil respiration measurements. The eddy covariance theory is described in detail and the standard EUROFLUX methodology is presented in Chapter 2. Soil respiration methods are discussed in Chapter 3, soil processes being an important component of the carbon balance of an ecosystem.

The sections that follow (Chaps. 4 to 7) deal with the current understanding of biospheric exchanges and the ecological factors controlling them, the chapters are organized according to biome: beech, spruce, and pine and evergreen oak forest. These are the main European forest types and emphasis is given to the efficiency of their biogeochemical cycles and to the comparison across sites of the same species growing in different climates and geographical locations.

The other sections contain new findings coming from across-biome comparisons using EUROFLUX data. In particular. Chapter 8 addresses the use of models and their validation with EUROFLUX datasets. In Chapter 9 the use of EUROFLUX data in upscaling carbon sequestration data at European and country-specific levels is shown, using a combination of GIS, remote sensing, and direct flux measurement data.

In Chapter 10, comparison of ecosystem energy balance and the way energy is partitioned at the land surface is discussed. In Chapter 11 an analogous discussion is carried out on carbon fluxes, showing that there is a latitude-based trend in the magnitude of carbon sinks of European forests and that ecosystem respiration is the main factor controlling this trend.

In Chapter 12 the respiration processes at soil level are discussed, and new insights on factors controlling such processes are presented. In Chapter 13 conclusions on the work reported in the book are presented.

Finally, the entire EUROFLUX dataset is available (http://carbodat.ei.jrc.it) with more than 300 Mbytes of data. Included are more than 50 site-specific sets of data, ranging from continuous half-hour fluxes, to meteorological information, to ecological factors related to aboveground biomass and soil parameters over 96–98 years

Further Reading

Keeling RF, Piper SC, Heimann M (1996) Global and hemispheric CO_2 sinks deduced from changes in atmospheric O_2 concentration. Nature 381:218–221

Tans PP, Fung IY, Takahashi T (1990) Observational constraints on the global atmospheric CO_2 budget. Science 247:1431–1438

Valentini R, Matteucci G, Dolman AJ, Schulze E-D, Rebmann C, Moors EJ, Granier A, Gross P, Jensen NO, Pilegaard K, Lindroth A, Grelle A, Bernhofer C, Grünwald T, Aubinet M, Ceulemans R, Kowalski AS, Vesala T, Rannik Ü, Berbigier P, Loustau D, Gudmundsson J, Thorgeirsson H, Ibrom A, Morgenstern K, Clement R, Moncrieff J, Montagnani L, Minerbi S, Jarvis PG (2000) Respiration as the main determinant of European forests carbon balance. Nature 404:861–865

2 Methodology for Data Acquisition, Storage, and Treatment

M. Aubinet, R. Clement, J.A. Elbers, T. Foken, A. Grelle, A. Ibrom, J. Moncrieff, K. Pilegaard, Ü. Rannik, C. Rebmann

2.1 Introduction

The computation of half-hourly fluxes is complex and requires the treatment of a large amount (on the order of 10^5) of instantaneous measurements. It requires several operations that may be performed in different ways, and experience has shown that the results were sensitive to the computation procedure. Before making any comparison between different sites, one must be assured that the fluxes are computed in the same way on each site. It is therefore necessary to define a methodology for measurement and flux computation to be used by all the network teams.

The fundamental equations describing the conservation equation of a scalar are recalled and discussed in the second section. In the third section, the measurement system is presented. All EUROFLUX teams used the same system. However, as several teams developed their own computer programs in order to retain flexibility and to be able to adapt the software for particular instrumentation configurations, there was no agreement on the choice of unique software for flux computation. A comparison performed between different software packages processing the same data set stressed significant differences between the calculated fluxes. The differences were generally due to (problem 1) individual errors or to (problem 2) the choice of different hypotheses. In order to avoid problem 1, it was decided that any flux computation software used in the framework of EUROFLUX had to be validated. To this end, a software validation set was constituted. It was made by a series of raw data files ("golden files") obtained from a standard eddy covariance system and by the fluxes resulting from the computation. These fluxes were obtained as the convergent results of, at least, three independent software packages. The software validation set is available at the internet address http://carbodat.ei.jrc.it. Technical details about the softwarc vailidation set are given in the Annex.

Ecological Studies, Vol. 163
R. Valentini (Ed.) Fluxes of Carbon, Water and Energy of European Forests

Problem 2 could be solved only by explicitly defining the hypotheses and relations that must be used for flux computation. One aim of this chapter is to present these relations. They are described in the fourth section.

Finally, some quality control procedures are described in the fifth section. More details about these different procedures can be found in Aubinet et al. (2000).

2.2 Theory

The conservation equation of a scalar is:

$$\frac{\partial \varrho_s}{\partial t}+u\frac{\partial \varrho_s}{\partial x}+v\frac{\partial \varrho_s}{\partial y}+w\frac{\partial \varrho_s}{\partial z}=S+D \tag{2.1}$$

where ϱ_s is the scalar density, u, v, and, w are the wind velocity components, respectively, in the directions of the mean (x), and lateral wind (y), and normal to the surface (z). S is the source/sink term and D is the molecular diffusion term. The lateral gradients and the molecular diffusion will be neglected afterwards. After application of the Reynolds decomposition ($u=\bar{u}+u', v=\bar{v}+v', w=\bar{w}+w', \varrho_s=\overline{\varrho_s}+\varrho_s'$ where the overbars characterize time averages and the primes fluctuations around the average), averaging, integration along z, and assumption of no horizontal eddy flux divergence, Eq. (1) becomes:

$$\underset{I}{\int_0^{h_m} S dz} = \underset{II}{\overline{w'\varrho_s'}} + \underset{III}{\int_0^{h_m} \frac{\partial \overline{\varrho_s}}{\partial t} dz} + \underset{IV}{\int_0^{h_m} \bar{u}\frac{\partial \overline{\varrho_s}}{\partial x} dz} + \underset{V}{\int_0^{h_m} \bar{w}\frac{\partial \overline{\varrho_s}}{\partial z} dz} \tag{2.2}$$

where I represents the scalar source/sink term that corresponds to the net ecosystem exchange when the scalar is CO_2 and to the ecosystem evapotranspiration when the scalar is water vapor; II represents the eddy flux at height h_m (the flux which is measured by eddy covariance systems); III represents the storage of the scalar below the measurement height; IV and V represent the fluxes by horizontal and vertical advection. Under conditions of atmospheric stationarity and horizontal homogeneity, the last three terms of the right-hand side of Eq. (2.2) disappear and the eddy flux equals the source/sink term. However, in forest systems, these conditions are not always met and both storage and advection may be significant, especially at night (Grace et al. 1996; Goulden et al. 1996; Lee 1998; Aubinet et al. 2000; Paw et al. 2000). This problem is analyzed in Section 7.

2.3 Material

The eddy covariance system used in EUROFLUX comprises a three-axis sonic anemometer (Solent 1012R2, Gill Instruments, Lymington, UK), a closed path infrared gas analyzer (IRGA) (LI-COR 6262, LI-COR, Lincoln, Neb., USA), a logging computer, and a suite of analysis software for real-time and post-processing analysis. Gas samples are taken next to the sonic path and ducted down a sample tube to the IRGA.

The system should be installed on a tower or on a mast. The measurement height must be chosen high enough to be above the canopy and to avoid any flow distortion by obstacles (isolated high trees, tower, or mast structure). On the other hand, it must be chosen low enough for the measurement be representative of the surface that is studied. The conditions of representativeness of the measurements will be analyzed below (Sect. 2.7). The tower and measurement heights at the different EUROFLUX sites are given in Table 2.1.

In order to avoid shadowing by the tower or mast structure, placement of the system at the top of the tower is recommended. In other cases, it should be mounted not closer than 1.5 times the largest lateral dimension of the tower (Kaimal and Finnigan 1994) and at the upwind side of the tower. Measurements taken downwind from the tower should be removed from the data set.

The sonic anemometer produces the values of the three wind components and the speed of sound at a rate of 20.8 times per second. It's accuracy is 1.5 % below 30 m s^{-1} and 3 % above this speed. The velocity offset is 0.02 m s^{-1}. It has a built-in five-channel analogue to digital (A/D) converter that permits simultaneous measurement of analogue sensors and their digitization and integration with the turbulence signals. The analogue channels are sampled at a rate of 10 Hz. In the EUROFLUX system, the analogue channels are used for water vapor and CO_2 concentration output from the IRGA.

Generally, measurement of the air temperature fluctuations is done by means of the speed-of-sound output of the sonic anemometer. At wind speeds above about 8–10 m s^{-1}, however, speed-of-sound data becomes noisier because of mechanical deformation of the probe head. The measurement of the covariance of the vertical velocity and temperature is impossible in those cases (Grelle and Lindroth 1996). Thus, additional fast thermometers (Pt resistances) are applied at sites with frequent high wind speeds.

The concentrations of water vapor and CO_2 are measured by an LI 6262 infrared gas analyzer (IRGA). This is a differential analyzer, which compares the absorption of infrared energy by water vapor and CO_2 in two different chambers within the optical bench. The calibration range is 0–3000 μmol mol^{-1} for CO_2 and 0–7.5 kPa for H_2O. The typical noise level is 0.3 μmol mol^{-1} at 350 μmol mol^{-1} for CO_2 and 0.002 kPa at 2 kPa for H_2O. Use of this instrument has been described in detail by Moncrieff et al. (1997). When using the fast response configuration, its 95 % time constant is 0.1 s (LI-COR 1991).

Table 2.1. Measurement height, surface parameters, fetch distance in the main wind direction at each EUROFLUX site. Computation of footprint function maximum, distance giving 80 % of the flux, and contribution to the flux of sources situated behind the fetch. Computations are made using the Haenel and Grünhage (1999) model

Site	Measurement height (h_m) (m)	Displacement height (d) (m)	Roughness length (z_o) (m)	Fetch in main wind direction	Footprint function maximum location (m)	80 % Cumulative footprint (m)	Relative contribution of sources behind fetch
IT1	32	17.9	2.8	900	19.7	204.6	95.9
IT2	18	9.8	1.3		16.6	157.7	
FR1	22	9.1	1.7		30.4	280.2	
FR2	25.5	13	1.8	600	26.8	250.6	91.8
DK1	43	20	2	800	79.2	684.0	82.8
SW1[a]	35	18	2		44.9	404.8	
	70				296.6	2451.1	
	100				575.5	4716.1	
SW2	14	5.4	.8		27.8	244.1	
GE1[a]	22	12	2		14	143.7	
	32				61.1	539.7	
NL1	27	8.1	1.5	1500	69.6	597.7	92.1
UK1	13	4.8	.6		31.9	272.3	
GE2	42	22.5	1.85	500	61.1	539.7	78.5
BE1	40	19	2.7	1500	51.6	469.9	94.0
BE2	42	18	2.4	500	73.7	647.8	74.6
IC1	2.5	0.4	0.1		13.9	110.6	
FI1	23	9.8	1.4	250	38.5	340.5	73.3

[a] Eddy fluxes are collected at three different heights on site SW1 and at two different heights on site GE1

In EUROFLUX, the analyzer is used in absolute mode, i.e., the reference chamber is filled with a gas scrubbed of water and CO_2. For this purpose a closed, pump-driven, air circuit (with magnesium perchlorate desiccant and CO_2-absorbing chemicals) through the reference cell may be used. An alternative method is to flush nitrogen gas from a cylinder to the chamber, maintaining a flow rate of about 20 ml/min. This avoids possible damage to the instrument by the desiccant.

The calibration of the IRGA is carried out periodically (at least fortnightly). Lower calibration frequencies may affect the mean CO_2 concentration measurement, as the LI-COR 6262 is subjected to a significant zero drift. However, as the slope drift is very low, the impact of the calibration frequency on the fluxes is limited.

The air flows from the sampling point to the IRGA analyzer chamber through a system of tubes, pumps, and filters. The concentration sampling point should be placed close to the sonic anemometer sampling volume since spatial separation between the two systems causes a flux underestimation (Sect, 2.4.3). On the other hand, the sampling tube cannot be placed so as to disturb the wind pattern in the sonic anemometer sampling volume. A good compromise is to place the concentration sampling point just below or just above the sonic anemometer sampling volume.

Different air flow configurations are used in EUROFLUX. Three typical configurations are described in Aubinet et al. (2000). Two of them are presented in Fig. 2.1. In all cases, the system must be designed so as to maintain a stabilized flow with a high Reynolds number (in order to minimize the frequency losses; Leuning and Moncrieff 1990), to generate pressure and temperature conditions in the IRGA chamber that comply with its range of operational parameters and to avoid pressure fluctuations, air contamination caused by the pump, and condensation within tubes and analyzer.

Most commonly, the analyzer is placed on the tower, some meters away from the sonic. Some groups use long sample tubes with the analyzer on the ground surface as this permits easy access for its servicing.

Turbulent airflow through the tube is sometimes chosen (Moore 1986) in order to minimize frequency losses, but this usually requires powerful pumps with a high power consumption. Laminar flow in the tube is maintained by less powerful pumps and the extra loss in frequency response can be accounted for adequately (Moncrieff et al. 1997).

To prevent condensation, in most cases the air is sucked through the tube and the analyzer (the air pump is the last link in the air transport chain). The data are then corrected for the effect of under-pressure within the analyzer's sample cell (Fig. 2.1a;Moncrieff et al. 1997). A sensor (LI-COR 6262–03, LI-COR) that measures the IRGA chamber pressure at 1 Hz may be used to that end. Another possibility is to blow the air through the analyzer, giving high flow rates and a reduced number of corrections. With a suitable air pump and appropriate heating, pump-induced distortions and condensation can be prevented effectively (Grelle 1997). A third possibility is to transport the air at a high flow rate close to the measurement point with subsampling at a lower flow rate for H_2O and CO_2. (Fig. 2.1b). This methods produces very high flow rates without creating too large an under-pressure in the measuring chamber, but it requires two pumps. It is used in long tube systems.

In addition, it is necessary to place filters in every system (ACRO 50 PTFE 1 μm, Gelman, Ann Arbor, Mich., USA) upstream to the analyzer in order to prevent contamination of the IRGA chamber. Except for sites with very clean air, at least two filters in series are recommended. The first filter should be placed at the inlet of the tube and replaced every fortnight (more frequent changes are necessary in winter at some sites when reduced mixing conditions causes accelerated filter contamination). The second filter (ACRO 50 PTFE 1 μm, or Spiral cap 0.2 μm, Gelman) may be placed close to the analyzer and must be changed annually.

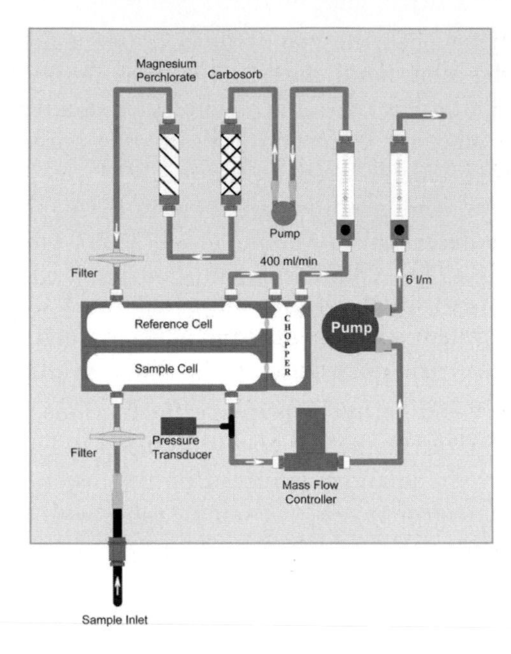

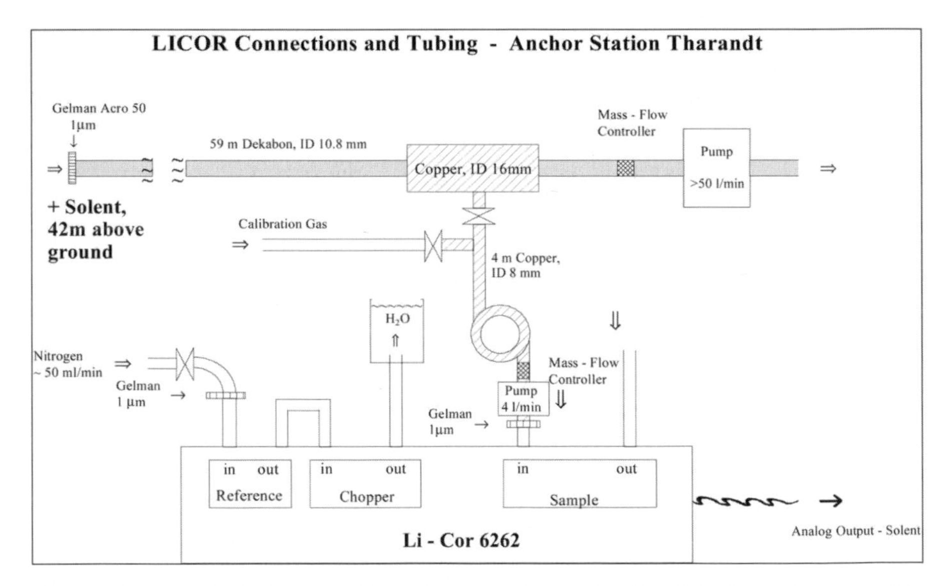

Fig. 2.1. Two typical flow configurations used in EUROFLUX. *Above* System with through flow in the sample circuit and chemicals in the reference circuit (Moncrieff et al. 1997). *Below* System with secondary flow in the sample circuit and nitrogen in the reference

The time lag that is introduced in a ducted system is taken into account by the software. It varies significantly from site to site according to tube length, flow rate(Leuning and Moncrieff 1990), and filter number and type. For a given system, the lag is fairly constant and its value can be specified in the software within an expected range. The exact time lag cannot be determined theoretically as it is difficult to estimate the effect of the filters. However, it can be computed by finding the maximum in the covariances (McMillen 1988; Grelle and Lindroth 1996; Moncrieff et al. 1997).

Given the aim of operating flux systems for extended periods of time, some protection of the system components from adverse weather is essential. Due to the large climate variability between the different EUROFLUX sites, the technical requirements for appropriate protection differ. In northern latitudes, the air analysis system is placed in thermally insulated boxes that are

Table 2.2. List of the meteorological variables measured in the EUROFLUX sites

Symbol	Unit	Variable	Instrument	
Radiation				
R_g	W m^{-2}	Global radiation	Pyranometer	O[a]
R_n	W m^{-2}	Net radiation	Net radiometer	O
PPFD	µmol m^{-2} s^{-1}	Photosynthetic photon flux density	Photodiode	O
R_{ref}	W m^{-2}	Reflected radiation	Pyranometer	F
R_d	W m^{-2}	Diffuse radiation	Pyranometer+screen	F
APAR		Light interception	Photodiodes	F
Temperature				
T_a	°C	Air temperatures (profile)	Resistance or thermocouple	O
T_{bole}	°C	Bole temperature	Resistance or thermocouple	O
T_s	°C	Soil temperature (profile 5–30 cm)	Resistance or thermocouple	O
T_c	°C	Canopy radiative temperature	Infra-red sensor	F
G	W m^{-2}	Soil heat flux density	Heat flux plates	O
Hydrology				
P	mm	Precipitation	Rain gauge	O
RH	%	Relative humidity profile	Resistance, capacitance, or psychrometer	O
SWC	% by volume	Soil water content (0–30 40–70 80–110)	TDR or theta probe (15 days)	O
SF	mm	Stem flow	Tree collectors (15 days)	F
T_f	mm	Throughfall	Pluviometer (15 days)	F
SNOWD	mm	Snow depth	Sensor (15 days)	F
Miscellaneous				
P_a	kPa	Pressure	Barometer	O

[a] O, Obligatory; F, voluntary.

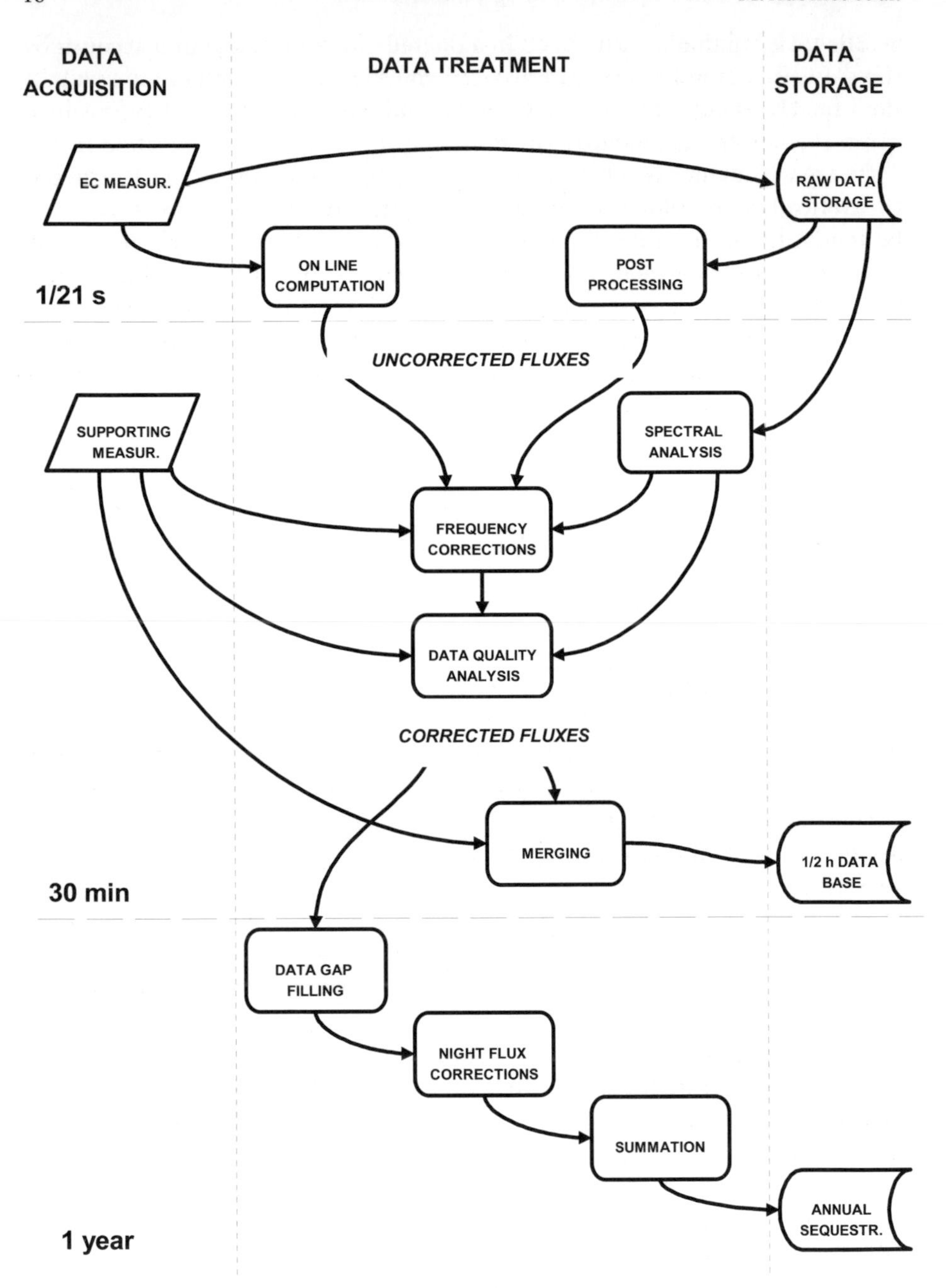

Fig. 2.2. Schematic of data acquisition, processing, and storage in EUROFLUX systems

ventilated by filtered air and heated to keep the internal temperature constant (Grelle 1997). In some cases, the sonic probes are heated to prevent riming. In the Mediterranean region, it is generally sufficient to keep the system in a clean and dry environment.

At each EUROFLUX site, the eddy covariance measurements are supported by a set of meteorological measurements that are also collected every half hour. The list of the variables that are measured is given in Table 2.2. These measurements allow characterization of the meteorological conditions under which the fluxes occur and provide a set of input data for soil vegetation atmosphere transfer (SVAT) models that can be used for calibration as well as for validation. In addition, they are necessary for eddy covariance flux correction, quality control, and gap filling.

2.4 Data Acquisition, Computation, and Corrections

2.4.1 Procedure Outline

The general procedure followed for collecting and processing the data is presented in Fig. 2.2. At intervals of 1/20.8 s, instantaneous measurements of the three components of the wind velocity, the speed of sound, and the molar fractions of CO_2 and water vapor are collected and stored. Software computes the mean values, the variances, the covariances and the so-called "uncorrected fluxes" (because, at this step, the corrections for high frequency losses are not performed).

The amount of raw data collected at these sample rates is about 600 Mb/month and is typically stored on tapes, CDs, or removable drives for post-processing, e.g., further analysis of the time series, data quality analysis, or spectral analysis. The supporting meteorological measurements are also averaged over intervals of a half hour. In particular, energy storage in the biomass and air (or possibly in the soil) or CO_2 storage in the air are computed at this step. Meanwhile, the corrections of fluxes are applied. These include corrections for latent heat or lateral momentum fluxes (Schotanus et al. 1983) and corrections for high frequency losses (Moore 1986; Leuning and Judd 1996; Moncrieff et al. 1997). There is no need for air density corrections of the CO_2 flux (Webb et al. 1980) as will be shown later.

The eddy covariance fluxes are then submitted to a series of tests that eliminate data that fails certain criteria. Different tests (based on statistical analysis, stationarity or similarity criteria, energy balance closure) are presented below. After correction and quality analysis, the eddy covariance and supporting measurements are merged and constitute the input from each group to the EUROFLUX database (cf. http://carbodat.ei.jrc.it)).

At a yearly time scale, the fluxes are summed in order to provide annual carbon sequestration estimation. This sum requires a complete data series, i.e., a procedure to fill in the data gaps and to correct them for the night flux error. Such procedures are described in other publications (Aubinet et al. 2000; Falge et al. 2001).

2.4.2 Computation of the Uncorrected Fluxes

The input data are the instantaneous values of the three components of the velocity (u, v, w [m s^{-1}]), of the sound speed (U_{son} [m s^{-1}]), and of the CO_2 and H_2O mole fractions [c, (μmol mol^{-1}), h (mmol mol^{-1})].

For two arbitrary variables ξ and η, the mean of ξ and the second moment of ξ and η are computed as:

$$\overline{\xi} = \frac{1}{n_s} \sum_{k=1}^{n_s} \xi_k \tag{2.3}$$

and

$$\overline{\xi'\eta'} = \frac{1}{n_s} \sum_{k=1}^{n_s} \xi'_k \eta'_k \tag{2.4}$$

where n_s is the number of samples and the fluctuations (ξ_k', η_k') around the mean at the step k $\left(\overline{\xi_k}, \overline{\eta_k}\right)$ are computed as:

$$\begin{aligned} \xi_k' &= \xi_k - \overline{\xi_k} \\ \eta_k' &= \eta_k - \overline{\eta_k} \end{aligned} \tag{2.5}$$

where $\overline{\xi_k}$ and $\overline{\eta_k}$ are computed either using a running mean (McMillen 1986, 1988; Baldocchi et al. 1988; Kaimal and Finnigan 1994) or a linear detrending algorithm (Gash and Culf 1996). The sonic temperature T_{son} [K] is deduced from sound velocity (U_{son}) by Schotanus et al. (1983) and Kaimal and Gaynor (1991) as:

$$T_{son} = \frac{U_{son}^{\ 2} M_d}{\gamma \Re} = \frac{U_{son}^{\ 2}}{403} \tag{2.6}$$

where M_d is the molar mass of dry air (0.028965 kg mol^{-1}), γ is the ratio of the air specific heat at constant pressure, and volume (1.4) and $\Re$ is the gas constant (8.314 J K^{-1} mol^{-1}).

For the computation of $\overline{w'c'}$ or $\overline{w'h'}$ the time series of w_k' must be delayed for synchronization with c_k' or h_k' to take into account the time taken for the air to travel down the sample tube.

Coordinate rotations are applied on the raw means and second moments. The two first rotations align u parallel to the mean wind velocity and nullify v and w. Their aim is to eliminate errors due to sensor tilt relative to the terrain surface or aerodynamic shadow due to the sensor or to the tower structure. However, they also suppress the vertical velocity component. This is not always appropriate as non-zero vertical velocity components may appear above tall vegetation when flux divergence or convergence occur (Lee 1998). The third rotation is performed around the x-axis in order to nullify the lateral momentum flux density ($\overline{v'w'}$ covariance). Indeed, this term is zero over plane surfaces and is likely to be very small over gentle hills. A complete justification on this point is given by Kaimal and Finnigan (1994). McMillen (1988) stresses that this rotation is not well defined in low speed conditions and recommends its application with care and, in any event, to limit it to 10°.

The eddy covariance fluxes of CO_2 (F_c, μmol m^{-2} s^{-1}) or H_2O (F_h, mmol m^{-2} s^{-1}) are directly deduced from the rotated covariances as:

$$F_c = \frac{\overline{P_a}}{\Re \overline{T}_{son}} \overline{w'c'} \qquad \text{and} \qquad F_h = \frac{\overline{P_a}}{\Re \overline{T}_{son}} \overline{w'h'} \tag{2.7}$$

where P_a is the atmospheric pressure.

The latent heat flux is defined as $\lambda M_w F_h/1000$ where M_w is the molar mass of water (0.0180153 kg mol^{-1}) and λ is the latent heat of vaporization of water (2441.78 J kg^{-1} at 25 °C)

Finally, the sensible heat flux, H, is given as:

$$H = \varrho_m C_m \overline{w'T_a'} = \frac{\overline{P_a} M_d}{\Re \overline{T}_{son}} C_d \left(\overline{w'T_{son}'} - 3.210^{-4} \overline{T}_{son} \overline{w'h'} + \frac{2U\overline{u'w'}}{403} \right) \tag{2.8}$$

where ϱ_m and C_m are the density and specific heat of the moist air (which is practically the same for humid and dry air, i.e., less than 0.5 % difference in the meteorological range). The second term in parenthesis accounts for the difference between sonic and real temperature, the third term corrects the sound speed for lateral momentum flux perturbations (Schotanus et al. 1983; Kaimal and Gaynor 1991). The relation (2.8) is rigorous only when the speed of sound is measured along a vertical sound path. Thus, in the case of the SOLENT, where the axis is tilted with respect to the vertical, this relation is not strictly valid. Exact relations are proposed by Liu et al. (2001). However, the departure from this relation is small. When the temperature fluctuations are measured with a fast thermometer (platinum resistance wire), these corrections are not necessary.

Corrections in order to remove air density fluctuations due either to temperature or to water vapor concentration fluctuations (Webb et al. 1980) are not necessary here. Successive analyses by Leuning and Moncrieff (1990), Leuning and King (1992), and Leuning and Judd (1996) showed that, for closed path systems, the temperature fluctuations were damped and became negligible provided that a minimum tube length of the order of 1000 times the tube inner diameter is used for the air sampling. Rannik et al. (1997) showed that this length also does not depend on the tube material. In addition, humidity fluctuations are automatically corrected by the LI-COR 6262 software.

2.4.3 Correction for Frequency Response Losses

The turbulent flow in the atmospheric boundary layer may be considered as a superposition of eddies of different sizes. At one measurement point they generate velocity and scalar concentration fluctuations of different frequencies. It is common to describe the frequency repartitions of the fluctuations by introducing the cospectral density. In the case of turbulent fluxes of a scalar, the cospectral density C_{ws} is related to the covariance (Stull 1988, Kaimal and Finnigan 1994) as:

$$\overline{w's'} = \int_0^\infty C_{ws}(f)df \tag{2.9}$$

where f is the cyclic frequency. In practice the integral range is limited at low frequency by the duration of observation and/or the autoregressive high-pass filtering, and at high frequency by the instrument response and tube effects. Consequently, the measured turbulent fluxes of a scalar s may be thought of as the integration over frequencies of the cospectral density ($C_{ws}(f)$) multiplied by a transfer function ($TF(f)$) characterizing the measurement process:

$$\overline{w's'}_{meas} = \int_0^\infty TF(f)C_{ws}(f)df \tag{2.10}$$

The fluxes must therefore be corrected so as to take these effects into account. A correction factor is introduced which is the ratio of the real flux (2.9) (i.e., the flux that would be measured by an ideal system) and the measured flux (2.10):

$$CF = \frac{\int_0^\infty C_{ws}(f)df}{\int_0^\infty TF(f)\ C_{ws}(f)df} = \frac{\int_0^\infty C_{ws}(f)df}{\int_0^\infty TF_{HF}(f)\,TF_{LF}(f)\,C_{ws}(f)df} \tag{2.11}$$

where the low (TF_{LF}) and high (TF_{HF}) frequency parts of the transfer function are separated. They will be discussed independently.

2.4.4 Effect of Low Frequency Cut on Fluxes

High pass filtering appears because of finite duration of the measurement period and use of recursive filters to compute the fluctuations. Two filter algorithms are generally used to this end: one is based on linear detrending (LD), the other on a running mean (RM). They are given in detail in Aubinet et al. (2000) as are the equations of the corresponding transfer functions for variances and co-variances. The transfer function depends only on the sampling duration (one half hour in this instance) for the LD case. In the RM case, it depends on the time constant of the filter that can be adjusted.

It is important to note that the transfer function associated with block averaging,

$$TF_{LF}^{BA}(f) = 1 - \frac{\sin^2(\pi fT)}{(\pi fT)^2} \tag{2.12}$$

applies when derivation of fluctuating components includes subtraction of mean values over the averaging period. When this is not done, the finite averaging time does not lead to underestimation of fluxes in case of RM filtering, and the transfer function $TF^{BA}{}_{LF}$ must not be applied together with the transfer function of the RM case. In case of LD, the transfer function $TF^{BA}{}_{LF}$ is a natural part of the corresponding transfer function (see Aubinet et al. 2000). The attenuation of low frequencies in EC flux calculation and application of corresponding transfer functions is discussed in detail in Rannik (2001) and Massman (2001).

The LD transfer function is displayed in Fig.2 3 as well as two RM transfer functions with time constants 200 s (RM200) and 1000 s (RM1000) (Kristensen 1998). The LD transfer function decays more sharply than the RM transfer functions and its filtering effect is intermediate between those of RM200 and RM1000. A more quantitative estimation of the filtering effect of each function may be estimated by combining it with the atmospheric surface layer cospectra at the neutral limit (Horst 1997). They are 0.7, 1.5, and 0.2 % for LD, RM200, and RM1000, respectively, when the measurement height (h_m–d) to wind speed ratio is 1 s. It increases up to 6.1, 11.4, and 1.8 % when this ratio is 10 s. Due to spectral shift in turbulence energy, the filtering impact at low frequencies is less under stable conditions and greater under unstable conditions; as a result, a relatively small error in short-period fluxes can translate into a big error in long-term averages.

In order to confirm these results experimentally, a series of raw data (one month of data in June 1999 at site BE1) was treated three times using the

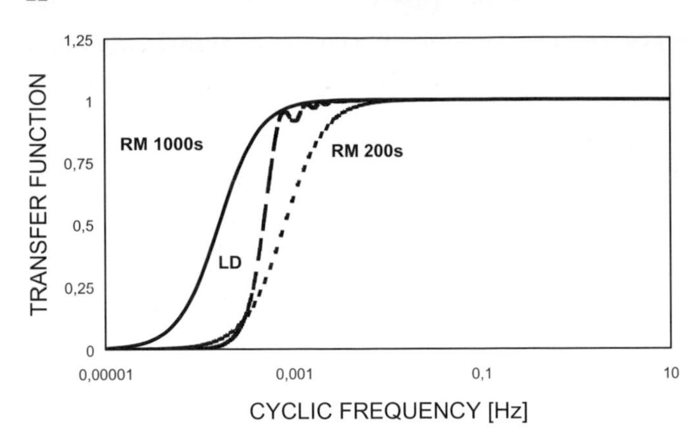

Fig. 2.3. Transfer functions corresponding to the high pass filters: *LD* Linear detrend; *RM 200s* running mean with time constant 200 s; *RM1000s* running mean with time constant 1000 s

same procedure and parameters, only varying the high pass filtering algorithm from one treatment to another. The three series of covariances (noted respectively $\overline{w'\eta'}_{RM200}$, $\overline{w'\eta'}_{LD}$ and $\overline{w'\eta'}_{RM1000}$ were compared. In spite of some spread due to the random error, it was clear that, for all scalars, $\overline{w'\eta'}_{RM200} < \overline{w'\eta'}_{LD} < \overline{w'\eta'}_{RM1000}$.

This is illustrated in particular for $\overline{w'c'}$ in Fig. 2.4a,b. The differences between $\overline{w'\eta'}_{RM200}$ and $\overline{w'\eta'}_{RM1000}$ were 7.5 % for $\overline{w'c'}$, 5.6% for $\overline{w'u'}$, 8.0% for $\overline{w'T'}$, and 13.4% for $\overline{w'q'}$. The differences between $\overline{w'\eta'}_{RM200}$ and $\overline{w'\eta'}_{LD}$ were 5.7, 4.4, 5.8, and 9.7 %, respectively. In addition, these differences were found to increase with the height to wind speed ratio. In particular, the averaged ratio $\dfrac{\overline{w'c'}_{RM1000}}{\overline{w'c'}_{RM200}}$ (shown in Fig. 2.5) is close to 1 at $(h_m-d)U^{-1}$ <5 s and increases to 1.1 for higher $(h_m-d)U^{-1}$ values.

Consequently, in order to observe negligible systematic errors in fluxes, the RM has to be applied with moderately long time constants, but this leads to systematic overestimation of variances in the periods of non-stationarity of first moments (Shuttleworth 1988). In addition, it was shown by Rannik and Vesala (1999) that fluxes are subject to increased random errors during episodes of non-stationarity. Thus, it seems justified to apply the method with a moderate time constant to avoid these problems at the expense of small systematic errors in fluxes, which should be then accounted for. In the case of atmospheric surface layer measurements, reasonable values of the time constant of the RM seem to be between 200 and 1000 s, probably around 500 s, being dependent on the measurement height (h_m-d).

Application of the LD method with commonly accepted averaging times does not generally lead to overestimation of variances and increases in ran-

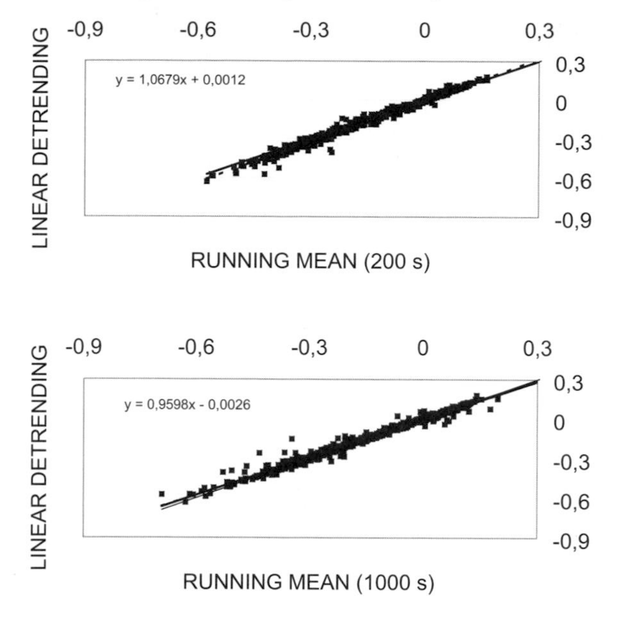

Fig. 2.4. *Above* Comparison between the covariances of wind velocity and CO_2 concentrations computed using 200-s time constant running mean and linear detrending. *Below* Comparison between the covariances of wind velocity and CO_2 concentrations computed using 1000-s time constant running mean and linear detrending

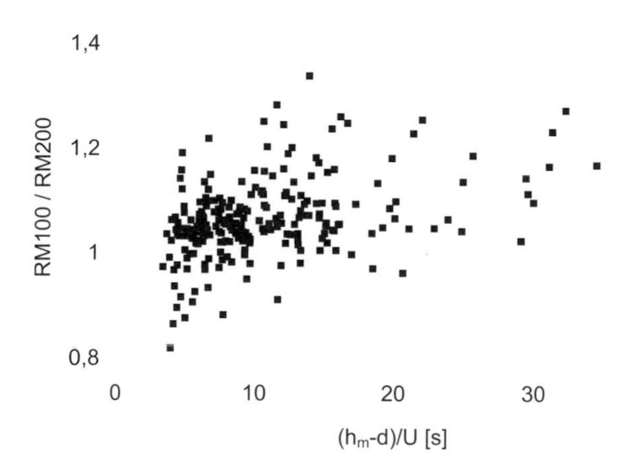

Fig. 2.5. Evolution with the wind velocity of the ratio of the covariances computed with 1000- and 200-s running mean time constants

dom errors, to the advantage of the method. However, the systematic errors in fluxes due to LD are not negligible under certain experimental conditions.

2.4.5 Effects of High Frequency Cut on Fluxes

Low pass filtering results from the inability of the system to resolve fluctuations associated with small eddies and induces an underestimation of the measured turbulent flux. The transfer function characterizing the high frequency losses (TF_{HF}) is 1 at low frequencies, decays to zero at high frequencies and may be characterized by its cut-off frequency (f_{co}), that is, the frequency at which the transfer function equals $2^{-1/2}$. When the cut-off frequency is known, the transfer function may be combined with the high pass transfer function and model cospectra as defined by Kaimal et al. (1972) or Horst (1997) in order to estimate the correction factor. Moncrieff et al. (1997) showed in this way that, for a given eddy covariance system, the correction factor is a function of wind speed (U) and measurement height above the displacement height (h_m–d). The relation between the correction factor and the cut-off frequency is given for five different (h_m–d)/U values in Fig. 2.6. The transfer function and its cut-off frequency may be estimated either theoretically or experimentally.

The theoretical approach (Moore 1986; Leuning and Moncrieff 1990; Leuning and King 1992; Leuning and Judd 1996; Moncrieff et al. 1997) describes the transfer function of an eddy covariance system as the product of six individual functions, each describing a particular instrumental effect: the dynamic frequency responses of the sonic anemometer and of the IRGA, the sensor response mismatch, the scalar path averaging, the sensor separation and, in closed path systems, the attenuation of the concentration fluctuations down the sampling tube.

The first four functions depend only on the sonic and IRGA characteristics and on the wind speed and are thus the same for all the EUROFLUX sites. The impact on flux measurements of these four effects was computed by Aubinet et al. (2000) for the EUROFLUX setup. They showed that, except for very low measurement heights and wind speeds, they were negligible. The last two transfer functions are site-specific. They depend on the separation distance between the sonic and the inlet of the sampling tube of the IRGA, the tube length and radius, the flow in the tube, the number of filters, and the wind velocity. The correction factor depends on all these parameters and also, under stable conditions, on the measurement height and on the Obukhov

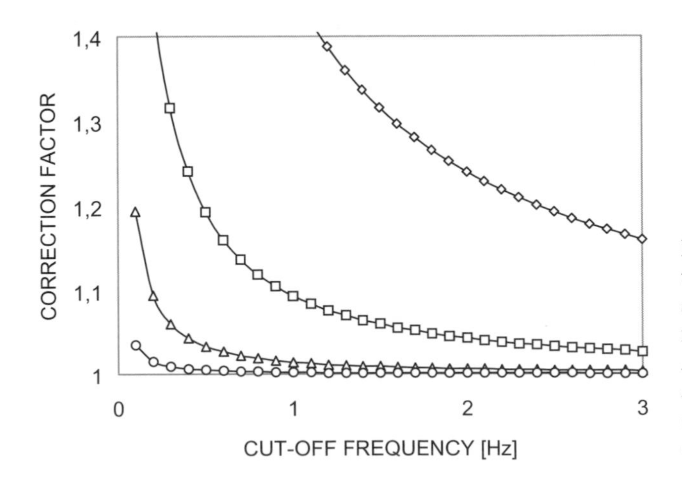

Fig. 2.6. Evolution of the correction factor according to the cut-off frequency. *Lozenges* h_m–d/U=0.2 s; *squares* (h_m–d)/U=1 s; *triangles* (h_m–d/U=5 s; *circles* (h_m–d)/U=25 s

length. It was shown by Aubinet et al. (2000, 2001) that the cut off frequency deduced by the theory was overestimated and the correction underestimated. This is probably true because the theoretical approach does not take the fluctuation damping by the filters into consideration.

For these reasons, the experimental approach is preferred. It consists of computing the heat and scalar co-spectra over a sufficiently long period (in order to reduce the uncertainties on the low frequency part of the cospectra) with good conditions (sunny and that meet both fetch and stationarity criteria). The experimental transfer function is then deduced as the ratio of the normalized cospectral densities:

$$TF_{HF}^{\exp}(f) = \frac{N_T}{N_s} \frac{C_{ws}(f)}{C_{wT}(f)} \tag{2.13}$$

where N_T and N_s are normalization factors. This function may then be fitted by an exponential regression, which may be introduced in Eq. (2.11):

$$TF_{HF}(f) = \exp\left\{ -\ln(\sqrt{2}) \left(\frac{f}{f_{co}} \right)^2 \right\} \tag{2.14}$$

The normalization factors N_T and N_r should be the real covariances (respectively $\overline{w'T'}$ or $\overline{w's'}$) or, if similarity between heat and scalar transport is assumed, the real standard deviation (respectively, σ_T and σ_s). However, the measured covariances and standard deviations are lower than the real ones, as affected by the high frequency attenuation, too. To avoid this bias, the normalization factors may be computed as the integrals of the cospectral densities between zero and a limit frequency, chosen to be low enough for the attenuation be negligible in the integrals but high enough for the number of points used to estimate the integrals to be sufficient and the uncertainty on the normalization factors to be low.

The cospectra for sensible heat are obtained experimentally and are therefore affected by the dynamic response of the sonic [or the high frequency (HF) thermometer], the scalar path averaging, and the sensor response mismatch. As the corresponding functions are not the same for the sonic or the HF thermometer and the IRGA, these effects are not correctly taken into account by the experimental transfer function. However, as these effects are small compared to the tubing or to the sensor separation effect, they could be neglected.

It was shown earlier that the transfer function only depends on the set-up. Consequently, it is not necessary to compute it every time. An estimation every month in order to follow the wear of the pumps (if the flow is not controlled) is sufficient. The correction factor may be expressed as a function of the wind velocity.

2.5 Quality Control

Direct validation of eddy covariance measurements is not possible since there is no alternative method for measuring the fluxes at the same spatial and temporal scale. In these conditions, assessment of the quality of the data becomes a primordial step in the measurement process. The procedure chosen here is to investigate empirically whether the fluxes meet certain plausibility criteria. Four criteria have been investigated that concern respectively the statistical characteristics of the raw instantaneous measurements, the stationarity of the measuring process, boundary layer similarity, and energy balance closure.

2.5.1 Raw Data Analysis

Tests on the raw data may be performed in order to detect anomalies. The first test to perform is the spike detection. Spikes can be caused by random electronic spikes or sonic transducer blockage (during precipitation, for example). Their detection and removal may be carried out by algorithms based on local averaging scale and point-to-point autocorrelation (Højstrup 1993). A warning may be noted (flagged) when the percentage of replaced data exceeds a threshold. Next, data are further required to fall within absolute limits and normal ranges of skewness and kurtosis. Discontinuities in the mean and variance of a time series (coherent on a local averaging scale) may be detected using the Haar transform (Mahrt 1991). Finally, each variable must be tested for a lower limit in absolute variance and a variance ratio for consecutive local windows. The software package that performs these tests was described by Vickers and Mahrt (1997). The FORTRAN package is freely available and is intended as a safety net to identify instrument and data-logging errors prior to data analysis.

2.5.2 Stationarity Test

One of the fundamental hypotheses that underlies the equivalence between the eddy covariance flux and the net ecosystem exchange is the stationarity of the measuring process. This assumption has then to be fulfilled for using the eddy covariance method.

The nonstationarity test has been used since the 1970s by Russian scientists (Gurjanov et al. 1984, see Foken and Wichura 1996). It is performed as follows: the measurement period is split into several (4–8) adjacent intervals of equal length. A partial covariance is computed on each interval using a relation similar to Eq. (2.4). The averaged partial covariance is then computed as the mean of the partial co-variances of all the intervals. It is compared to the co-variance directly computed on the whole measurement period. If there is a difference of less than 30 % between the two variables, then the measurement is considered to be stationary. For practical use: All data with differences <30 % have a high quality, and those with differences between 30 and 60 % have an acceptable quality (Foken et al. 1997).

2.5.3 Integral Turbulence Characteristics Test

Another test may be performed by analyzing the integral characteristics (flux-variance similarity) of the vertical wind and the temperature. Relations between these characteristics and the stability parameter were extensively investigated by Wichura and Foken (1995). By definition, the integral characteristics are basic similarity characteristics of the atmospheric turbulence (Obukhov 1960; Wyngaard et al. 1971). They characterize whether or not the turbulence is well developed according to the similarity theory of turbulent fluctuations. It is possible to discover some typical effects of non-homogeneous terrain. Firstly, if there is additional mechanical turbulence caused by obstacles or generated by the measuring device itself, the measured values of integral characteristics are significantly higher than predicted by the model. Secondly, the measured values of integral characteristics are significantly higher than the model for terrain with an inhomogeneity in surface temperature and moisture conditions, but not for inhomogeneities in surface roughness. This was found by De Bruin et al. (1991) and confirmed by Wichura and Foken (1995).

The data quality is good if the difference between the measured integral characteristics and the calculated value differs by not more than 20–30 %. According to Foken (2000), this test can also be used for stable stratification. For neutral stratification, this test cannot be used for scalar fluxes.

2.5.4 Energy Balance Closure

The closure of the energy balance is a useful parameter to check the plausibility of data sets obtained at different sites. In this approach, the sum of turbulent heat fluxes is compared with the available energy flux (the net radiative flux density less the storage flux densities in the observed ecosystem, including soil, air, and biomass). According to the first law of thermodynamics, the two terms should balance, i.e., the regression of turbulent fluxes according to the available energy flux should have a slope of 1 and an intercept of zero. However, non closure of the energy balance was observed on all EUROFLUX sites (Aubinet et al. 2000). The slope of the regression was found always lower than one (from 0.726 to 0.997) and the intercept was found always negative (from –4 to –32 W m^{-2}, the lower intercept corresponding to the higher slope).

This shows that the turbulent fluxes were systematically lower than the other energy fluxes deduced from radiation and heat storage measurements. The evolution of the slope with stability and with friction velocity were also investigated. It was shown to be greater under conditions of unstable stratification and less at stable stratification. In addition, at all sites, the slope was found to increase with increasing u^* in the range between zero and 0.4 m s^{-1} at stable stratification.

It was shown by Wilson et al. (in prep.) that all the FLUXNET sites behave similarly. Nonclosure of the energy balance can be for different reasons. In a detailed analysis of the possible causes of energy balance nonclosure, Wilson et al., (in prep.) identified five main primary error sources: (1) low and high frequency losses of turbulent fluxes, (2) neglected energy sinks, (3) footprint mismatch, (4) systematic instrument bias, (5) mean advection of heat and/or water vapor. Error 1 can be avoided by applying an appropriate correction as described above. For error source 2, Wilson et al. cite energy storage in the soil above heat flux plates, latent heat losses below the heat flux plates, sensible and latent heat between the soil, and the measurement point. All these terms are usually considered in the energy budgets of the EUROFLUX sites. Besides, some processes like melting, freezing, or heat conductance to cold rain are not considered in the budget. Error 3 results from the differences between the source areas of the different instruments measuring the energy fluxes (Schmid 1997). The source area of the radiometer is limited to a circle whose radius depends on sensor height and that is constant with time, while the source area of the eddy covariance system is approximated by an ellipse whose shape and position vary with wind and stability conditions. In these conditions, systematic difference between the radiation and the turbulent fluxes may appear in heterogeneous sites. The systematic instrument bias (error 4) can affect the eddy covariance system and also the heat storage measurement systems and radiation instruments. In a field intercomparison within the BOREAS project, Smith et al. (1997) observed deviations up to 16 % from their standard instrument, due to different calibrations of the sensors. Such comparison was not made between the EUROFLUX sites. Neglecting horizontal and vertical advection fluxes (error 5) could be one major reason for the lack of energy balance, in particular at night, during stable periods. As advection fluxes concern heat as well as water vapor and carbon dioxide, the lack of energy balance here could be an indicator of an underestimation of the carbon dioxide flux during this period. This point will be discussed below (Sect, 2.7).

The degree of closure of the energy balance may be used as a criterion to determine if the vertical turbulent fluxes do represent the total fluxes of a scalar or not. In particular, perturbation of the flux by the tower in some directions may be expressed by a lower degree of closure. On the other hand, the energy balance may serve as a tool for better analyzing night processes.

2.6 Spatial Representativeness of the Fluxes

To cover as many forest types as possible, the flux network must consider sites that are not necessarily ideal for flux measurements, in particular, heterogeneous forests. One main problem in this case is to know to what extent the fluxes (and the annual sum) are representative of the real ecosystem flux. To know this, it is necessary to locate the flux sources and to establish the distribution of the frequencies at which they influence the flux measurements. A first rule of thumb (Baldocchi et al. 1988) was to consider that the boundary layer grows with an approximate angle of 1/100, which means that the source is located within a distance equal to 100 times the measurement height above the effective surface. It was shown later (Leclerc and Thurtell 1990) that this rule was too drastic for eddy measurements, scalar flux source areas being smaller than scalar concentration areas by approximately one order of magnitude (Schmid 1994, 1997). Footprint models may help to refine this criterion.

Footprint analysis was developed in order to estimate the source location. It is based on Lagrangian analysis and relates the time-averaged vertical flux of a quantity at the measurement point to its turbulent diffusion from sources located upwind from the measurement point. The extent to which an upwind source located a certain distance from the measurement point contributes to observed flux has been termed the source weight function or flux footprint (Schmid 1994). In footprint analysis, the relationship between the surface sources/sink and the measured flux is studied. Footprint analysis may be used to:

- evaluate the adequacy of the fetch of a homogeneous stand in a certain wind direction and under certain stability conditions;
- estimate the fraction of flux contributed by the sources within the homogeneous fetch assuming uniform surface sources;
- estimate the upwind distance giving a significant fraction of flux.

Different analytical models were presented that differ in their complexity. They were based on the analytical diffusion theory (e.g., Schmid and Oke 1990; Schuepp et al. 1990; Horst and Weil 1992, 1994; Schmid 1994) or by Lagrangian stochastic simulation (Leclerc and Thurtell 1990; Flesch 1996; Rannik et al. 2000). All these models showed that the source area size and upwind distance increased with the measurement height above the displacement height and with thermal stability, and decreased with surface roughness and thermal instability. By combining a simple footprint model with a site map, Aubinet et al. (2001) showed that, owing to the modifications of the flux source area with stability conditions, flux measurements made in heterogeneous forests could describe ecosystems that differ completely between night and day.

In order to give an indication about the representativeness of EUROFLUX sites, the measurements height (h_m) stand characteristics (displacement height, d, roughness length, z_o) and fetch (D_f) in the main wind direction are given in Table 2.1. In addition, we give the footprint maximum location (D_{max}), the upwind distance giving 80 % of the flux (D_{80}), and the relative contribution of sources situated behind the fetch (F_{crel}). The footprint calculations were made according to the analytical model by Haenel and Grünhage (1999), which is a version of the Horst and Weil (1994) model modified to ensure correct asymptotic behavior of the footprint function. Estimations are for neutral stratification conditions. Consequently, the variables obtained by footprint calculation are subjected to great variations with thermal stratification. Indeed, both D_{max} and D_{80} increase under stable conditions and decrease under unstable conditions, while F_{crel} exhibits the opposite behavior.

2.7 Night Flux Corrections

It is likely that, during stable night-time conditions, the CO_2 exchange is underestimated by the eddy covariance measurements. This is supported by different arguments. First, a correlation between u^* (used as a measure for turbulent mixing) and the CO_2 efflux is observed under low u^* on all sites (Aubinet et al. 2000). As the night flux results from ecosystem respiration which is controlled by temperature and soil water content and is independent from turbulence, it is likely that such correlation is linked to the measurement process. Secondly, the increase with increasing u^* of the slope of the energy balance regression (Sect. 2.5.4) between zero and 0.4 m s^{-1} at stable stratification confirms that the observed underestimation of the CO_2 fluxes at night comes from a lack of turbulence rather than from a varying source strength. Finally, comparisons made at some sites with soil chamber measurements (Aubinet et al. 2001) showed that the night eddy flux measurements clearly underestimate the soil CO_2 fluxes at night. Underestimation of the night-time CO_2 fluxes is an example of selective systematic error and as such can have an important impact on long term budgets estimations (Moncrieff et al. 1996). As it appears systematically at night when the forest produces CO_2, it leads to an overestimation of the forest carbon sequestration.

The night flux underestimation could be due to technical limitations of the eddy covariance system (the concentrations variations may be too rapid or of too low amplitude to be detected by the system, Goulden et al. 1996) or, more likely, to the presence at night of alternative transport processes that are not taken into account by the eddy covariance system. In particular, storage of the CO_2 in the layer below the eddy flux system [term III in Eq. (2.2)] and mass flow by horizontal or vertical advection [terms IV and V in Eq. (2.2)] may be significant at night. CO_2 storage becomes significant at night, especially under stable conditions when it may become the most important term of the carbon balance (Berbigier et al. 2001). However, it does not affect the long-term carbon budget as long as the morning flush is captured by the measurement. Consequently, it is not necessary to apply a correction to the night fluxes if the underestimation results only from CO_2 storage.

The transport by advection takes place in presence of a concentration gradient and of a non-zero velocity in the same direction. Above homogeneous terrains the vertical velocity and the horizontal concentration gradients are supposed to be zero and advection flows are not considered in the carbon budget. In the present methodology, the second rotation applied to the velocity vector (Sect. 2.5.2) systematically sets the vertical component of the mean wind speed, *w*, and thus the mean vertical advection to zero. However, non-zero vertical advection is not unrealistic and may be induced by horizontal heterogeneity and mesoscale movements. It is important to note that if CO_2 leaves the forest by advection, contrary to the storage, it will be lost to the eddy system. Consequently, if the night flux underestimation results from advection, it needs to be corrected.

Lee (1998) and Finnigan (1999) proposed an alternative way to perform axis rotation that allows the evaluation of these flows. Some first estimations of the mass flow exchanged by advection using this procedure were made by Paw et al. (2000) and Grelle (2001). Research on this topics is still in progress.

In the absence of a direct estimation of the advection fluxes, the more usual way to correct for flux underestimation during stable nights is to replace the measured fluxes by the simulated efflux parameterized by a temperature function that has been derived during well-mixed conditions. Two general problems with this approach are the sensitivity of the CO_2 budget with the threshold value of u^* that is used to distinguish between stable and well-mixed conditions, and the risk of double counting of the flux. Aubinet et al. (2000) showed that the threshold varied from site to site but was typically about 0.4 m s^{-1}. The double counting of the flux arises in presence of CO_2 storage: indeed, under these conditions, flux would be taken into account for the first time when applying the parameterization correction to the measurement taken during stable periods, and for a second time when it would be measured during the morning flush. To avoid this duplication, it is necessary to correct the eddy flux for storage effects when the parameterization correction is not applied. Aubinet et al. (Aubinet et al. 2002) showed that the error introduced by double counting could lead to an underestimation of the forest carbon sequestration that could be as important as the night flux error itself.

Annex: Technical Details About the Software Validation Set

The raw data set comprises 19 successive 30-min digital files containing raw time series data from the sonic anemometer, viz., the three velocity components, speed of sound (all in m s^{-1}), CO_2 (μmol mol^{-1}), and H_2O concentrations [mmol mol^{-1}] sampled at 20.8 Hz. The data cover a range of meteorological conditions, at night and during the day. The data were obtained at the Vielsalm site (Belgium) on 1 June 1997, between 12:00 P.M. and 9:00 P.M. A description of this site is given by Aubinet et al. (2001).

The computation results used as reference are obtained independently by six different software packages. The computation involves the following steps: estimation of the temperature from the sound speed, computation of the mean values, computation of the fluctuations (using a high pass filter), variances and covariances (including an estimation of the time lag for CO_2 and H_2O fluxes), three axes rotation, and conversion of the eddy covariance in fluxes. Although recommended, the Schotanus correction [Eq. (2.9)] was not applied in this set. As the EUROFLUX software packages are designed to treat data issued from closed chamber IRGA, no correction for density fluctuation is introduced. As stated, the results are based on the data obtained by six independent software packages. Three used a recursive digital high pass with a 200-s time constant: *EdiSol* (Edinburgh University, Scotland), *UIA* (University of Antwerpen, Belgium), *SMEAR Solent* (University of Helsinki, Finland). Three used a linear detrending algorithm: *EDDYMEAS* (University of Bayreuth, Germany), *RISOE* (RISOE National Laboratory, Denmark), *IBK* (Georg August University of Göttingen, Germany). All of them output the mean wind velocity (three components), temperature, water vapor and carbon dioxide concentrations, and the fluxes of momentum (or the friction velocity), sensible heat, water vapor (or latent heat), and carbon dioxide. The agreement among all software packages has been found to be better than 1 % for the wind velocity, 0.01 % for CO_2 concentration, and 0.03 % for the water vapor concentration. The agreement among linear detrend or running mean software packages was better than 2 % for the latent heat and better than 1 % for the other fluxes. These results as well as the tolerance (maximum difference among software packages) are given at the internet address http://carbodat.ei.jrc.it.

References

Aubinet M, Chermanne B, Vandenhaute M, Longdoz B, Yernaux M, Laitat E (2001) Long term measurements of water vapour and carbon dioxide fluxes above a mixed forest in Ardenne's region. Agric For Meteorol 108:293–315

Aubinet M, Heinesch B, Longdoz B, (2002) Estimation of the carbon sequestration by a heterogeneous forest: night flux corrections, heterogeneity of the site and inter-annual variability. Global Change Biol 8:1053–1071

Baldocchi D, Hicks BB, Meyers TD (1988) Measuring biosphere-atmosphere exchanges of biologically related gases with micrometeorological methods. Ecology 69:1331–1340

Berbigier P, Bonnefond JM, Mellmann P (2001) CO2 and water vapour fluxes for 2 years above Euroflux forest site. Agric For Meteorol 108:183–197

De Bruin HAR, Bink NJ, Kroon LJM (1991) Fluxes in the surface layer under advective conditions. In: Schmugge TJ, André JC (eds) Workshop on land surface evaporation measurement and parameterization. Springer, Berlin Heidelberg New York, pp 157–169

Falge E, Baldocchi D, Olson R, Anthoni P, Aubinet M, Clement R, Granier A, Bernhofer C, Hollinger D, Lai C.T, Kowalski A, Meyers T, Moors EJ, Munger JW, Pilegaard K, Rannik Ü, Rebmann C, Verma S, Law B, Moncrieff J, Grünwald T, Katul G, Wofsy S, Jensen NO, Vesala T, Tenhunen J, Suyker A, Wilson K (2001) Gap filling strategies for defensible annual sums of net ecosystem exchange. Agric For Meteorol 107:43–69

Finnigan J (1999) A comment on the paper by Lee (1998): "On micrometeorological observations of surface-air exchange over tall vegetation." Agric For Meteorol 97:55–64

Flesch TK (1996) The footprint for flux measurements, from backward lagrangian stochastic models. Boundary Layer Meteorol 78:399–404

Foken T (2000) The turbulence experiment FINTUREX at the Neumayer-Station/ Antarctica. Bericht des Deutschen Wetterdienstes (in press)

Foken T, Wichura B (1996) Tools for quality assessment of surface-based flux measurements. Agric For Meteorol 78:83–105

Foken T, Jegede OO, Weisensee U, Richter SH, Handorf D, Görsdorf U, Vogel G, Schubert U, Kirzel H-J, Thiermann V (1997) Results of the LINEX-96/2 experiment. Deutscher Wetterdienst, Geschäftsbereich Forschung und Entwicklung, Arbeitsergebnisse 48, 75 pp

Gash JHC, Culf AD (1996) Applying linear detrend to eddy correlation data in real time. Boundary Layer Meteorol 79:301–306

Goulden ML, Munger JW, Fan S-M, Daube BC, Wofsy SC (1996) Measurements of carbon sequestration by long-term eddy covariance: methods and a critical evaluation of accuracy. Global Change Biol 2:159–168

Grace J, Malhi Y, Lloyd J, McIntyre J, Miranda AC, Meir P, Miranda HS (1996) The use of eddy covariance to infer the net carbon dioxide uptake of Brazilian rain forest. Global Change Biol 2:209–217

Grelle A (1997) Long-term water and carbon dioxide fluxes from a boreal forest: methods and applications. Doctoral Thesis. Silvestria 28, Acta Universitatis Agriculturae Sueciae, 80 pp

Grelle A (2001) Flux experiments in a Boreal Forest Stand within the Framework of EUROFLUX. In: Shimizu H (ed) Carbon dioxide and vegetation: advanced approaches for absorption of CO_2 and responses to CO_2. National Institute for Environmental Studies, Tsukuba, Japan, pp 17–27

Grelle A, Lindroth A (1996) Eddy-correlation system for long term monitoring of fluxes of heat, water vapour, and CO_2. Global Change Biol 2:297–307

Gurjanov AA, Zubkovskii SL, Fedorov MM (1984) Mnogoknal'naja avtomatizirovannaja sistema obrabotki signalov no baze EVM. Geod Geophys Veröff R II(26):17–20
Haenel HD, Grunhage L (1999) A closed analytical solution based on height-dependent profiles of wind speed and eddy viscosity. Boundary Layer Meteorol 93: 395–409
Højstrup J (1993) A statistical data screening procedure. Meas Sci Technol 48:472–492
Horst TW (1997) A simple formula for attenuation of eddy fluxes measured with first order response scalar sensors. Boundary Layer Meteorol 82:219–233
Horst TW, Weil JC (1992): Footprint estimation for scalar flux measurements in the atmospheric surface layer. Boundary Layer Meteorol 59:279–296
Horst TW, Weil JC (1994) How far is far enough? The fetch requirements for micrometeorological measurement of surface fluxes. J Atmos Oceanic Technol 11:1018–1025
Kaimal JC, Finnigan JJ (1994) Atmospheric boundary layer flows: their structure and measurement. Oxford Univ Press, Oxford, 289 pp
Kaimal JC, Gaynor JE (1991) Another look at sonic thermometry. Boundary Layer Meteorol 56:401–410
Kaimal JC, Wyngaard JC, Izumi Y, Cote OR (1972) Spectral characteristics of surface-layer turbulence. Q J R Meteorol Soc 98:563–589
Kristensen L (1998) Time series analysis. Dealing with imperfect data. Risø National Laboratory, Riso-I-1228(EN), 31 pp
Leclerc MY, Thurtell GW (1990) Footprint prediction of scalar fluxes using a Markovian analysis. Boundary Layer Meteorol 52:247–258
Lee X (1998) On micrometeorological observations of surface-air exchange over tall vegetation. Agric For Meteorol 91:39–49
Leuning R, Judd MJ (1996) The relative merits of open- and closed-path analysers for measurements of eddy fluxes. Global Change Biol 2:241–254
Leuning R, King KM (1992) Comparison of eddy-covariance measurements of CO_2 fluxes by open-and-closed-path CO_2 analysers. Boundary Layer Meteorol 59:297–311
Leuning R, Moncrieff J (1990) Eddy-covariance CO_2 measurements using open- and closed-path CO_2 analysers: corrections for analyser water vapor sensitivity and damping of fluctuations in air sampling tubes. Boundary Layer Meteorol 53:63–76
LI-COR (1991) LI-6262 CO_2/H_2O analyser instruction manual. LI-COR, Lincoln, NE., 91 pp
Liu H, Peters G, Foken T (2001) New equations for sonic temperature variance and buoyancy heat flux with an omnidirectional sonic anemometer. Boundary Layer Meteorol 100(3):459–468
Mahrt L (1991) Eddy asymmetry in the sheared heated boundary layer. J Atmos Sci 4:153–157
Massman WJ (2001) Reply to comment by Rannik on 'A simple method for estimating frequency response corrections for eddy covariance systems'. Agric For Meteorol 107:247–251
McMillen RT (1986) A BASIC program for eddy correlation in non simple terrain. NOAA Technical Memorandum, ERL ART-147, NOAA, Silver Spring, MD
McMillen RT (1988) An eddy correlation technique with extended applicability to non-simple terrain. Boundary Layer Meteorol 43:231–245
Moncrieff JB, Mahli Y, Leuning R (1996) The propagation of errors in long-term measurements of land-atmosphere fluxes of carbon and water. Global Change Biol 2:231–240
Moncrieff JB, Massheder JM, de Bruin H, Elbers J, Friborg T, Heusinkveld B, Kabat P, Scott S, Soegaard H, Verhoef A (1997) A system to measure surface fluxes of momentum, sensible heat, water vapour and carbon dioxide. J Hydrol 188/189:589–611
Moore CJ (1986) Frequency response corrections for eddy correlation systems. Boundary Layer Meteorol 37:17–35

Obukhov AM (1960) O strukture temperaturnogo polja i polja skorostej v uslovijach konvekcii. Izv AN SSSR Ser Geofiz:1392–1396

Paw UKT, Baldocchi DD, Meyers TP, Wilson KB (2000) Correction of eddy covariance measurements incorporating both advective effects and density fluxes. Boundary Layer Meteorol 97:487–511

Rannik Ü (2001) A comment on the paper by W-J. Massman, 'A simple method for estimating frequency response corrections for eddy covariance systems'. Agric For Meteorol 107:241–245

Rannik Ü, Vesala T (1999) Autoregressive filtering versus linear detrending in estimation of fluxes by the eddy covariance method. Boundary Layer Meteorol 91:259–280

Rannik Ü, Vesala T, Keskinen R (1997) On the damping of temperature fluctuations in a circular tube relevant to the eddy covariance measurement technique. J Geophys Res 102:12789–12794

Rannik Ü, Aubinet M, Kurbanmuradov O, Sabelfeld T, Markkanen T, Vesala T (2000) Footprint analysis for measurements over a heterogeneous forest. Boundary Layer Meteorol 97:137–166

Schmid HP (1994) Source areas for scalars and scalar fluxes. Boundary Layer Meteorol 67:293–318

Schmid HP (1997) Experimental design for flux measurements: matching scales of observations and fluxes. Agric For Meteorol 87:179–200

Schmid HP, Oke TR (1990) A model to estimate the source area contributing to turbulent exchange in the surface layer over patchy terrain. Q J R Meteorol Soc 116:965–988

Schotanus P, Nieuwstadt FTM, de Bruin HAR (1983) Temperature measurement with a sonic anemometer and its application to heat and moisture flux. Boundary Layer Meteorol 26:81–93

Schuepp PH, Leclerc MY, MacPherson JI, Desjardins RL (1990) Footprint prediction of scalar fluxes from analytical solutions of the diffusion equation. Boundary Layer Meteorol 50:355–373

Shuttleworth WJ (1988) Corrections for the effect of background concentration change and sensor drift in real-time eddy correlation systems. Boundary Layer Meteorol 42:167–180

Smith EA, Hodges GB, Bacrania M, Cooper HJ, Owens MA, Chappell R, Kincannon W (1997) BOREAS net radiometer engineering study, final report. NASA, Grant NAG5-2447

Stull RB (1988) An introduction to boundary layer meteorol. Kluwer, Dordrecht, 666 pp

Vickers D, Mahrt L (1997) Quality control and flux sampling problems for tower and aircraft data. J Atmos Oceanic Technol 14:512–526

Webb EK, Pearman GI, Leuning R (1980) Correction of flux measurements for density effects due to heat and water vapour transfer. Q J R Meteorol Soc 106:85–100

Wichura B, Foken T (1995) Anwendung integraler Turbulenzcharakteristiken zur Bestimmung von Beimengungen in der Bodenschicht der Atmosphäre. DWD, Abteilung Forschung, Arbeitsergebnisse no 29, 52 pp

Wilson KB, Goldstein A, Falge E, Aubinet M, Baldocchi DD, Bernhofer C, Ceulemans R, Dolman DH, Field C, Grelle A, Law B, Loustau D, Meyers T, Moncrieff J, Monson R, Oechel W, Tenhunen J, Valentini R, Verma S. Energy balance closure at FLUXNET sites (submitted)

Wyngaard JC, Coté OR, Izumi Y (1971) Local free convection, similarity and the budgets of shear stress and heat flux. J Atmos Sci 28:1171–1182

3 Measurement of Soil Respiration

H. Lankreijer, I.A. Janssens, N. Buchmann, B. Longdoz,
D. Epron, S. Dore

3.1 Introduction

Within terrestrial ecosystems, the soil CO_2 efflux is one of the largest carbon flux components. The global efflux of carbon from the soil is estimated between 50 and 75 Gt C year^{-1} and makes up 20–40 % of the total annual input of carbon dioxide into the atmosphere (Houghton and Woodwell 1989; Raich and Schlesinger 1992; Schimel 1995). The magnitude of the soil flux is similar to that of the net primary productivity (Houghton and Woodwell 1989). It has been suggested that as global temperature rises, enhanced decomposition of the large soil carbon stock (1580×10^{15} g; Schimel 1995), especially in the high northern latitudes, might increase the input of carbon into the atmosphere (Gordon et al. 1987; Kirschbaum 1995; Trumbore et al. 1996; Zimov et al. 1996). However, the effect of a temperature increase on the decomposition rate is still unsolved and a point of discussion. Others suggest that decomposition rates in forest soils are not controlled by temperature (Liski et al. 1999; Giardina and Ryan 2000). Besides the potential temperature-induced feedback, changes in land use and forest management, which do affect the storage of carbon in the soil, were important points of discussion throughout the negotiations of the Kyoto protocol. Therefore, understanding the processes underlying the exchange of carbon from and into the soil is needed to make "management of the net carbon budget" possible (IGBP Terrestrial Carbon Working Group 1998).

The efflux of CO_2 from the soil originates from different sources. Decomposition of organic matter (heterotrophic respiration) and respiration by living roots (autotrophic respiration) are the two main sources, but chemical oxidation and carbonate dissolution may also contribute to the total flux (Burton and Beauchamp 1994). The contribution of the diverse sources to the total flux is difficult to obtain. Reported estimates of the contribution of root respiration to the total soil CO_2 efflux in forests range from 10 to 90 % (Tate et al. 1993; Thierron and Laudelout 1996; Hanson et al. 2000) and an average of 45 % is given by Landsberg and Gower (1997). However, part of the observed

Ecological Studies, Vol. 163
R. Valentini (Ed.) Fluxes of Carbon,
Water and Energy of European Forests

variability is related to the use of different methodologies (see also Chap. 12, this Vol.).

The efflux of CO_2 from the soil is very heterogeneous both in time and space. The high (spatial) variability introduces an uncertainty in the estimation of a mean or a total annual value and is caused by the heterogeneity in soil structure, temperature, soil moisture, bacterial, fungal, and root density distributions, and soil organic matter content. Also, the variability in the transport processes of CO_2 from deeper layers to the surface (soil diffusivity), and the turbulence and pressure patterns above the soil contribute to the heterogeneity in the soil CO_2 efflux.

Measurements of the rate of CO_2 efflux from the soil performed with an hourly or daily time step show high correlation with soil temperature and/or soil moisture content (Janssens et al., Chap. 12). Soil respiration is limited under low and very high soil water contents. Low soil water content may lead to lower quantities of dissolved organic carbon (DOC), which is an important substrate for heterotrophic soil respiration (Billings et al. 1998). Under water-saturated conditions respiration depends strongly on the transport of dissolved gases and can be limited by poor aeration (Freijer and Leffelaar 1996). Heterotrophic respiration is mainly determined by temperature and water content, but also by substrate quality expressed by the concentration of lignin and nitrogen (Ågren and Bosatta 1996b; Ryan et al. 1997). Boone et al. (1998) found that root respiration has even a higher sensitivity to temperature than heterotrophic respiration. They suggested that when plants increase their allocation to the roots under elevated atmospheric CO_2 concentration, elevated temperatures lead to lower sequestration of carbon in the soil due to the higher root respiration. This means that autotrophic respiration is also influenced by allocation of carbon assimilates to the roots and the fast turnover of these assimilates. The exact influence of mycorrhizae on soil respiration is not yet known and may be related to the activity of the trees. Taking into account the significant export of carbon assimilates [up to 25% of net primary production, (NPP)] to the prolific mycorrhizae (100–800 km of living hyphae may be found per gram of soil), the contribution of the mycorrhizae to the soil carbon efflux cannot be neglected (R. Finlay, pers. comm.). Root respiration might be sensitive to the soil CO_2 concentrations, and high CO_2 concentrations in the soil atmosphere have been described as inhibiting root respiratory activity (Qi et al. 1994; Burton et al. 1997).

The soil CO_2 efflux on the short time scales of minutes to days depends not only on the production of CO_2 by roots and soil organisms, but also on the transport from the subsurface upwards (Fang and Moncrieff 1999a). In the unsaturated soil layers the transport can occur in both the liquid and the gas phase. Diffusion – driven by concentration gradients – is considered to be the primary mechanism, but transport by convection and dispersion may also occur, especially in water-saturated soils (Simùnek and Suarez 1993a; Freijer and Leffelaar 1996). Precipitation, pressure differences, and turbulence above the soil surface can influence the efflux (Baldocchi and Meyers 1991; Hanson et al. 1993). It is expected that the latter effect is more profound in soil with a thick litter layer and less so in less porous, bare soils. Carbon dioxide stored in the (porous) litter layer exchange faster under conditions of turbulence.

As soil respiration is mainly determined by temperature, its seasonal variability usually tracks the temperature trend over the whole year (Boone et al. 1998). Soil water content might change this picture by limiting soil respiration under dry circumstances, while the contribution of root respiration may also differ strongly throughout the year, depending on growth rates and allocation patterns. In winter, soil respiration rates are usually expected to be low due to low temperatures. However, recent studies report a consistent CO_2 flux from forest tundra of 89 g C m^{-2} s^{-1} (Zimov et al. 1996) and 2–69 g C m^{-2} s^{-1} from tussock tundra during winter (Oechel et al. 1997; Fahnestock et al. 1998). The sources and the control of soil respiration during winter in arctic ecosystems are not well understood (Grogan et al. 2001). The occurrence of soil frost and/or a snow layer may lead to release of carbon in pulses when temperature rises again above zero. Flush of carbon from the decomposition of killed microbes, stimulation of the microbes by higher temperatures, release of CO_2 trapped in ice, or accumulation of CO_2 under the snow are considered causing those often observed flushes of carbon dioxide (Billings et al. 1998).

3.1.1 Measuring Soil Respiration

Considering the complexity of processes behind the CO_2 efflux from the soil, its heterogeneity in both space and time, and the interactions with the forest canopy above the soil, an estimate of the total CO_2 efflux from the soil and its components is not easy to obtain.

Many commercially available or self-made systems are used to measure soil respiration rates directly at the soil level (Norman et al. 1997; Janssens et al. 2000), and several different systems were applied at the EUROFLUX sites. The most common technique is to place a chamber on the soil surface and measure the change in CO_2 concentration in it. An advantage of the chamber system is the relative easy application and straightforward approach. Soil CO_2 efflux can also be estimated from measurements of the CO_2 concentration profile in or above the soil (e.g., Zimov et al. 1996). The advantage of profile measurements in the soil is the possibility of thus estimating the source depth of the flux, but the disadvantage is the difficulty in estimating soil and air diffusivity. The aboveground profile measurements are easier to perform, but when the difference in CO_2 concentration along the vertical axis is small, errors are large.

The eddy covariance method applied directly above the forest soil has several advantages over chamber-based methods and is probably the most suitable method: (1) the soil surface and soil microclimate are not disturbed, (2) measurements are performed under "natural" turbulent conditions, and (3) a larger surface area is covered. The technique requires sufficient turbulence below the canopy (Baldocchi et al. 1997) and no other sources and sinks between the soil surface and the sensor. Above-canopy eddy flux measurements also include, besides soil respiration, the respiration and photosynthesis of the vegetation. Total ecosystem respiration from both the soil and the vegetation can be derived from eddy flux measurements above the canopy from night-time flux extrapolation or by analysis of the daytime measurements (see Chap. 8, Falge et al.). However, distinction between respiration from the soil and from the vegetation above the ground is not possible with these methods, without using empirical estimates, and when both storage during stable conditions and advection of the carbon flux exist on the site, correction of the night-time fluxes is needed.

3.1.2 Modeling

Estimates of soil carbon fluxes by simulation models have been used in numerous studies. Many simulation studies focus, however, on the decomposition of organic matter in the soil (Ågren and Bosatta 1987, 1996a; Jenkinson 1990, 1991; Liski 1997) and describe the change in the storage of carbon in the soil. In spite of its importance, the simulation of the soil carbon efflux by process-based models including both heterotrophic and autotrophic respiration has been rather limited (Simùnek and Suarez 1993a; Freijer and Leffelaar 1996; Fang and Moncrieff 1999a). In general, predictive models have used regression functions fitting the CO_2 flux to environmental parameters (Hanson et al. 1993; Lloyd and Taylor 1994; Peterjohn et al. 1994; Lavigne et al. 1997). Regression of the CO_2 efflux by soil temperature and humidity typically results in r^2 values above 0.7, but still does not explain the total variance in the efflux (see Chap. 12, this Vol.). This shortcoming can be partly attributed to the lack of a detailed description of the production and transport processes as well as to the use of inaccurate techniques for measuring soil CO_2 efflux. Factors such as root/mycorrhizal activity, atmospheric turbulence, substrate "quality" (Ågren and Bosatta 1996b), soil structure, and diffusivity might be important, but are more difficult to assess. Seasonal variations can be described and simulated quite well, whereas the reasons for spatial variations are still poorly understood.

Using only average air temperature may be sufficient for simulating long-term decomposition of soil organic matter, but is less suited for analyzing short-term processes. The simulation of long-term carbon efflux (on monthly or longer time scales) based on the decomposition rates has been performed by, e.g., Hyvönen et al. (1996) and Liski (1997). Models based on the concept of humus quality show promising results (Bosatta and Ågren 1999; Joffre et al. 2001).

3.2 Soil CO_2 Efflux Measurements in the EUROFLUX Project

To obtain and analyze the total gain and losses of carbon from forest ecosystems and its compartments, soil respiration was measured at all EUROFLUX sites. Results of the soil respiration measurements within EUROFLUX are presented and discussed in Chapter 12.

The applied techniques for estimation of the soil respiration differ between sites. If the estimation of the soil respiration from eddy flux measurements above the canopy is not considered here, there were 13 different systems for measuring soil respiration (Table 3.1). Most of the systems used a chamber placed at the soil surface. All systems measured the soil efflux without making specific distinction between decomposition and biomass respiration. Several systems included the photosynthesis and respiration of the forest floor vegetation. Roughly, the measurements could be divided into either "continuous" or "periodic" with a certain time interval. The continuous systems measured with a time resolution of 10–30 min, and were used at ten sites. At four of those sites, measurements were performed during the entire year with two to five chambers, while at the other sites the continuous measurements were performed during campaigns. The eddy covariance and profile techniques were used during campaigns at five sites. At 14 of the 18 EUROFLUX sites, soil respiration was measured at intervals with a mobile chamber system. Data collected with the periodic systems represent point measurements with a time resolution varying between 8 and 45 days. At some sites the data are limited to just a few days or nights, but at several sites multiple years of data are available (Table 3.1).

3.2.1 Chamber Systems

Depending on the presence or absence of air circulation through chamber and analyzer, chamber techniques have been categorized as either *static* or *dynamic* (Witkamp and Frank 1969). Static chamber techniques are based either on enrichment or absorption of CO_2 in the headspace. The alkali solution method (Lundegårdh 1927) is probably the oldest method, while the soda lime method (Monteith et al. 1964; Howard 1966; Edwards 1982; Grogan 1998) is probably the most frequently used technique because it is inexpensive, easy to use, and particularly suitable where spatial variation is large (Kleber and Stahr 1995; Keith et al. 1997; Janssens and Ceulemans 1998). However, static techniques tend to be less accurate than dynamic systems due to effects on the diffusion process (Nay et al. 1994; Janssens et al. 2000), and are therefore often regarded as inferior to dynamic chamber systems (Norman et al. 1992). Measurements can be improved if they are compared to other measurements (Janssens and Ceulemans 1998).

Table 3.1. Soil respiration measurements. See text for the description of the systems. The frequency gives the interval between the measurements, while the number represents the number of locations measured. Under remarks the number of days with data is given, or how often a reading was taken during a day

Site	System	Frequency	Number	Period	Remarks	References
Periodic						
IT1	PP	8–45 days	15–30	96/5–98/11	34 days	Matteucci et al. (2000)
IT2	PP	15–20 days	30	96–98/6		Dore (1999)
FR1	L2	14–28 days	72	96/6–98/10	28 days	Epron et al. (1999a); Le Dantec et al. (1999)
SW1	L2	1 × year	36	97–98	Summer	
SW2	L2	14 days	24	97/5–99/10		Widén and Majdi (2000)
GE1	L4	1 × month	20	98/3–98/10	1–2 x day	Buchman (2000)
GE2	OG	1–2 × month	3	98/6–98/10		
NE1	PP	2 nights	20	97		
BE1	LH	14 days	15	97/8–98/8	40 days	Longdoz et al. (2000)
BE2	PP			96/4–98/2	Combined with SL	Janssens et al. (2000)
BE2	SL	3–4 weeks	47	96/4–98/2		Janssens and Ceulemans (1998)
FI1	SC, SP	1–2 weeks	3	97–99		Ilvesniemi and Pumpanen (1997); Pumpanen et al. (2001)
FI1	SC	3 ×	10	97–99	Summer	
EX1	PP	1 × month	10	98/3–98/12	4 x day	
EX3	LH	1 × month	4	99/5–99/10		
IS1	L2	4–5 ×	44; 48	96; 97	Summer	
Continuous						
FR2	EC			2 months 1997 16 days 1998		
DK1	DC	12 min	5 (10)	96/6; 96/9; 97/5	1–6 weeks	
DK1	DC	12 min	5 (10)	98/4–99	Year-round	
SW1	OS	10 min	1 (2)	97–98	3–4 days	
SW2	OS	10 min	3 (20)	95–01	Year-round	Iritz et al. (1997); Widén and Lindroth (2002)
FI1	OC	30 min	2	97/10–99/5	Year-round	
EX2	P, OC	15 min	2	96/6–99/7	Year-round	
NE1	P	30 min		96/6–present	Year-round	
BE2	EC	30 min		98/7	11 days; 6 nights	
GE1	EC	30 min				

Dynamic chamber systems typically use an infrared gas analyzer (IRGA). Two approaches can be distinguished: *closed* and *open* dynamic systems. In *closed* chamber IRGA systems, air circulates in a loop between the chamber and an external IRGA, and the change in CO_2 concentration over time is measured (Parkinson 1981; Norman et al. 1992; Goulden and Crill 1997; Rochette et al. 1997). In *open* systems, air does not circulate in a loop but is vented to the atmosphere. *Open* chamber systems have a constant airflow through the chamber, and the difference in CO_2 concentrations of the ambient and internal air at the inlet and the outlet are continuously monitored (Witkamp and Frank 1969; Edwards and Sollins 1973; Kanemasu et al. 1974; Schwartzkopf 1978; Denmead 1979; Fang and Moncrieff 1996; Iritz et al. 1997; Rayment and Jarvis 1997).

The classic closed-static, soda lime technique (SL) was applied at the Belgian site Brasschaat (BE2). Another closed-static technique, which made use of manual syringe sampling from a closed soil chamber (SC), was used at the Finish site Hyytiäla (FI1). According to Janssens et al. (2000), the corrected soda lime measurements agreed well with measurements acquired with the portable closed-dynamic system of PP systems (Hitchin, UK) (PP*). This last system consists of a CIRAS-1 or EGM-1 infrared gas analyzer and a cylindrical soil chamber (the SRC-1). The PP-system soil respiration set (Parkinson 1981) was used at five sites. The comparable closed-dynamic system by Li-Cor, the Li-Cor 6200 and Li-Cor 6400 gas analyzers (Li-Cor; Li-Cor 1993) combined with the Li 6000-09 or Li 6400-09 soil chambers (L2, L4), was also used at five sites, while the sites at Gembloux (BE1) and Bílý Kríž (EX3) used the same type of IRGA, but with home made chambers (LH) based on the same technique, as described by Norman et al. (1992). Use of the portable closed-dynamic systems at a total of 11 sites makes it the most commonly used technique within EUROFLUX for direct measurement of the soil CO_2 efflux.

The portable, closed-dynamic systems usually resulted in periodic measurements with a long time interval, but with a relative high number of spa-

* The abbreviations are used in Table 3.2

Table 3.2. Range of measured fluxes in μmol m^{-2} s^{-1}. Measurements were performed over 3 days and averages are shown for each day, if available

System	Range	Average day 1	Average day 2	Average day 3
LH BE1	2–6	4.8	3.0	
L2 SE2	1.5–6	2.5	3.0	
PP BE2	3–11	5.5	4.0	
GE-open	1–3			1.6
UK-open	1.6–3.4	2.6	1.9	
SE-Lab	0.5–1.5 and 2–3			
OS SE2	1–50			

tially distributed measurements. Based on the same principle as the closed-dynamic system, automatic soil chambers (OC) were used at the Sollingen site in Germany (EX2) and Hyytiäla in Finland (FI). Rain and temperature fluctuations reached the soil normally since the chambers were open between the measurements. Readings were almost continuous, however, with a low spatial distribution.

The open-dynamic chambers were all non-commercial home-built systems, mainly developed to obtain continuous readings of the soil carbon efflux. At the Danish site Lille Boegeskov the system consisted of five simultaneously operating chambers (DC). Each chamber covered an area of 28 × 28 cm. The chambers were closed for 25 min of each hour, but were left open during rainfall. The Swedish open system (OS) consisted of a tunnel-shaped chamber, covering an area of 200 × 30 cm, and having a continuous flow of air through the system by a fan. The CO_2 flux was obtained by measuring the difference between the inlet and outlet concentration of the air with a gas analyzer (IRGA LI-6262, Li-Cor). At the German site Tharandt, the system (OG) consisted of three chambers of 30 × 35 cm where the change in CO_2 concentration was determined non-differentially and compared to a reference reading of ambient air just before the measurements.

3.2.2 Meteorological and Profile Gradient Systems

3.2.2.1 Eddy Correlation Below the Canopy (EC)

At the French site (FR2) a 3-D Gill anemometer type R2, and an "open path" IRGA was used (Advanet E009, the "Otahki" devices), both 6 m above ground surface. Corrections according to Webb et al. (1980) for open path systems were made on turbulent fluxes. At the Belgian site in Brasschaat (BE2) the equipment consisted of a sonic anemometer (USAT-3, Metek, Germany) and a Li-6262 (Janssens et al. 2000). The anemometer was mounted at a height of 1.65 m above the forest floor.

3.2.2.2 Flux Profile Measurements (P)

At the Dutch site Loobos the net carbon exchange was estimated from below canopy CO_2 concentration profiles measured with a gas analyzer of PP systems. The profile had five levels and measurements were averaged over 5 min at each level, resulting in a profile for every 30 min. At the German site in Solling (EX2), both the soil chambers and the profile measurements were connected to the same gas analyzer (Li-Cor 6251). Each measurement cycle of 15 min consisted therefore of measured fluxes from two chamber-plots and a profile from several levels within the forest canopy. At the French site Le Bray

(FR2) profile measurements were made at ten levels between 0.1 and 25 m, with a particular emphasis on the lower levels to catch the night-time storage (levels 0.1, 0.2, 0.6, 0.9, and 2.0 m).

3.2.2.3 CO_2 Profile in the Soil (SP)

At the Finnish site Hyytiäla the CO_2 profile in the soil was measured by manual sampling from tubes installed at different depths in the soil (Ilvesniemi and Pumpanen 1997), while in the winter of 1997/1998 the CO_2 concentration profile of the snow was measured once a month.

3.2.2.4 Additional Soil Measurements

Soil temperature, soil water content, and soil texture/density were measured at all sites, generally at the same time that soil respiration rates were measured. Further ancillary data included measurements of C and N content of the upper horizons, litter layer thickness, root-biomass distribution, litter decomposition rates, soil temperature profiles, air temperature and humidity just above the soil surface, radiation at the soil level, and atmospheric pressure. These measurements differed in frequency and methodology among the sites and a description of the different techniques is not included here.

3.3 Comparison of Systems

Results from in situ comparisons of systems are still limited (Norman et al. 1997). Within the EUROFLUX framework direct comparison of different systems was performed at several sites (Le Dantec et al. 1999; Janssens et al. 2000; Longdoz et al. 2000). Comparison of seven systems was performed during a special organized workshop in Uppsala (Sweden) in June 1997. The systems involved in this experiment were the PP Systems (PP), Li-6200 (L2 and LH), and the Swedish open chamber (OS). Those four systems, applied at the EUROFLUX sites, were also compared to three techniques that were not used at the EUROFLUX sites: two automatic open-dynamic chamber systems, abbreviated as UK-open (Fang and Moncrieff 1996) and GE-open (Kutsch 1996), and one based on CO_2 accumulation rates from soil samples incubated at constant temperature (Persson et al. 1989). To create quasi-controlled conditions a container (4 × 4 m) was filled with a 30-cm-thick layer of (bare) organic soil. However, the spatial variability of the flux of CO_2 was still large. During a measurement session prior to the site comparisons, the flux showed an average of 2.87 $\mu mol\ m^{-2}\ s^{-1}$ and an SD of 0.77 $\mu mol\ m^{-2}\ s^{-1}$ (CV=26.5 %).

Thus direct comparison of systems required either measurement at the same location, one after another, or a high number of measurements by each system spread over the entire container. Direct comparison of the portable chamber systems on exactly the same spot revealed the problem of soil disturbance when placing the chambers, either by such placement on the soil or by movement of the collar. This observation made it clear that field measurements with portable chambers, even if using prefixed collars, also have to be performed carefully to avoid pulses of high CO_2. Further comparison of systems was therefore restricted to the daily means obtained by each system. The ranges of measured effluxes from 2.5 days are given in Table 3.2.

In general, the open systems from the UK and Germany (UK-open and GE-open) gave lower fluxes than the closed chamber systems (PP, L2, and LH). Of the three closed chamber systems, the PP system systematically gave the highest average value, supporting results reported by Le Dantec et al. (1999) and Janssens et al. (2000). The ventilation fan inside this system might be the reason for the higher measured flux. Direct comparison of both systems, under field and laboratory conditions, showed that the internal wind speed in the chamber as well as the difference between inside and outside wind speed are important factors (Le Dantec et al. 1999). Another reason behind the overestimation of the flux was the use of the EGM-1 analyzer. At that time, this analyzer did not separate IR absorption by CO_2 and water vapor, and because of the rather wet soil, both the CO_2 and water efflux was measured. As could be expected from a static technique, fluxes estimated with the accumulation technique (SE-Lab) resulted in the lowest values. The Swedish chamber system (OS) showed a high range with values up to five times higher than the other systems. This high efflux of CO_2 may be related to the fact that this system uses a transparent chamber; solar radiation could heat up the soil surface. When the chamber was covered by dark plastic, the measured flux rates decreased considerably. Testing the portable Li-6200 (L2) with transparent collars of 10 cm height also showed higher flux rates. Selection of chamber design (transparent or opaque) should be carefully considered and is an important factor when soil respiration measurements of different sites and systems are compared with each other.

Differences between methodologies for measuring soil CO_2 were discussed at the LESC workshop 6–8 April 2000 in Edinburgh, Scotland (Rayment 2000). The workshop resulted in a list with guidelines and recommendations concerning measurements with the different (chamber) systems. Although none of the systems were rejected, and each system has its advantages, all methodologies have to be cross-calibrated and carefully applied. If chambers are used, the open dynamic system is assumed to be the most reliable system. Chambers should be either removed between the measurements or opened between the readings to limit the alteration of the soil.

3.4 Discussion

Based on experience with the large number of different systems applied within EUROFLUX, a number of general problems with the interpretation of the soil respiration estimates can be identified. The main problem is that most systems are not "cross-calibrated" (Rayment and Jarvis 1997). Such calibrations have been performed up to now on a limited scale and, to date, there is no standard method for measurement of soil respiration (Nay et al. 1994; Conen and Smith 2000; Widén and Lindroth 2002). Janssens et al. (2000) found that measurements performed with the PP system and the Li-Cor chamber systems showed a high correlation, indicating that calibration against a standard system is possible.

Each technique has its specific time and space resolution. Integration of measured fluxes over large areas is hampered by the high heterogeneity in the soil, resulting in highly varying CO_2 efflux rates. For chamber techniques this means that a high number of replicates at different spots is required. In order to analyze the processes underlying the flux and to separate the total flux into different components, multiple techniques are needed.

A disadvantage of all systems, except for the meteorological techniques, is that they enclose a part of the soil surface and exclude the effect of turbulence and pressure fluctuations on the soil CO_2 efflux. The effect of turbulence is probably the most underestimated factor, since strong gusts of wind as well as long undisturbed stable conditions exist inside the canopy. Thus, how the chamber influences the efflux of carbon dioxide from the soil and its internal flow makes the interpretation of the measurements complicated. On the other hand, fans inside chambers generating the necessary air-mixing may induce an unnatural turbulence, which might result in an increased efflux of CO_2 from the soil that can be sustained by enhanced lateral diffusion (Le Dantec et al. 1999; Janssens et al. 2000). Only for the closed-dynamic chamber used at the EX2 site in Sollingen, was a correction term mentioned as rectifying the possible error.

Open chamber systems are extremely sensitive to pressure differences between the chamber and the atmosphere (Kanemasu et al. 1974; Fang and Moncrieff 1996; Rayment and Jarvis 1997; Lund et al. 1999). Several approaches have been suggested to minimize these pressure differences, such as simultaneously blowing and drawing air through the chamber (Fang and Moncrieff 1996), and the use of very large air inlet apertures (Iritz et al. 1997; Rayment and Jarvis 1997), but elimination of pressure gradients is still a problem with today's systems. In closed systems, pressure equilibration between the chamber and the atmosphere can be achieved with a properly designed venting tube (Hutchinson and Mosier 1981; Norman et al. 1992), through which leakage can be restricted to a minimum.

All chamber techniques have the potential problem of disturbance of the respiratory processes by the technique itself, i.e., the chamber (Nay et al. 1994;

Lund et al. 1999; Conen and Smith 2000). When forest floor vegetation is enclosed in the chamber, plant respiration is included in the measurements, which makes distinction of the fluxes difficult. Further, when the chamber is transparent for light, the measured flux can include the uptake of CO_2 by photosynthesis.

The eddy covariance technique applied below the canopy is probably a very suitable method for measuring the CO_2 from the soil as the natural distribution of the vertical pressure gradient, the horizontal air velocity, and the vertical CO_2 concentration gradient are not disturbed (Longdoz et al. 2000). However, the conditions for using this technique are not always suitable – for example, in young and low forest stands when a sink of carbon exists between the soil and the sensor. Eddy flux measurements cover a relative large area under the canopy. Comparison of below canopy eddy flux measurements with chamber measurements has to take these conditions in to account, but have shown good agreement (Law et al. 1999; Matteucci et al. 2000; Janssens et al. 2001).

Major advantages of continuous chamber systems are availability of series of data over a long period of time and measurement under relatively undisturbed conditions. However, usually the number of monitored locations with continuous measuring systems is low, thus limiting their potential for scaling in space. In addition, with time, the conditions within some of the continuous systems might differ strongly from the surroundings, again limiting their use for extrapolation.

Advantages of the mobile chamber systems are (1) no permanent power requirements, and (2) a potential for covering large areas and accounting for great spatial variations. Chamber measurements are more useful for distinguishing the spatial distribution and contribution of different sources at the soil surface. However, portable chamber measurements have to be performed carefully, considering the potential disturbances when no preinstalled collars are used; and when collars are used, it is often difficult to determine to what depth they can be inserted without disturbing roots.

Separation of heterotrophic and autotrophic soil respiration, and in some cases respiration of the aboveground biomass, is not possible with any of the systems described above. Within the EUROFLUX project, root respiration was estimated by comparing efflux measurements from root-free plots and control plots (Epron et al. 1999b; Janssens 1999). Other techniques for separating root from microbial respiration, which have been applied elsewhere, are using root cuvettes in the field (e.g. Gansert 1994; Ryan et al. 1996), excavating and directly measuring in a closed chamber in the field (Widén and Majdi 2000), excising roots in the laboratory (Burton et al. 1998), trenching (Bowden et al. 1993; Fisher and Gosz 1986; Boone et al. 1998; Hart and Sollins 1998), girdling of trees (Högberg et al. 2001), performing ^{14}C, ^{13}C, or ^{18}O studies (Horwath et al. 1994; Swinnen et al. 1994; Högberg and Ekblad 1996; Lin et al. 1999; Högberg et al. 2001), inhibiting one respiratory component with specific inhibitors or herbicides (Helal and Sauerbeck 1991; Nakane et al. 1996), applying a controlled accumulation technique in the laboratory (Persson et al. 1989), and enhancing one component over the other (Bowden et al. 1993). Lin et al. (1999) used stable isotopes, but their system was strongly influenced by the tank CO_2 with a very different carbon isotopic composition compared to ambient CO_2. Isotopes have to be measured frequently if partitioning between microbial and root respiration is the objective, since activities change so fast seasonally.

3.5 Conclusions

Despite its long history of measurements, the process of soil respiration remains difficult to assess and to interpret. As there is not yet one preferable system applicable and suitable for all environmental conditions and ecosystems, a careful comparison with other techniques and a thorough analysis of potential effects of the applied technique on the flux itself is needed whenever soil respiration is measured. Based on the experiences within EUROFLUX, a system that causes no or only small changes in the environmental conditions (inside the chambers) has to be used for a correct assessment of the actual soil respiratory fluxes. Spatial and temporal variability has to be accounted for by an adequate sampling design.

In order to be able to explain the measured flux, determination of the soil temperature, soil water content, soil carbon/organic matter content and distribution, fine root biomass and distribution, soil texture, and litter-layer thickness and nutrient content need to be included in the measurement program.

References

Ågren GI, Bosatta E (1987) Theoretical analysis of long-term dynamics of carbon and nitrogen in soils. Ecology 68:1181–1189

Ågren GI, Bosatta E (1996a) Theoretical ecosystem ecology; understanding element cycles. Cambridge Univ Press, Cambridge, 234 pp

Ågren GI, Bosatta E (1996b) Quality: a bridge between theory and experiment in soil organic matter studies. Oikos 76:522–528

Aubinet M, Grelle A, Ibrom A, Rannik U, Moncrieff J, Foken T, Kowalski AS, Martin P, Berbigier P, Bernhofer C, Clement R, Elbers I, Granier A, Gruenwald T, Morgenstern K, Pilegaard K, Rebmann C, Snijders W, Valentini R, Vesala T (2000) Estimates of the annual net carbon and water exchange of European forests: the EUROFLUX methodology. Adv Ecol Res 30:113–175

Baldocchi DD, Meyers TP (1991) Trace gas exchange above the floor of a deciduous forest. 1. Evaporation and CO_2 efflux. J Geophys Res 96(D4):7271–7285

Baldocchi DD, Vogel CA, Hall B (1997) Seasonal variation of carbon dioxide exchange rate above and below a boreal jack pine forest. Agric For Meteorol 83:147–170
Billings SA, Richter DD, Yarie J (1998) Soil carbon dioxide fluxes and profile concentrations in two boreal forests. Can J For Res 28:1773–1783
Bonan GB (1992) Physiological controls of the carbon balance of boreal forest eceosystems. Can J For Res 23:1453–1471
Boone RD, Nadelhoffer KJ, Canary JD, Kaye JP (1998) Roots exert a strong influence on the temperature sensitivity of soil respiration. Nature 396:570–572
Bosatta E, Ågren GI (1999) Soil organic matter quality interpreted thermodynamically. Soil Biol Biochem 31:1889–1891
Bowden RD, Nadelhoffer KJ, Boone RD, Melillo JM, Garrison JB (1993) Contributions of aboveground litter, belowground litter, and root respiration to total soil respiration in a temperate mixed hardwood forest. Can J For Res 23:1402–1407
Buchmann N (2000) Biotic and abiotic factors modulating soil respiration rates in Picea abies stands. Soil Biol Biochem 32:1625–1635
Burton DL, Beauchamp EG (1994) Profile nitrous oxide and carbon dioxide concentrations in a soil subject to freezing. Soil Sci Soc Am J 58:115–122
Burton AJ, Zogg GP, Pregitzer KS, Zak DR (1997) Effect of measurement CO_2 concentration on sugar maple root respiration. Tree Physiol 17:421–427
Burton AJ, Pregitzer KS, Zogg GP, Zak DR (1998) Drought reduces root respiration in sugar maple forests. Ecol Appl 8:771–778
Conen F, Smith KA (2000) An explanation of linear increases in gas concentration under closed chamber used to measure gas exchange between soil and the atmosphere. Eur J Soil Sci 51:111–117
Denmead OT (1979) Chamber systems for measuring nitrous oxide emission from soils in the field. Soil Sci Soc Am J 43:89–95Dore S (1999) Functional characteristics of carbon balance and energy use in forest ecosystems: comparison between three study cases. PhD Thesis. Universita degli Studi di Padova (in Italian)
Edwards NT (1982) The use of soda-lime for measuring respiration rates in terrestrial ecosystems. Pedobiologia 23:231–330
Edwards NT, Sollins P (1973) Continuous measurement of carbon dioxide evolution from partitioned forest floor components. Ecology 54:406–412
Epron D, Farque L, Lucot E, Badot P-M (1999a) Soil CO_2 efflux in a beech forest: the dependence on soil temperature and soil water content. Ann For Sci 56:221–226
Epron D, Farque L, Lucot E, Badot P-M (1999b) Soil CO_2 efflux in a beech forest: the contribution of root respiration. Ann For Sci 56:289–295
Fahnestock JT, Jones MH, Brooks PD, Walker DA, Welker JM (1998) Winter and early spring CO_2 efflux from tundra communities of northern Alaska. J Geophys Res 103:29023–29027
Fang C, Moncrieff JB (1996) An improved dynamic chamber technique for measuring CO_2 efflux from the surface of soil. Funct Ecol 10(2):297–305
Fang C, Moncrieff JB (1999a) A model for soil CO_2 production and transport 1: model development. Agric For Meteorol 95:225–236
Fang C, Moncrieff JB (1999b) A model for soil CO_2 production and transport 2: application to a Florida *Pinus elliotte* plantation. Agric For Meteorol 95:237–256
Fisher FM, Gosz JR (1986) Effects of trenching on soil processes and properties in a New Mexico mixed-conifer forest. Biol Fertil Soils 2:35–42
Freijer JI, Leffelaar PA (1996) Adapted Fick's law applied to soil respiration. Water Resour Res 32(4):791–800
Gansert D (1994) Root respiration and its importance for the carbon balance of beech seedlings (*Fagus sylvatica* L.) in a montane beech forest. Plant Soil 167:109–119

Giardina CP, Ryan MG (2000) Evidence that decomposition rates of organic carbon in mineral soil do not vary with temperature. Nature 404:858–861
Gordon AM, Schlentner RE, Van Cleve K (1987) Seasonal patterns of soil respiration and CO_2 evolution following harvesting in the white spruce forests of interior Alaska. Can J For Res 17:304–310
Goulden ML, Crill PM (1997) Automated measurements of CO_2 exchange at the moss surface of a black spruce forest. Tree Physiol 17:537–542
Goulden ML, Munger JW, Fan S-M, Daube BC, Wofsy SC (1996) Exchange of carbon dioxide by deciduous forest: response to inter-annual climate variability. Science 271:1576–1578
Grogan P (1998) CO_2 flux measurements using soda lime: correction for water formed during CO_2 absorption. Ecology 79:1467–1468
Grogan P, Illeris L, Michelsen A, Johasson S (2001) Respiration of recently-fixed plant carbon dominates mid-winter ecosystem CO_2 production in sub-arctic heath tundra. Climatic Change 50:129–142
Hart SC, Sollins P (1998) Soil carbon and nitrogen pools and processes in an old-growth conifer forest 13 years after trenching. Can J For Res 28:1261–1265
Hanson PJ, Wullschleger SD, Bohlman SA, Todd DE (1993) Seasonal and topographic patterns of forest floor CO_2 efflux from an upland oak forest. Tree Physiol 13:1–15
Hanson PJ, Edwards NT, Garten CT, Andrews JA (2000) Separating root and soil microbial contributions to soil respiration: a review of methods and observations. Biogeochemistry 48:115–146
Helal HM, Sauerbeck D (1991) Short-term determination of the actual respiration rate of intact plant roots. In: McMichael BF, Persson H (eds) Plant roots and their environment. Elsevier, London, pp 88–92
Högberg P, Ekblad A (1996) Substrate-induced respiration measured in situ in a C3-plant ecosystem using additions of C4-sucrose. Soil Biol Biochem 28:1131–1138
Högberg P, Nordgren A, Buchmann N, Taylor AFS, Ekblad A, Högberg MN, Nyberg G, Ottosson-Löfvenius M, Read DH (2001) Large-scale forest girdling shows that current photosynthesis drives soil respiration. Nature 411:789–792
Horwath WR, Pregitzer KS, Paul EA (1994) ^{14}C allocation in tree-soil systems. Tree Physiol 14:1163–1176
Houghton RA, Woodwell GM (1989) Global climatic change. Sci Am 260:18–26
Howard PJA (1966) A method for the estimation of carbon dioxide evolved from the surface of soil in the field. Oikos 17:267–271
Hutchinson GL, Mosier AR (1981) Improved soil cover method for field measurements of nitrous oxide fluxes. Soil Sci Soc Am J 45:311–316
Hyvönen R, Ågren GI, Andrén O (1996) Modelling long-term carbon and nitrogen dynamics in an arable soil receiving organic matter. Ecol Appl 6(4):1345–1354
IGBP Terrestrial Carbon Working Group (1998) The terrestrial carbon cycle: implications for the Kyoto protocol. Science 280:1393–1394
Ilvesniemi H, Pumpanen J (1997) SMEARII station for measuring forest ecosystem-atmosphere relation. In: Haataja J, Vesala T (eds) University of Helsinki Department of Forest Ecology Publications 17. University of Helsinki, Helsinki, pp 30–37Iritz Z, Lindroth A, Gärdenäs A (1997) Open ventilated chamber system for measurements of H_2O and CO_2 fluxes from the soil surface. Soil Technol 10:169–184
Janssens IA (1999) Soil CO_2 efflux in a mixed forest ecosystem in the Antwerp Campine region. PhD Thesis, Universiteit Antwerpen (UIA)
Janssens IA, Ceulemans R (1998) Spatial variability in forest soil CO_2 efflux assessed with a calibrated soda lime technique. Ecol Lett 1:95–98
Janssens IA, Kowalski AS, Longdoz B, Ceulemans R (2000) Assessing forest soil CO_2 efflux: an in situ comparison of four techniques. Tree Physiol 20:23–32

Janssens I, Kowalski A, Ceulemans R (2001) Forest floor CO_2 fluxes estimated by eddy covariance and chamber-based model. Agric For Meteorol 106:61–69

Jenkinson DS (1990) The turnover of organic carbon and nitrogen in soil. Philos Trans R Soc Lond B 329:361–368

Jenkinson DS (1991) Model estimates of CO_2 emissions from soil in response to global warming. Nature 351:304–306

Joffre R, Ågren GI, Gillon D, Bosatta E (2001) Organic matter quality in ecological studies: theory meets experiment. Oikos 93:451–458

Kanemasu ET, Powers WL, Sij JW (1974) Field chamber measurements of CO_2 flux from soil surface. Soil Sci 118:233–237

Keith H, Jacobsen KL, Raison RJ (1997) Effects of soil phosphorus availability, temperature and moisture on soil respiration in *Eucalyptus pauciflora* forest. Plant Soil 190:127–141

Kirschbaum MUF (1995) The temperature dependence of soil organic matter decomposition, and the effect of global warming on soil organic C storage. Soil Biol Biochem 27(6):753–760

Kleber M, Stahr K (1995) Soil carbon turnover in subalpine systems and its dependence on climate. In: Zwerver S, van Rompaey RSAR, Kok MTJ, Berk MM (eds) Climate change research: evaluation and policy implications. Elsevier, Amsterdam, pp 561–566

Kutsch WL (1996) Untersuchungen zur Bodenatmung zweier Ackerstandorte im Bereich der Bornhöveder Seenkette. EcoSyst Suppl 16:1–125

Landsberg JJ, Gower ST (1997) Applications of physiological ecology to forest management. Academic Press, San Diego, 354 pp

Lavigne MB, Ryan MG, Anderson DE, Baldocchi DD, Crill PM, Fitzjarrald DR, Goulden ML, Gower ST, Massheder JM, McCaughey JH, Rayment M, Striegl RG (1997) Comparing nocturnal eddy covariance measurements to estimates of ecosystem respiration made by scaling chamber measurements at six coniferous boreal sites. J Geophys Res 102(D24):28,977–28,985

Law B, Baldocchi D and Anthoni P (1999) Below-canopy and soil CO_2 fluxes in a ponderosa pine forest. Agric For Meteorol, 94:171–188

Le Dantec V, Epron D, Dufrêne E (1999) Soil CO_2 efflux in a beech forest: comparison of two closed dynamic systems. Plant Soil 214:125–132

Li-Cor (1993) LI-6000-09 Soil respiration chamber, instruction manual. Publication no 9311-69. Li-Cor, Inc Lincoln, NE, USA

Lin G, Ehleringer JR, Rygiewicz PT, Johnson MG, Tingey DT (1999) Elevated CO_2 and temperature impacts on different components of soil CO_2 efflux in Douglas-fir terracosms. Global Change Biol 5:157–168

Lindroth A, Grelle A, Morén AS (1998) Long-term measurements of boreal forest carbon balance reveal large temperature sensitivity. Global Change Biol 4:443–450

Liski J (1997) Carbon storage of forest soils in Finland. University of Helsinki Department of Forest Ecology Publication 16. University of Helsinki, HelsinkiLiski J, Ilvesniemi H, Mäkelä A, Westman CJ (1999) CO_2 emissions from soil in response to climate warming are overestimated – the decomposition of old organic matter is tolerant to temperature. Ambio 28:171–174

Lloyd J, Taylor JA (1994) On the temperature dependence of soil respiration. Funct Ecol 8:315–323

Longdoz B, Yernaux M, Aubinet M (2000) Soil CO_2 efflux measurements in a mixed forest: impact of chamber disturbances, spatial variability and seasonal evolution. Global Change Biol 6:907–917

Lund CP, Riley WJ, Pierce LL, Field CB (1999) The effects of chamber pressurization on soil-surface CO_2 flux and the implications for NEE measurements under elevated CO_2. Global Change Biol 5:269–281

Lundegårdh H (1927) Carbon dioxide evolution of soil and crop growth. Soil Sci 23:417–453
Matteucci G, Dore S, Stivanello S, Rebmann C, Buchmann N (2000) Soil respiration in beech and spruce forest in Europe: trends, controlling factors, annual budgets and implications for the ecosystem carbon balance. In: Schulze E-D (ed) Carbon and nitrogen cycling in European forest ecosystem. Ecological studies 142. Springer, Berlin Heidelberg New York, pp 217–236
Moncrieff JB, Mahli Y, Leuning R (1996) The propagation of errors in long-term measurements of land-atmosphere fluxes of carbon and water. Global Change Biol 2:231–240
Monteith JL, Szeicz G, Yabuki K (1964) Crop photosynthesis and the flux of carbon dioxide below the canopy. J Appl Ecol 1:321–337
Morén AS, Lindroth A (1998) Field measurements of water vapour and carbon dioxide fluxes – chamber system and climatic monitoring by an automatic station. Report 4. Department for Production Ecology, Faculty of Forestry, SLU, Uppsala, Sweden
Nay SM, Mattson KG, Bormann BT (1994) Biases of chamber methods for measuring soil CO_2 efflux demonstrated with a laboratory apparatus. Ecology 75(8):2460–2463
Nakane K, Kohno T, Horikoshi T (1996) Root respiration rate before and just after clear-felling in a mauture, deciduous, broad-leaved forest. Ecol Res 11:111–119
Norman JM, Garcia R, Verma SB (1992) Soil surface CO_2 fluxes and the carbon budget of a grassland. J Geophys Res 97:18845–18853
Norman JM, Kucharik CJ, Gower ST, Baldocchi DD, Crill PM, Rayment M, Savage K, Striegl RG (1997) A comparison of six methods for measuring soil-surface carbon dioxide fluxes. J Geophys Res 102:28771–28777
Oechel WC, Vourlitis G, Hastings SJ (1997) Cold season CO_2 emission from arctic soils. Global Biogeochem Cycles 11:163–172
Parkinson KJ (1981) An improved method for measuring soil respiration in the field. J Appl Ecol 18:221–228
Persson T, Lundkvist H, Wirén A, Hyvönen R, Wessén B (1989) Effects of acidification and liming on carbon and nitrogen mineralization and soil organisms in mor humus. Water Air Soil Pollut 44:77–96
Peterjohn WT, Melillo JM, Steudler PA, Newkirk KM (1994) Responses of trace gas fluxes and N availability to experimentally elevated soil temperatures. Ecol Appl 4(3):617–625
Pumpanen J, Ilvesniemi H, Keronen P, Nissinen A, Pohja T, Vesala T, Hari P (2001) An open chamber system for measuring soil surface CO_2 efflux: analysis of error sources related to the chamber system. J Geophys Res Atmos 106(D8):7985–7992
Qi J, Marshall JD, Mattson KG (1994) High soil carbon dioxide concentrations inhibit root respiration of Douglas fir. New Phytol 128:435–442
Raich JW, Schlesinger WH (1992) The global carbon dioxide flux in soil respiration and its relationship to vegetation and climate. Tellus 44B:81–99
Rayment MB (2000) Investigating the role of soils in terrestrial carbon balance – harmonising methods for measuring soil CO_2 efflux. LESC Exploratory workshop, Edinburgh, 6–8 April. European Science Foundation. http://www.esf.org/generic/163/2073 aappitem5.1a.pdf
Rayment MB, Jarvis PG (1997) An improved open chamber system for measuring soil CO_2 effluxes in the field. J Geophys Res 102(D24):28,779–28,784
Rochette P, Ellert B, Gregorich EG, Desjardins RL, Pattey E, Lessard R, Johnson BG (1997) Description of a dynamic closed chamber for measuring soil respiration and its comparison with other techniques. Can J Soil Sci 77:195–203
Ryan MG, Hubbard RM, Pongracic S, Raison RJ, McMurtrie RE (1996) Foliage, fine-root, woody-tissue and stand respiration in *Pinus radiata* in relation to nitrogen status. Tree Physiol 16:333–343

Ryan MG, Lavigne MB, Gower ST (1997) Annual carbon cost of autotrophic respiration in boreal forest ecosystems in relation to species and climate. J Geophys Res 102 (D24):28.871–28.883

Schimel DS (1995) Terrestrial ecosystems and the carbon cycle. Global Change Biol 1:77–91

Simùnek J, Suarez DL (1993a) Modelling of carbon dioxide transport and production in soil. 1. Model development. Water Resour Res 29(2):487–497

Simùnek J, Suarez DL (1993b) Modelling of carbon dioxide transport and production in soil. 2. Parameter selection, sensitivity analysis, and comparison of model predictions to field data. Water Resour Res 29(2):499–513

Schwartzkopf SH (1978) An open chamber technique for the measurement of carbon dioxide evolution from soils. Ecology 59:1062–1068

Swinnen J, van Veen JA, Merckx R (1994) Rhizosphere carbon fluxes in field-grown spring wheat: model calculations based on 14C partitioning after pulse-labelling. Soil Biol Biochem 26:171–182

Tate KR, Ross DJ, O'Brien BJ, Kelliher FM (1993) Carbon storage and turnover, and respiratory activity, in the litter and soil of an old-growth southern beech (*Nothofagus*) forest. Soil Biol Biochem 25:1601–1612

Thierron V, Laudelout H (1996) Contribution of root respiration to total CO_2 efflux from the soil of a deciduous forest. Can J For Res 26:1142–1148

Trumbore SE, Chadwick OA, Amundson R (1996) Rapid exchange between soil carbon and atmospheric carbon dioxide driven by temperature change. Science 272(5260): 393–396

Webb EK, Pearman GI, Leuning R (1980) Correction of flux measurements for density effects due to heat and water vapour transfer. Q J R Meteorol 106:85–100

Widén B, Lindroth A (2002) A calibration system for soil CO_2 efflux chamber systems: description and application. Soil Sci Soc Am J (in press)

Widén B, Majdi H (2000) Soil CO2 efflux and root respiration at three sites in a mixed pine and spruce forest: seasonal and diurnal variation. Can J For Res 31:786–796

Witkamp M, Frank ML (1969) Evolution of CO_2 from litter, humus and subsoil of a pine stand. Pedobiologia 9:358–365

Zimov SA, Davidov SP, Voropaev V, Prosiannikov SF, Semiletov IP, Chapin MC, Chapin FS (1996) Siberian CO_2 efflux in winter as a CO_2 source and cause of seasonality in atmospheric CO_2. Climatic Change 33:111–120

4 Deciduous Forests: Carbon and Water Fluxes, Balances and Ecophysiological Determinants

A. Granier, M. Aubinet, D. Epron, E. Falge, J. Gudmundsson, N.O. Jensen, B. Köstner, G. Matteucci, K. Pilegaard, M. Schmidt, J. Tenhunen

4.1 Introduction

Among the five EUROFLUX broad-leaved sites, four are beech stands and one is a poplar plantation. These sites are spread over a wide range of climatic conditions: mid-oceanic (France), continental (Germany) and subpolar (Iceland), and the elevation ranges from plain to mountain. Furthermore, these stands are of various ages: the poplar stand is a young plantation, the beech stand of Hesse is a young natural regeneration, while the other beech stands included in this study are much older (70 to 140 years).

Beech (*Fagus sylvatica* L.) covers a large area in Europe, estimated at 120,000 km^2. It extends from southern England, France, and northern Spain in the west, to Poland and Romania in the east. Latitudinal extension ranges from southern Sweden to northern Italy and Greece. In its northern area, beech is mainly found at low elevation, while in the southern part of Europe it is located at higher elevations (Tessier du Cros 1981). This species plays an important economic and recreative role. From the ecological point of view, beech is being more widely used, in particular as an alternative tree species to conifers such as spruce. Therefore, there is an increasing need for information on carbon and water cycles for this species.

This chapter compares hourly and annual water and carbon dioxide fluxes among the broad-leaved EUROFLUX sites. Some measurements obtained in other projects are included in this present synthesis (Tables 4.1 and 4.2). One goal will be to examine, among those beech and poplar stands, to what extent water and carbon fluxes differ or show comparable response to environmental factors.

Ecological Studies, Vol. 163
R. Valentini (Ed.) Fluxes of Carbon, Water and Energy of European Forests

Table 4.1. Site characteristics: climate. Season is defined here as the period from bud break to the date of net carbon flux inversion

Site	Elevation (m)	Average air temperature (°C)	Average annual air temperature (°C)	Average season Pi (mm)	Average annual Pi (mm)	Cumulated season PAR (mol m^{-2})	Season (days)
Hesse (F)	300	9.0	15.0	820	350	5600	157
Soroe (DK)	40	8.5	15.3	510	174	5682	147
Collelongo (IT)	1550	6.2	12.6	1180	419	4655	135
Vielsalm (B)	450	7.5	13.5	1000	297	4750	175
Kiel (D)		8.1		697			
Steigerwald (D)	440	7.5	14.5	795	385	5060	163
Karlsruhe (D)	220	9.5	15.9	771	416	5944	183
Gunnarsholt (IC)	78	4.3	9.7	1044	288	2065	144
Aubure (F)	1100	6.0	13.0	1350	372	3550	121

Table 4.2. Site characteristics: stand structure

Site	Age years	Density (n ha^{-1})	Height (m)	Ground area (m^2 ha^{-2})	LAI (m^2 m^{-2})	Wood biomass (t ha^{-1})	Remarks	Project[a]
Hesse (F)	30	3800	14	19.6	5.7	88		E-C-W
Soroe (DK)	80	430	25	22.0	4.8	247		E-C-W
Collelongo (IT)	105	885	21	34.3	5.0	268		E-C-W
Vielsalm (B)	90	230	27	12.9	5.0	245	Mixed	E-C-W
Gunnarsholt (IC)	7	10 000	1.0	3.8	2.6			E-C-W
Kiel (D)	100	150	29		4.5	128		O-C-W
Steigerwald (D)	140	358	31	28.0	6.9		Mixed	O-W
Karlsruhe (D)	74		28	34.0			Mixed	O-W
Aubure (F)	120	429	22.5	43.3	5.7			O-W

[a] E, EUROFLUX site; O, other project; C, carbon; W, water.

4.2 Water Transfer and Water Balance

4.2.1 Sap Flow in Beech, Tree Transpiration

Sap flow measurements performed at some of the sites presented in Tables 4.1 and 4.2 attempt to: (1) give precise information on the spatial and temporal variations of sap flow at the tree scale, (2) estimate stand transpiration and compare it to total evapotranspiration (from eddy covariance measurements).

Beech is a diffuse-porous species in which sapwood cannot be distinguished visually from heartwood; in general, a transverse section of wood has an even color. Nevertheless, in some experiments, a decrease in wood water content towards the center of the tree was observed (Granier et al. 2000b) that could be linked to a decrease in its water transfer capabilities.

4.2.2 Sap Flow Profile in Trunks

Sap flow measurements made at increasing depth below the bark showed an exponential decrease, the maximum rate being found in the outer 0–20 mm annulus. This pattern as well as the water distribution mentioned above, plus computer tomography observations (Habermehl 1982; see also Raschi et al. 1995), reveal that beech does not exhibit a sharp transition between sapwood and heartwood, contrary to the common observation in either ring-porous or coniferous species. Accordingly, biochemical markers indicate a transition zone in beech where histological sapwood cells were found up to 60 annual rings below the cambium (Magel et al. 1997).

Figure 4.1 shows radial patterns of sap flux density obtained in different beech stems, from the Steigerwald and the Hesse forests. Besides the large scatter in the data when expressed to the relative depth below cambium (absolute depth of sapwood up to 150 mm), similar relationships were

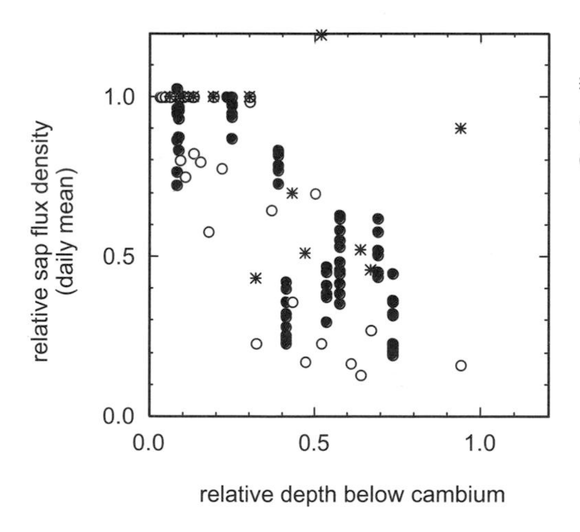

Fig. 4.1. Variation of sap flux density within the trunks of beech of various ages (30–120 years) as a function of depth below cambium. Sap flux density is expressed in percent of maximum observed in the outer part of the wood. Two techniques were used: heat pulse velocity with Greenspan sensors (Pausch, Dawson and Köstner, unpubl.; for method see Pausch et al. 2000), and Granier-type radial sap flowmeters

observed in the trees from two sites, irrespective of tree size. At the Karlsruhe site (Lang 1999), the sap flux decrease according to the depth was less pronounced (data not shown).

4.2.3 Range of Sap Flow and Among-Tree Variation

Total sap flow circulating in a tree is equal to the product of sap flux density by the sapwood area at the measurement level in the trunk. Therefore, under a given climatic demand, sap flow is higher in large trees than in small trees. In a young beech stand (Hesse, 30 years old), total daily sap flow ranged between 2 and 25 kg day^{-1} during bright days in summer, according to tree circumference. In a 74-year-old beech stand at Karlsruhe, the daily sap flow amounted 104 to 182 kg day^{-1}. In older stands, (site Steinkreuz/Steigerwald, 140 years old), maximum flux rates ranged between 10 and 400 kg day^{-1} in trees of 12 and 60 cm dbh, respectively.

Coefficient of variance of sap flow (CV) depends upon the stand structure: when the stand is homogeneous, especially in the older stands, CV is low. At the Karlsruhe site, CV equaled 14 % (Lang 1999). At Steigerwald, on bright days, CV in the outer sapwood reached 45 % in subcanopy trees and 17 % in trees of the main canopy. In the young stand of Hesse, where competition among trees is greater, CV was much higher, reaching 61 % due to very low sap flux densities in intermediate and in suppressed trees whose crowns are less illuminated. In this stand, an increasing relationship between sap flux density and tree circumference was observed.

4.2.4 Canopy Conductance for Water Vapor: Effect of Radiation, Vapor Pressure Deficit, Temperature, Leaf Area, and Drought

Tree canopy conductance ($g_{c,}$, cm s^{-1}) can be derived from sap flow and from climate measurements, using the inverted Penman-Monteith equation, as in Gash et al. (1989). Sap flow data measured on a sample of trees are scaled up to the stand taking into account the stem distribution (Köstner et al. 1998; Granier et al. 2000b).

We used the multiplicative-type functions proposed by Jarvis (1976) and Stewart (1988) relating canopy conductance to environmental and structural factors:

$$g_c = g_{cmax}(R,D) \, . \, f(t) \, . \, g(REW) \, . \, h(LAI) \quad (4.1)$$

where
R is the global radiation (W m^{-2}),
D is the air vapor pressure deficit (kPa),

f, g, h are the limiting functions [0..1],
t is air temperature (°C),
REW is the relative extractable water in the soil defined as extractable soil water/maximum extractable water,
LAI is the leaf area index.

The function g_{cmax} represents the maximum canopy conductance observed when temperature, soil water deficit, and LAI are not limiting tree transpiration. This function depends on incident radiation and on air vapor pressure deficit (see Fig. 4.2).

The following function was applied at Hesse:

$$g_{cmax}=[R/(R+282)]\ [7.019/(1+3.44\ D)]\quad r^2=0.86 \tag{4.2}$$

Figure 4.2 shows the comparison of the g_{cmax} functions obtained in four beech sites: young (Hesse) and old (Kiel) stands growing in plain, and old stands in montane areas(Aubure and Steigerwald). In three stands a strong similarity in the response of g_{cmax} to radiation and to vapor pressure deficit was found, despite differences in age and site conditions. This is probably due to the close values of LAI among the beech sites studied here. Beech is sensitive to atmospheric drought, canopy conductance decreasing sharply when air vapor pressure deficit (D) increases. Steigerwald data show a similar pat-

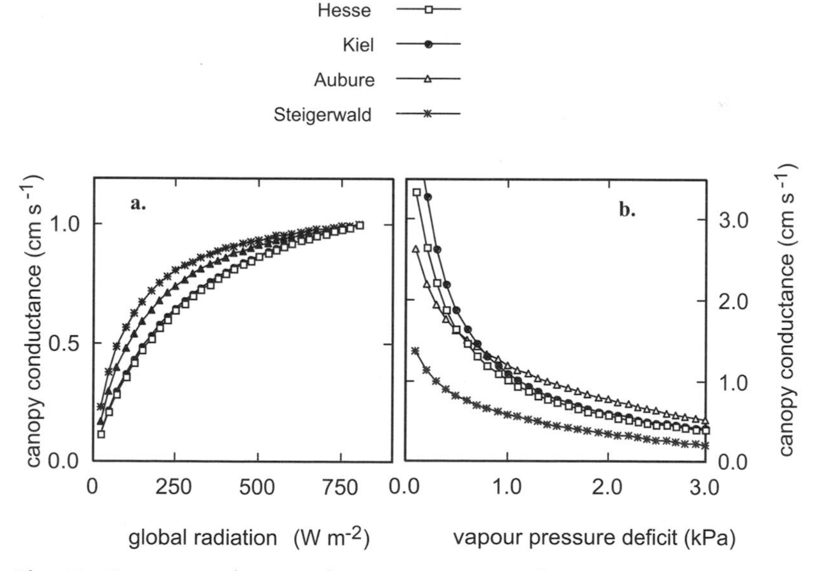

Fig. 4.2. Canopy conductance for water vapor as a function of **a** global radiation, and of **b** air vapor deficit, under saturating global radiation (R>500 W m^{-2}) in four beech stands. *Kiel* data are from Herbst (1995)

tern of g_{cmax} variation to D, but values were much lower, probably due to an unknown limiting factor. The functions f, g, h of temperature, soil water content, and LAI were calibrated at Hesse (Granier et al. 2000b). Canopy conductance (g_c) in the spring was strongly limited by temperature, when it dropped below a threshold of ca. 17 °C. This sensitivity to temperature was not observed at the end of the leafy period (September–October). An effect of increasing soil water deficit was observed at Hesse during the summer.

4.2.5 Annual Water Balance

The annual amounts of the main water balance components are shown in Table 4.3. Rainfall interception was within a narrow range of 98–127 mm $year^{-1}$ among the stands, despite a high variability in rainfall through the years and among the sites. No correlation was found between rainfall interception and LAI. Transpiration varied to a much larger extent, 218–421 mm $year^{-1}$, probably linked to the large variation in the climatic demand and to soil water deficit that developed in some sites (Hesse). Nevertheless, the ratio of transpiration to total evapotranspiration was rather stable, and equaled 0.73 on average.

Table 4.3. Cumulated values of rainfall (R), net interception (In), transpiration (T), and total evapotranspiration (Et). T was computed from sap flow measurements; Et was measured over the stand by eddy correlation technique. Data are expressed in mm

Year	Site	R (annual)	R (season)	In (season)	T (season)	Et (season)	T/Et (season)
1996	Hesse	672	325	116	256	338	0.76
1997	Hesse	853	440	125	253	351	0.72
1998	Hesse	974	453	107	257	313	0.82
1999	Hesse	1073	490	122	333	384	0.86
1992	Kiel	756		98	421	575	0.73
1993	Kiel	838		127	326	515	0.63
1994	Kiel	977		108	389	517	0.75
1995	Kiel	687		108	419	577	0.73
1998	Karlsruhe	677	249		310	423	0.73
1995	Aubure	1255	436	102	218		

Note: data from Kiel are outputs of a model (Herbst et al. 1999)
Aubure data are from Biron (1994)

4.3 Carbon Fluxes and Carbon Balance

Net ecosystem exchange (NEE, measured by eddy covariance) is equal to the algebraic sum of the gross assimilation flux (GPP) and the total ecosystem respiration R_{eco}(the sum of aerial and soil respiration). The assumption that R_{eco} as measured during the night can be extrapolated to daylight periods is questionable when leaves are taken into consideration, because mitochondrial respiration may be depressed by 40–90 % during the day (Brooks and Farquhar 1985). This factor could probably lead to an overestimation of total ecosystem respiration.

4.3.1 Net Ecosystem Exchange

Two typical time courses of NEE are shown in Fig. 4.3. In broad-leaved stands, variation of NEE within the year is strongly dependent on tree phenology. After bud break, as is true for stand transpiration, the rapid increase in carbon uptake by the forest is roughly parallel to the increase in LAI. The maximum carbon fixation rate varies between –5 and –8 g C m^{-2} day^{-1}, according to the sites and to the period.

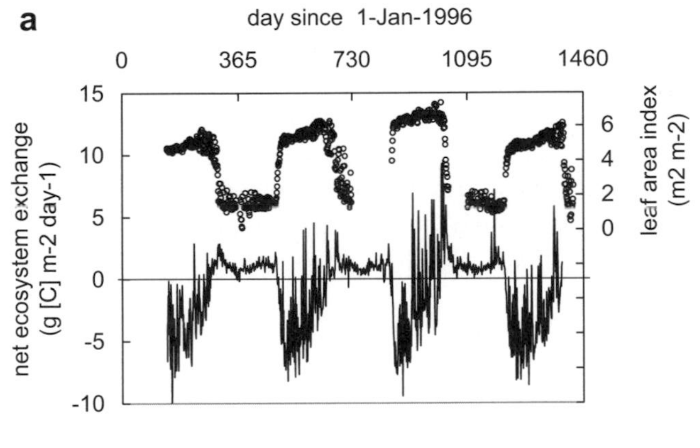

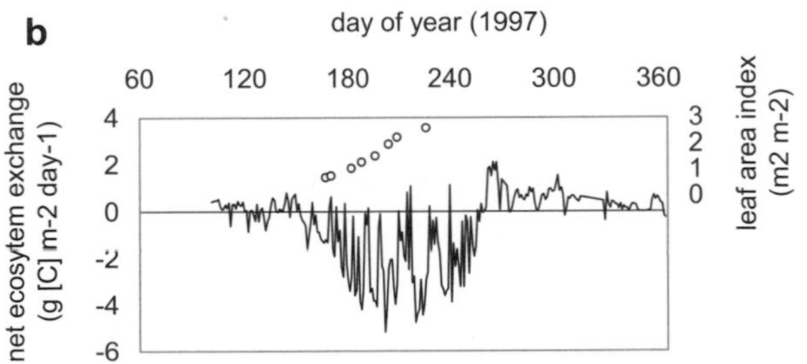

Fig. 4.3. Annual variation of NEE at two sites, and the variation of LAI as estimated from intercepted global radiation. **a** Hesse forest (France), **b** Gunnarsholt (Iceland)

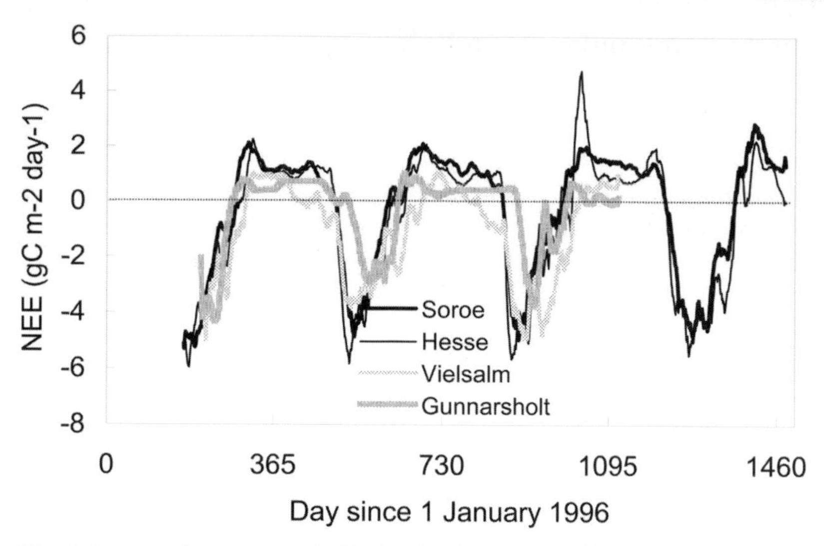

Fig. 4.4. Annual variation of NEE in the four EUROFLUX beech stands from 1996 to 1999. Data are smoothed using a moving average over 21 days. Gaps were filled according to Falge et al. (2001)

During the leafed periods, short-term (day-to-day) fluctuations in NEE can be observed, corresponding to conditions of low photosynthesis and/or high ecosystem respiration (low incident radiation, rainfall events, and high air temperature). During the period when trees are without leaves, NEE variation is less variable; its time course depends mainly on temperature (see below). Moving averaged time courses of NEE are presented in Fig. 4.4. Except for Gunnarsholt, where bud break occurs later due to lower temperatures, there is a good synchronism in the seasonal variation, especially between Hesse and Soroe. In Vielsalm, a significant carbon uptake can be observed earlier in the spring and later in the fall. This is probably due to the presence of coniferous species (*Pseudotsuga menziesii*, *Abies alba*, *Pinus sylvestris*) mixed with the beech.

4.3.2 Ecosystem Respiration

Daily ecosystem respiration (R_{eco}) in the EUROFLUX sites was calculated as the sum of measured fluxes during the night plus the extrapolated daytime respiration. R_{eco} shows an exponential increase with temperature. In all the studied sites, these relationships were rather poor due to considerable scattering of the data (see Fig. 4.5). We attempted to separate between leafed and non-leafed phases, as shown in Fig. 4.5. However, no clearly different behavior between the phases was found, the two regression lines showing no significant difference. We applied Arrhenius-type equations to the data. The rela-

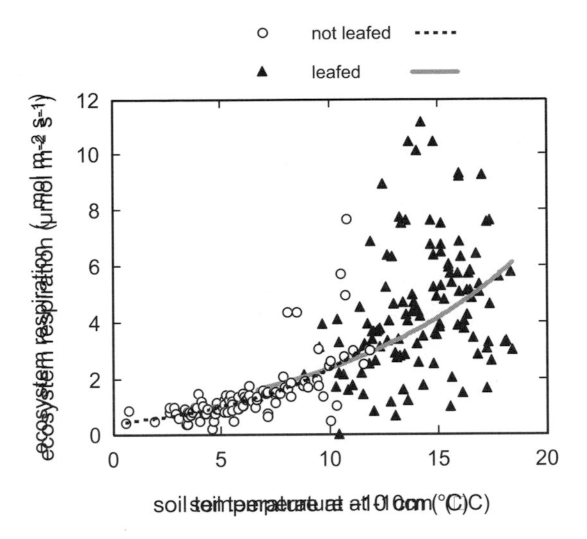

Fig. 4.5. Ecosystem respiration as a function of soil temperature in Hesse. Data are 5-day means

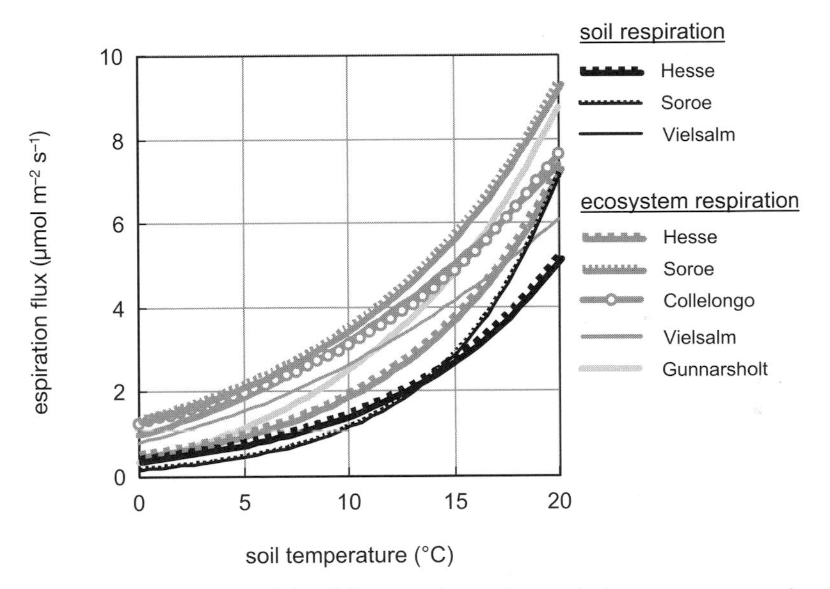

Fig. 4.6. Comparison of the different relationships relating ecosystem and soil respiration to soil temperature measured at –5 or –10 cm according to the sites

tionships obtained in the five EUROFLUX sites are shown in Fig. 4.6. On average, R_{eco} equaled 2.0–3.7 µmol m^{-2} s^{-1} at 10 °C, and 7.3–9.2 µmol m^{-2} s^{-1} at 20 °C. It can be seen that very close relationships were found at Vielsalm and Collelongo. Higher ecosystem respiration rates were found at Soroe, independent of temperature. At low temperature (0 to 10 °C), the lowest R_{eco} rate was observed at the Gunnarsholt site, but it increased more sharply than at the other sites with rising soil temperature.

4.3.3 Soil Respiration

Soil chambers were used to measure CO_2 efflux from the soil surface (Epron et al. 1999a,b; Le Dantec et al. 1999; Janssens et al. 2001) at three of the beech sites. As with R_{eco}, temperature-dependent relationships were found (see Fig 4.6). Soil temperature measured at –5 or –10 cm, depending on the site, was used as the reference temperature. Functions applied at Hesse and Soroe were very close in the range 0–15 °C, while soil respiration was much higher at Vielsalm. The effect of decreasing soil moisture was only observed at Hesse (Epron et al. 1999a,b), where a summer drought developed.

At Hesse, the following relationship was obtained:

$$R_{soil}=1.13\ \theta_v\ e^{0.136\ t_{-10}}\quad r^2=0.86 \tag{4.3}$$

where

θ_v is the soil water content in the 0–10 cm zone ($m^3\ m^{-3}$),

t_{-10} is soil temperature (°C) measured at –10 cm.

The ratio of R_{soil} to R_{eco} ranged from 0.38 to 0.82 depending on the sites and years, with an average value of 0.65. This agrees with data from other sites showing that soil respiration represents 60–80 % of total ecosystem respiration (Wofsy et al. 1993; Goulden et al. 1996; Davidson et al. 1998).

At Hesse, root respiration was separated from total soil respiration by comparing control plots to trench plots in which living roots were excluded, and by using an estimate of dead root decomposition within the trench plots. On average, roots contributed to ca. 60 % of the soil respiration (Epron et al. 1999b).

4.3.4 Daily Net Ecosystem Exchange

During the full leaf expansion period, the time course of daylight NEE is well related to incident photosynthetically active radiation (PAR) (Fig. 4.7). In beech, contrary to observations of net assimilation as measured at the leaf level, NEE did not show a saturation when increasing PAR. Temperature, vapor pressure deficit, and the proportion of diffuse to direct radiation also play a role in NEE variation (data not shown), but to a lesser extent.

The same model was used at the four beech sites:

$$NEE=Rd + aPAR / (b + PAR) \tag{4.4}$$

In the Gunnarsholt poplar site (data of July), the best fit was obtained with the following function:

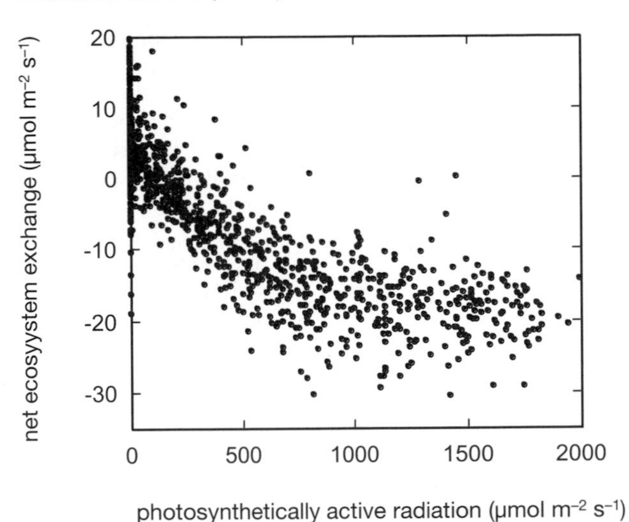

Fig. 4.7. NEE as a function of PAR. Half-hourly data of June 1997 during the daylight hours (Hesse forest)

Table 4.4. Value of the three parameters in Eq. (4.4) for the June period

	Collelongo (IT)	Soroe (DK)	Vielsalm (B)	Hesse (F)
a	−42.5	−38.2	−24.7	−33.3
b	859	955	465	746
Rd	3.0	4.9	3.3	3.9

$$NEE=-12.69+15.798 \exp(-0.002276*PAR) \quad (4.5)$$

The fitted coefficients of Eq. (4) are given in Table 4.4. The functions are shown in Fig. 4.8. For beech, two stands (Soroe and Hesse) exhibited very close NEE(PAR) relationships, while the Collelongo forest showed highest CO_2 fixation rate, and the Vielsalm forest the lowest. The higher assimilation rate in Collelongo, whatever the PAR, can be related to the much lower ecosystem respiration at high temperatures. Carbon uptake was lower at Gunnarsholt than in the beech stands, due to its much lower LAI.

The seasonal variations in those relationships were analyzed in some sites. Except for Gunnarsholt, where LAI reached its maximum in late August, the maximum fixation rates were measured in June, when LAI and radiation are maximum, while soil water content is still high. Few data are yet available, however, on the effects of drought on carbon fixation. Water stress was reported at two beech sites (Collelongo 1993, in Matteucci 1998; Hesse 1996, 1998). In those stands, soil water content depletion induced a significant decrease in NEE when relative extractable water (REW) in the soil (see Fig. 4.9) dropped below about 40–50 % of field capacity.

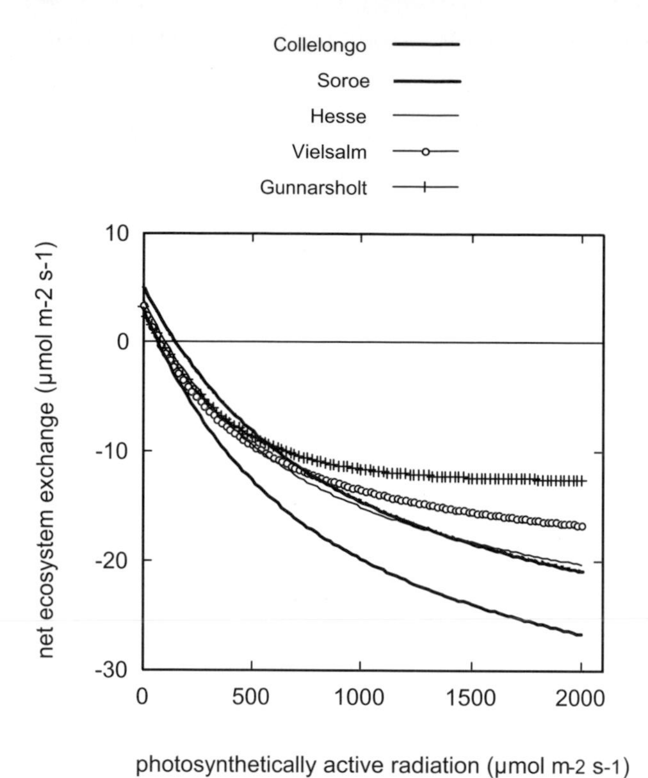

Fig. 4.8. Comparison of NEE as a function of incident PAR in the four beech and poplar stands. For the beech stands, the functions were fitted on data of June, corresponding to the maximum net assimilation. For the poplar stand, data of July were used (because much lower LAI and temperature were observed in June)

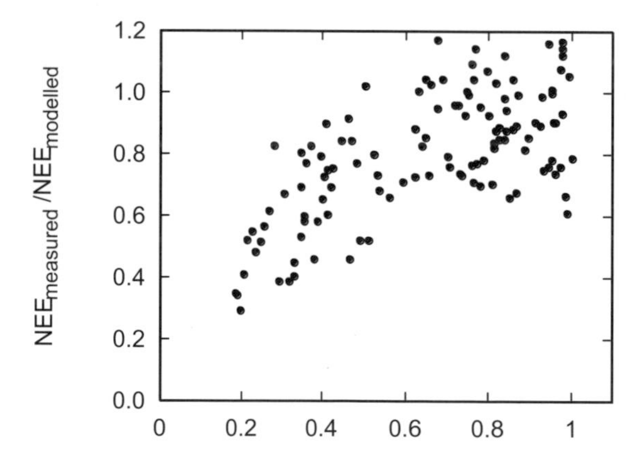

Fig. 4.9. Decrease in daily cumulated NEE as a function of REW in the soil during 1997 for the period of full leaf expansion in the beech forest of Hesse (France)

Table 4.5. Annual carbon fluxes over and within the EUROFLUX broad-leaved stands. Values are in g C m^{-2} $year^{-1}$

Site	Year	NEE	NPP	GPP	R_{eco}	R_{soil}	Biomass increment
Hesse (F)	1996	−218	−402	−1011	793	575	383
	1997	−257	−556	−1245	988	663	472
	1998	−79	−364	−1314	1235	713	355
	1999	−299		−1335	1036		482
Soroe (DK)	1996	−150		−1025	875	377	
	1997	−6		−1255	1249	432	
	1998	−27		−1214	1187	398	
	1999	−110		−1355	1245	457	
Collelongo (I)	1993	−472	−802	−1016	544		
	1997	−663	−822	−1302	636	682	
	1998	−631		−1339	703		
Vielsalm (B)	1996	−452		−1300	848	825	
	1997	−378		−1298	920	887	
	1998	−519		−1386	867	826	
Gunnarsholt (IC)	1997	−101		−708	607		
	1998	−121		−704	583		
Kiel (D)	1992–93	−242	−628	−1019	777	617	349

Note: At some sites, $R_{soil}>R_{eco}$. This is due to independent estimates of both fluxes through different techniques (EC and soil chambers)
Hesse and Soroe data from Granier et al. (2000a) and Pilegaard et al. (2001), repectively

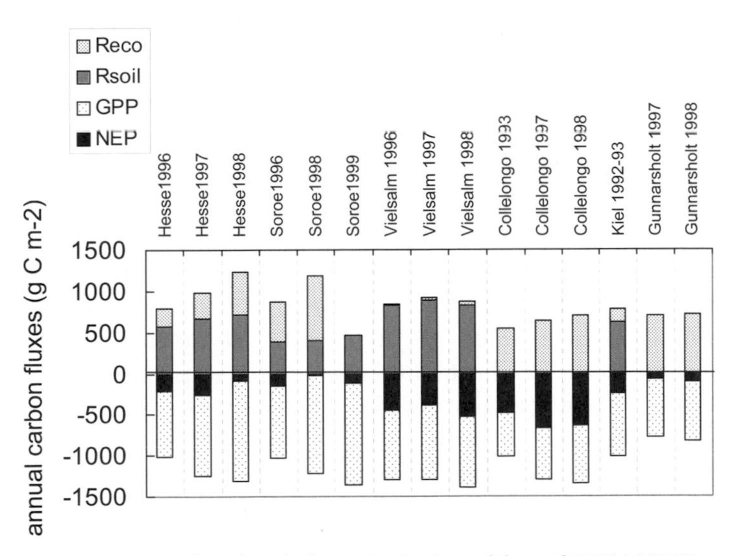

Fig. 4.10. Annual carbon balance in the broad-leaved EUROFLUX stands

4.3.5 Annual Carbon Balance, Intersite and Interannual Comparison

Annual fluxes are reported in Table 4.5. GPP was calculated here as the sum of NEE and estimated R_{eco}.

NEE shows large variations among the sites, ranging between –6 and –663 g C m^{-2} $year^{-1}$. As shown in Table 4.5 and in Fig 4.10, NEE also shows a large interannual variation. Among the beech stands, GPP variation is smaller than for NEE: GPP ranges between –1011 and –1474 g C m^{-2} $year^{-1}$. Much lower GPP was estimated in Iceland, probably due to limiting air temperature and radiation, and to the low LAI in the poplar stand. In contrast, much larger differences are found for R_{eco}, with minimum and maximum values of 544 and 1245 g C m^{-2} $year^{-1}$, respectively. Therefore, interannual or intersite variation in R_{eco} causes large differences in annual NEE: the higher the R_{eco}, the lower the CO_2 sequestration.

4.4 Conclusion

The beech stands showed comparable transpiration rates, due to close values of LAI and to similar canopy conductance response to radiation and to vapor pressure deficit. Therefore, when maximum extractable soil water is available, water balance can be modeled with accuracy. The combination of both sap flow and eddy covariance methods in some of the studied sites allowed the separation of tree transpiration from total evapotranspiration. In the beech stands investigated here, under dry foliage conditions, most (ca. 80–90 %) of stand evapotranspiration originated from transpiration.

Although showing a large variation among sites and years, the annual NEE in the broad-leaved EUROFLUX sites was within the range of other data from Europe and North America. Still, the annual carbon fixation in the sites investigated here was significantly lower than for the high productive coniferous forests, like the Sitka spruce or the maritime pine stands studied in the EUROFLUX program.

We have pointed out that most of the NEE variation originates from ecosystem respiration. On the whole, when R_{eco} increases, net carbon fixation decreases. This observation could be of importance in the perspective of global warming. At sites where chamber measurements were performed, R_{soil} was estimated at ca. 50–80 % of R_{eco} (Janssens et al. 2001), playing therefore a major role in the forest ecosystem carbon balance. A way to analyze the causes of the site-to-site differences in annual NEE is to use simple models relating the carbon fluxes to climatic variables, and to run the models on a data set using the same climate. The above-described relationships for net assimilation and for ecosystem respiration were used to simulate the annual time course of NEE, GPP, and R_{eco} from half-hourly climatic data. When applied on a same climate data set, the models lead to different estimated NEE (see the chapter on modeling, this Vol.). It can be concluded that both climate and stand physiology explain the observed differences among sites.

Aerial and belowground heterotrophic respiration can also explain a part of R_{eco} and therefore of NEE differences observed among sites. Two important sources of heterotrophic respiration are the woody debris decomposition at the soil surface, and the variation of carbon content in the soil. Management practices, e.g., thinning residuals left on the ground, or soil disturbances, have an effect on heterotrophic respiration.

On the other hand, comparison between NEE and net annual increment in the biomass (see data from Hesse and Kiel in Table 4.5) shows a discrepancy between both estimates that can be explained by a variation in the carbon stock, not measured in the EUROFLUX sites. This heterotrophic source of carbon could be located either in the soil or in the aerial compartments of the ecosystem. At Hesse, woody debris decomposition on the soil surface is suspected of increasing ecosystem respiration and therefore decreasing NEE. Unfortunately, measurement of this term "stock variation" is difficult, since the annual variation is small as compared to the totality of the stock.

References

Biron P (1994) Le cycle de l'eau en forêt de moyenne montagne: flux de sève et bilans hydriques stationnels (bassin versant du Strengbach à Aubure Hautes Vosges). PhD Thesis, University of Strasbourg I, 244 pp

Brooks A, Farquhar GD (1985) Effect of temperature on the CO_2/O_2 specificity of ribulose-1,5-biphosphate carboxylase/oxygenase and the rate of respiration in the light. Planta 165:397–406

Davidson EA, Beck E, Boone RD (1998) Soil water content and temperature as independent or confounded factors controlling soil respiration in a temperate mixed hardwood forest. Global Change Biol 4:217–227

Epron D, Farque L, Lucot E, Badot PM (1999a) Soil CO_2 efflux in a beech forest: dependence on soil temperature and soil water content. Ann Sci For 56:221–226

Epron D, Farque L, Lucot E, Badot PM (1999b) Soil CO_2 efflux in a beech forest: the contribution of root respiration. Ann Sci For 56:289–295

Falge E, Baldocchi D, Olson RJ, Anthoni P, Aubinet M, Clement R, Granier A, Bernhofer C, Hollinger D, Ta Lai C, Kowalsky A, Meyers T, Moors EJ, Munger JW, Pilegaard K, Rannik U, Rebmann C, Verma S (2001) Gap filling strategies for defensible annual sums of net ecosystem exchange. Agric For Meteorol 107:1–27

Gash JHC, Shuttleworth WJ, Lloyd CR, André J-C, Goutorbe J-P, Gelpe J (1989) Micrometeorological measurements in Les Landes forest during HAPEX-MOBILHY. Agric For Meteorol 46:131–147

Goulden ML, Munger JW, Fan SM, Daube BC, Wofsy SC (1996) Measurements of carbon sequestration by long-term eddy covariance: methods and a critical evaluation of accuracy. Global Change Biol 2:162–182

Granier A, Ceschia E, Damesin C, Dufrêne E, Epron D, Gross P, Lebaube S, Le Dantec V, Le Goff N, Lemoine D, Lucot E, Ottorini JM, Pontailler JY, Saugier B (2000a) The carbon balance of a young beech forest. Funct Ecol 14:312–325

Granier A, Biron P, Lemoine D (2000b) Water balance, transpiration and canopy conductance in two beech stands. Agric For Meteorol 100:291–308

Habermehl A (1982) A new non-destructive method for determining internal wood condition and decay in living trees. I. Principles, method, and apparatus. Arboricult J 6:1–8

Herbst M (1995) Stomatal behaviour in a beech canopy: an analysis of Bowen ratio measurements compared with porometer data. Plant Cell Environ 18:1010–1018

Herbst M, Eschenbach C, Kappen L (1999) Water use in neighbouring stands of beech (*Fagus sylvatica* L.) and black alder (*Alnus glutinosa* (L.) Gaertn.). Ann Sci For 56:107–120

Janssens IA, Lankreijer H, Matteucci G, Kowalski AS, Buchmann N, Epron D, Pilegaard K, Kutsch W, Longdoz B, Grünwald T, Montagnani L, Dore S, Rebmann C, Moors EJ, Grelle A, Rannik Ü, Morgenstern K, Clement R, Gumundsson J, Minerbi S, Berbigier P, Ibrom A, Moncrieff J, Aubinet M, Bernhofer C, Jensen NO, Vesala T, Granier A, Schulze E-D, Lindroth A, Dolman AJ, Jarvis PG, Ceulemans R, Valentini R (2001) Productivity and disturbance overshadow temperature in determining soil and ecosystem respiration across European forests. Global Change Biol 7:269–278

Jarvis PG (1976) The interpretation of the variations in leaf water potential and stomatal conductance found in canopies in the field. Philos Trans R Soc Lond 273:593–610

Köstner B, Granier A, Cermák J (1998) Sap flow measurements in forest stands – methods and uncertainties. Ann Sci For 1/2:13–27

Kutsch WL, Eschenbach C, Dilly O, Middelhoff U, Steinborn W, Vanselow R, Weisheit K, Wötzel J, Kappen L (1999) 5. The carbon cycle of contrasting landscape elements in the Bornhöved Lake district. In: Lenz R, Hantschel R, Tenhunen JD (eds) Ecosystem properties and landscape function in central Europe. Ecological studies. Springer, Berlin Heidelberg New York

Lang S (1999) Ecophysiological and anatomical investigations of sapflow in different sapwood depths of *Fagus sylvatica* L. Karlsr Beitr Pflanzenphysiol 35:1–184

Le Dantec V, Epron D, Dufrêne E (1999) Soil CO_2 efflux in a beech forest: comparison of two closed dynamic systems. Plant Soil 214:125–132

Magel E, Hillinger C, Höll W, Ziegler H (1997) Biochemistry and physiology of heartwood formation: role of reserve substances. In: Rennenberg H, Eschrich W, Ziegler H (eds) Trees – contributions to modern tree physiology. Backhuys Publishers, Leiden, pp 477–506

Matteucci G (1998) Bilanco del carbonio in una faggetta dell'Italia centro-meridionale: determinanti ecophysiologici, integrazione a livello di copertura e simulazione dell'impatto dei cambiamenti ambientali. PhD Thesis, University of Padova

Pausch RC, Grote EE, Dawson TE (2000) Estimating water use by sugar maple trees: considerations when using heat-pulse methods in trees with deep functional sapwood. Tree Physiol 20:217–227

Pilegaard K, Hummelshøj P, Jensen NO, Chen Z (2001) Two years of continuous CO_2 eddy-flux measurements over a Danish beech forest. Agric For Meteorol 107:29–41

Raschi A, Tognetti R, Ridder H-W, Béres C (1995) Water in the stems of sessile oak (*Quercus petraea*) assessed by computer tomography with concurrent measurements of sap velocity and ultrasound emission. Plant Cell Environ 18:545–554

Stewart JB (1988) Modelling surface conductance of pine forest. Agric For Meteorol 43:19–35

Tessier du Cros E (1981) Le hêtre. INRA, Paris, 615 pp

Wofsy SC, Goulden ML, Munger JW, Fan SM, Bakwin PS, Daube BC, Bassow SL, Bazzaz FA (1993) Net exchange of CO_2 in a mid-latitude forest. Science 260:1314–1317

5 Coniferous Forests (Scots and Maritime Pine): Carbon and Water Fluxes, Balances, Ecological and Ecophysiological Determinants

R. Ceulemans, A.S. Kowalski, P. Berbigier, A.J. Dolman, A. Grelle, I.A. Janssens, A. Lindroth, E. Moors, U. Rannik, T. Vesala

5.1 Introduction

5.1.1 Aims and Objectives

This chapter represents a synoptic synthesis of the results from the five EUROFLUX sites considered representative of the *Pinus* genus. Data from the EUROFLUX database for each pine site in 1997 and (partially) 1998 have been compiled and examined in order to reveal similarities and differences across the pine transect. Short-term climatic conditions (temperature, precipitation, radiation) are considered and presented as potentially important factors determining the functional responses of the individual ecosystems, which are presented as the fluxes of carbon, water, and energy. It should be remembered that the flux results may vary also in relation to other important site-to-site differences such as speciation, age, soil properties, and forest management. *"Pinus is a remarkable genus with a very large distribution range in the northern hemisphere. Where they occur, pines usually form the dominant vegetation cover and are extremely important components of the ecosystems. They also provide a wide range of products for human use. Pines are now widely grown in commercial plantations, both within and outside their natural ranges."* (Richardson 1998).

5.1.2 Scots Pine and Maritime Pine as Representatives of the Genus *Pinus* in the EUROFLUX Network

There are more pines than any other tree genus growing in the Northern Hemisphere. The approximately 80 species come from all the northern temperate regions, occasionally from the warmer parts, and are sometimes found

Ecological Studies, Vol. 163
R. Valentini (Ed.) Fluxes of Carbon, Water and Energy of European Forests

as far north as the Arctic Circle and beyond. Some pine species will survive conditions which one might think would inhibit all plant growth, in places where a few decimeters down the soil is permanently frozen and where the surface, around the roots, thaws out only long enough to permit a growing season of 6–8 weeks. Pines are a fast-growing source of softwood and are grown in vast numbers (Gholz et al. 1994). In addition to their northern native hemisphere, *Pinus radiata* have been planted by the millions in South Africa, Australia, and New Zealand. In the five forest sites included in the European pine transect of EUROFLUX, two different dominant pine (*Pinus*) species were involved, i.e., Scots pine (*Pinus sylvestris* L.) and maritime pine (*Pinus pinaster* Ait). Although both species have quite a few characteristics in common, they also have a number of specific differences. A comprehensive comparison of both species is given in Table 5.1.

Scots pine (*Pinus sylvestris*) is naturally distributed in a zonal band over Europe and Asia, from the Atlantic Ocean to the Pacific Ocean. So, its natural distribution area goes from Scotland to eastern Siberia, and between the boreal and temperate vegetation zones as far south as Spain. In prehistoric times Scots pines (*Pinus sylvestris*) were native to maritime climatic regions in Europe such as the Netherlands and Belgium. After having completely disappeared, Scots pines were reintroduced in the Netherlands and Belgium around 1515. In regions with a more continental character, Scots pines did not disappear. At its southern extent, Scots pine naturally occupies sites in the mountains up to 2000 m elevation. In northern Scandinavia, Scots pine forms the alpine and arctic timber line against the tundra. Knowledge of ecology, genetics, and ecophysiology of Scots pine are mainly based on research conducted in Finland and Sweden, and may not apply universally to the numerous variants occurring in different parts of the geographic distribution of the species (Stenberg et al. 1994).

Table 5.1. Comparison of the two pine (*Pinus*) species included in the EUROFLUX measuring network with their common name, biogeographic region of origin, habitat and a number of selected morphological features. (After Gholz et al. 1994; Richardson 1998)

Characteristics	*Pinus pinaster* Ait.	*Pinus sylvestris* L.
Common name	Maritime pine	Scots pine
Biogeographical region	SW Europe and W. Mediterranean basin	Europe and central Asia
Habitat	Mediterranean coastal	Boreal forest, temperate, subalpine
Needle number	2	2
Needle longevity (years)	3	2–5
Cone length (cm)	10–22	3–6
Height (m)	20–35 (40)	30–35

Table 5.2. General description and site characteristics of the five forest sites along the pine transect in the EUROFLUX measuring network. Geographical location, site description, climate, stand characteristics, and ecological/biological features have been listed. Sites have been arranged along a N-S gradient

	Finland	Sweden	The Netherlands	Belgium	France
			Geographical location and climate		
Name of site	Hyytiälä	Norunda	Loobos	De Inslag	Le Bray, Les Landes
Nearest city	Tampere	Uppsala, Stockholm	Wageningen	Brasschaat, Antwerpen	Bordeaux
Geographical coordinates	61°5′N 24°17′E	60°5′N 17°29′E	52°10′N 5°44′E	51°18′N 4°31′E	44°5′N 0°5′E
Elevation (m)	170	45	25	16	60
Mean temp. (°C)	3.5	5.5	10.3	9.8	13.5
Mean precipitation (mm)	640	530	800	750	900
Nitrogen deposition (kg N ha^{-1} a^{-1})	3	4	80	60–90	8
			Stand characteristics		
LAI	3	5	1.9–2.2	2.4	2.6–3.1
Height (m)	12	25	15	23	17
Year of plantation	1962	1898	1909	1929	1970
Stand density (number ha^{-1})	2500	600	360	556	500
Wood increment or yield (m^3 ha^{-1} a^{-1})	10	5–6	6.3	7	17
Topography (slope)	Slight slope	Flat	Flat	Flat	Flat

Table 5.2. (*Continued*)

	Finland	Sweden	The Netherlands	Belgium	France
			Vegetation and species composition		
Species in overstory	Mono-species *Pinus sylvestris*, but with *Picea abies*, *Betula pubescens*, *Alnus incana*, and *Populus tremula* (together less than 1 %)	*Pinus sylvestris* and *Picea abies*	*Pinus sylvestris*	*Pinus sylvestris*, with *Quercus robur* (15 %)	Monospecific *Pinus pinaster*
Species in understory	*Vaccinium vitis-idaea*, *V. myrtillus*, *Calluna vulgaris* and *Dicranum undulatum* (moss)	*Picea abies*	*Deschampsia flexuosa*	*Prunus serotina*, *Rhododendron ponticum* and *Molinia caerulea*	*Molinia caerulea*, *Arrhenaterum thorei*, *Pteridium aquilinum*, *Calluna vulgaris*, *Ulex* spp.
			Eddy covariance flux measurements began		
	April 1996	May 1994	1995 (water) and 1996 (carbon)	April 1996	July 1996

There are both plantations and natural forests of maritime pine (*Pinus pinaster)* everywhere in the southwestern part of Europe (SW France, Portugal, and Spain) and in Morocco. Botanical dictionaries (Richardson 1998) mention its presence on the coasts and islands of the Tyrrhenian Sea (Italian and French coasts, Sicily, Sardinia, and Corsica) as well as a patch in the Aurès ridge in Algeria near the Tunisian border. In France, maritime pines were planted during the 19th century for three main reasons: (1) to drain the marshes that covered Les Landes, (2) to stabilize the coastal dunes, and (3) to produce natural resins. Presently, only the wood of both pine species is exploited.

Although pine is the dominant (overstory) species in each of the five experimental sites, it should be emphasized here that many other species affected the flux measurements, such as oak and beech in the forest in Belgium (De Inslag), spruce in Finland (Hyytiälä), as well as various understory species at each of the sites (see Table 5.2).

5.2 Description of Experimental Sites and Data Collection

For the pine transect of the EUROFLUX network, five experimental sites were monitored along a north–south gradient in Europe. A comparative description of the five experimental sites, including general stand characteristics, ecological and biological features is given in Table 5.2. Long-term mean annual precipitation ranged from 530 mm a^{-1} (Sweden) to 900 mm a^{-1} (France). In the boreal sites the ground might be solid frozen for a major part of the year. However, in Finland only the first few centimeters are generally frozen, while at a depth of 5 cm the soil temperature remains stable around the freezing point (0 to 1 °C) during the winter due to the thick and stable snowpack. In this chapter and in the figures, all five sites will always be referred to by country rather than by site name. Leaf area index (LAI) values of the five sites were quite similar, although different methods had been used to assess LAI. The LAI of the site in Belgium ranged from 2 to 3 during the year (Gond et al. 1999). The LAI of the forest in Les Landes, France, ranged from 2.6 in winter to 3.1 at the end of summer or beginning of autumn. Since the life span of a maritime pine needle is 3 years, this means an LAI of about one per needle cohort. For the forests in Belgium and the Netherlands the LAI-2000 (LiCor Inc.) instrument – that uses diffuse light, integrated over a variety of angles – was employed, and all measurements were corrected for clumping. In France, the LAI was measured with a Demon instrument (CSIRO, Canberra, Australia), which is an optical system using the interception of the direct solar beam. Under the assumptions that the needles were randomly distributed and since all measurements were systematically made around 33° solar elevation, the measured LAI is half the developed needle surface area (one-sided LAI). This optically measured LAI, accurate to one tenth, was validated by some destructive measurements (Berbigier et al. 2001).

The LAI of the understory vegetation (and of some other species in the overstory) differed significantly among the five experimental sites, and could in some sites not be neglected. Because of the overstory speciation in the forest site in Belgium, only some 50 % of the overstory vegetation was Scots pine, the other main species being oak (*Quercus robur* L.) and beech (*Fagus sylvatica* L.). The understory was very important in this Belgian site, where the overstory canopy was far from closed and the LAI of the understory (*Rhododendron ponticum* and *Prunus serotina*, in particular) exceeded that of the overstory in some places (de Pury and Ceulemans 1997). At the Norunda site (Sweden) the forest was a mixture of Scots pine and Norway spruce (*Picea abies*) (Table 5.2). The understory of the French site appeared as a continuous cover of mainly Gramineae, which is typical for the wet Landes. The dominant species of the understory were *Molinia caerulea* and tore oat (*Arrhenaterum thorei*), while also ferns, heather and *Ulex* species were present. The LAI (one-sided) of the understory grasses was 1.5 at maximum in summer.

Data collection and treatment recommendations for EUROFLUX are described in Aubinet et al. (1999). Individual data collection methodologies for each site are described elsewhere (Grelle and Lindroth 1996; Rannik 1998; Vesala et al. 1998; Kowalski et al. 1999; Berbigier et al. 2001; Janssens et al. 2001). The analyses presented here made use of the half-hourly results (fluxes and meteorological variables) provided to the EUROFLUX database by each team. Data are from 1997 and (partially) 1998. For the forest sites in France and Sweden only a half year of data for 1998 were used in the analysis of the results presented here. Rules for data formatting and submission to the EUROFLUX database are described on the Internet: http://www.daac.ornl.gov/FLUXNET/euro_db.html. From these data two types of averages were computed for the presentations in this chapter. Scatter plots and annual trends are of daily means computed by averaging the 48 half-hour values for each day. In the case of missing data, a mean over at least 46 half hours was required to constitute a good daily estimate. Diurnal trends have been computed as the average of available monthly data for each hour in the day. No net radiometer was present at the forest site in Finland and no photosynthetic photon flux density (PPFD) was measured at the site in the Netherlands.

5.3 Results and Observations

5.3.1 Climate

The five sites of the EUROFLUX pine transect differ in terms of their climate, as illustrated by air temperature and precipitation (Fig. 5.1; note that the precipitation data from Sweden are missing). In Fig. 5.1, daily averages of air temperature and precipitation were plotted versus day of the year 1997 (annual trend). The 30-year mean is presented numerically, and as a dashed line for each site. Mean annual temperature had a very clear N–S trend, with lowest temperature in Finland (3.5 °C) and highest in France (13.5 °C), i.e., a difference of 10 °C over the entire N–S transect. In 1997 the mean air temperature measured at every site was about 1 °C warmer than the 30-year mean, except in the Netherlands, which was about 0.5 °C cooler. In comparison with the long-term (30 years) precipitation record, the year 1997 was relatively dry in northern Europe, while France received above-normal precipitation. The precipitation of the year 1998 was quite normal in Finland, but SW France was relatively dry during the first half of the year. In north–central Europe the year 1998 was extremely wet, and in the Belgian Campine region the most precipitation on historical record was measured.

5.3.2 Radiation

The annual trend in incoming shortwave ("global") radiation (Rg), which drives canopy photosynthetic processes, is presented for the five sites in 1997 in Fig. 5.2. At first glance, the five sites appear to be quite similar. However, some important differences can be seen. For the northern European sites, mean daily global radiation did not reach the annual mean value of 10 MJ m^{-2} day^{-1} until early or mid-March. At the French site, however, this value was reached about 1 month earlier, and the annual mean was nearly 40 % higher. The decrease in radiation in late fall was quite similar; the radiation climate in France lagged that of northern Europe by about 1 month, and midwinter values were much higher than they were further north.

Net (shortwave + longwave) radiation was more or less 70 % of the incoming global radiation (Fig. 5.3) for all periods and for all sites, except Finland which lacked net radiation measurements. In this ratio of net radiation to global radiation very little variation among the sites was observed. The incoming photosynthetic photon flux density (PPFD) was always half (50 %) of the global radiation, and no variation was observed among three sites (Fig. 5.4, Table 5.4). In Table 5.3 the regression statistics of the net radiation as a function of incoming (global) shortwave radiation for each site are summarized, while in Table 5.4 the regressions of PPFD versus incoming global radi-

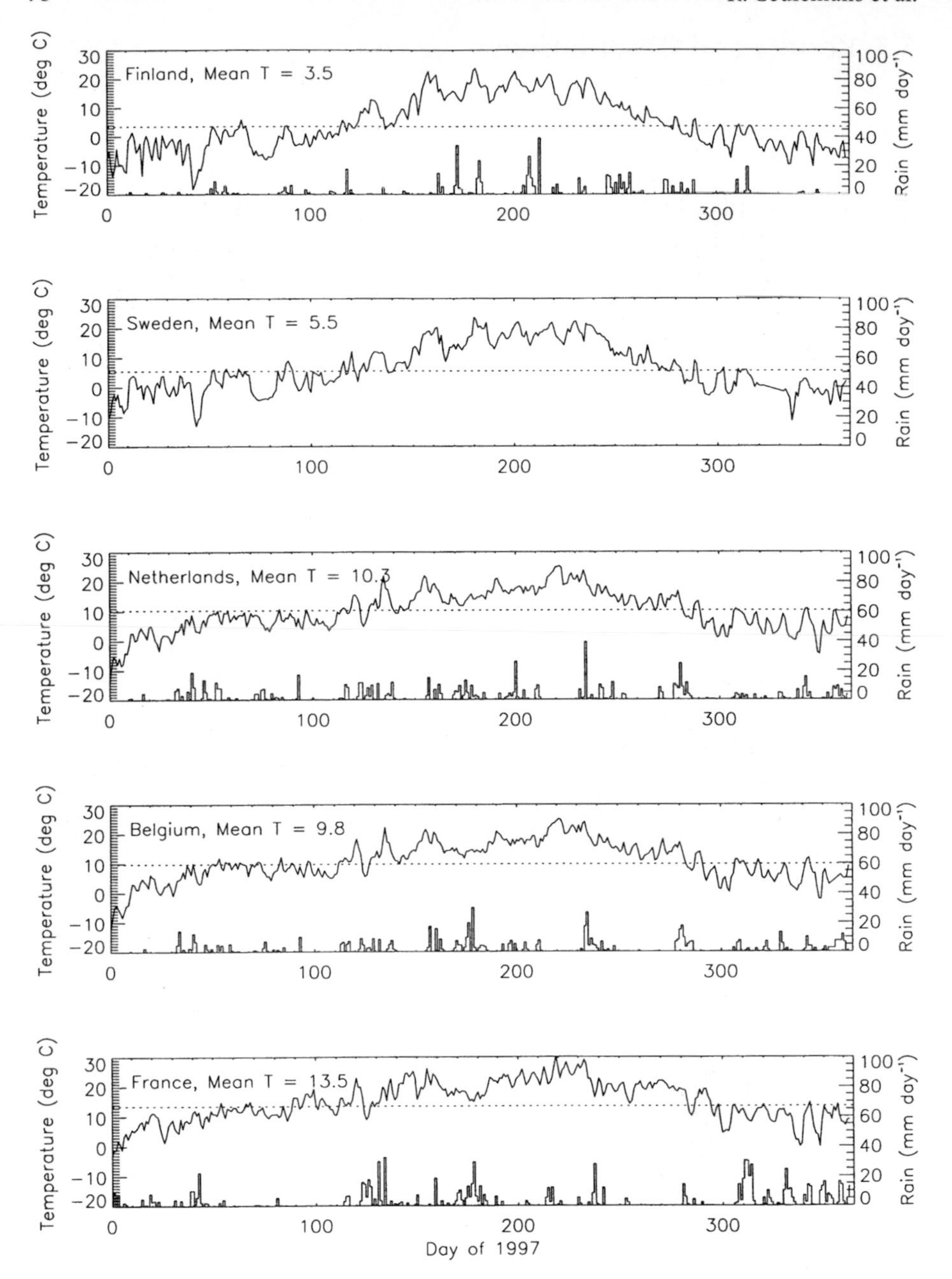

Fig. 5.1. Air temperature (°C) and precipitation (mm day^{-1}) during 1997 at the five sites of the EUROFLUX pine transect. The *dashed line* and the mean temperature represent the long-term (30-year) average temperature for each of the sites. The sites are presented along a N–S gradient. Each *point* represents a daily average

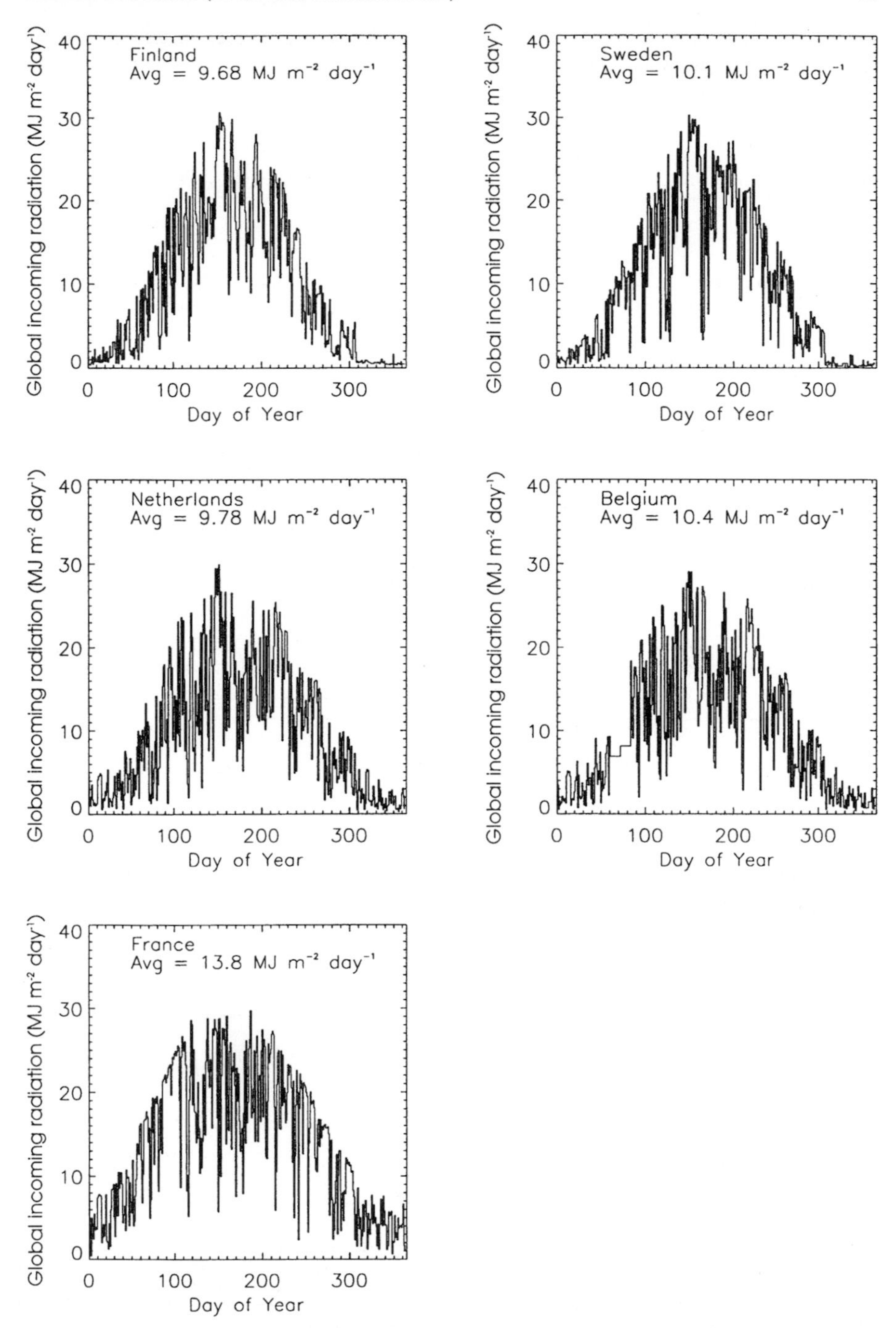

Fig. 5.2. Global incoming radiation (R_g, in MJ m^{-2} day^{-1}) for 1997 at the five sites of the EUROFLUX pine transect

Table 5.3. Regression statistics for the linear relationship (Fig. 5.3) of daily mean net radiation (R_n) depending on daily mean global radiation (R_g) for four of the five sites of the EUROFLUX pine transect

Name of site	Sweden Norunda	The Netherlands Loobos	Belgium De Inslag	France Le Bray, Les Landes
1997				
N	316	364	272	364
% Variance explained, linear model (100 × R^2)	93	92	95	96
Y-intercept (R_n for R_g=0; MJ m^{-2} day^{-1})	−2.5	−1.2	−1.5	−1.2
Dimensionless slope ($\Delta R_n/\Delta R_g$)	0.74	0.70	0.70	0.69
1998				
N	180	310	312	180
% Variance explained, linear model (100 × R^2)	88	92	93	95
Y-intercept (R_n for R_g=0; MJ m^{-2} day^{-1})	−1.7	−0.8	−1.8	−2.1
Dimensionless slope ($\Delta R_n/\Delta R_g$)	0.68	0.70	0.73	0.75

Table 5.4. Regression statistics for the linear relationship (Fig. 5.4) of daily mean photosynthetically active radiation (PPFD) depending on daily mean global radiation (R_g) for three of the five sites of the EUROFLUX pine transect

Name of site	Sweden Norunda	Belgium De Inslag	France Le Bray, Les Landes
1997			
N	349	140	364
% Variance explained, linear model (100 × R^2)	99	98	98
Y-intercept (PPFD for R_g=0; mol m^{-2} day^{-1})	0.1	−0.56	1.01
Slope (ΔPPFD/ΔR_g, mol MJ^{-1})	1.93	2.04	1.91
1998			
N	179	352	180
% Variance explained, linear model (100 × R^2)	99	99	98
Y-intercept (PPFD for R_g=0; mol m^{-2} day^{-1})	−0.1	−0.6	1.0
Slope (ΔPPFD/ΔR_g, mol MJ^{-1})	2.0	2.1	2.1

ation have been presented. For modeling purposes (see Table 5.4) the regression of PPFD versus incoming shortwave radiation is particularly useful (Jarvis et al. 1985).

5.3.3 Energy Budget

The energy budget for an ecosystem comprises terms associated with radiation, storage, metabolic activity, and fluxes to the soil and atmosphere. These fluxes are sensible heat and latent heat in the case of water vapor exchange. Under typical daytime conditions, the budget is dominated by net radiative heating, and cooling through exchange of sensible and latent heat with the atmosphere. Plants have some ability to control the balance between latent and sensible heat loss through stomatal activity. Here, energetics are considered both through the ability to close the energy budget, and through the Bowen ratio. The Bowen ratio is defined as the ratio of sensible to latent heat fluxes ($H/\lambda E$). The plots of energy budget closure (Fig. 5.5) not only illustrate the importance of the fluxes in relation to the other components, but also allow verification of the eddy covariance measurements (through the latent and sensible heat flux estimates). The month of April 1998 was exceptionally rainy in France – 179 mm. Rainfall was mainly concentrated within the days 92 to 107, which all had more than 5 mm day^{-1}, and eight of those days had rainfall above 10 mm day^{-1}. For this reason (i.e., the great atypicality), all of these days were excluded from Fig. 5.5.

The most interesting aspects here are believed to be the variation in Bowen ratio among the different sites (Fig. 5.6a,b). The plots of Fig. 5.6a,b demonstrate differences between sites for various times of the year (spring, summer, fall) in terms of water management strategy (Bowen ratio). A typical value for Bowen ratio over well-watered short grass or a wet soil surface is around 0.2. As the surface dries, the Bowen ratio becomes larger and approaches infinity for a dry surface. The Bowen ratio may become negative in arid areas as a result of advection (horizontal heat transport from surrounding areas) (Campbell 1997). To present the differences in Bowen ratio among the five different forest sites and among different time periods of the year, the months of April 1997 and July 1997 were selected (Fig. 5.6a,b). In April the Bowen ratio was larger than 1 at any time of the day and at all five sites, indicating a larger sensible heat loss than evaporative latent heat losses. At the Finnish and the Swedish sites evaporative heat losses were very small in April and hence resulted in far larger Bowen ratios (up to 4) as compared to the three other sites. At the Finnish site there was neither net carbon uptake nor transpiration in April. Evaporation from the surface of the snowpack – which disappeared around the beginning of May – was very small. These factors explain the large Bowen ratio in April in Finland. In July 1997 Bowen ratios were close to or smaller than unity, suggesting a larger proportion of latent heat losses during

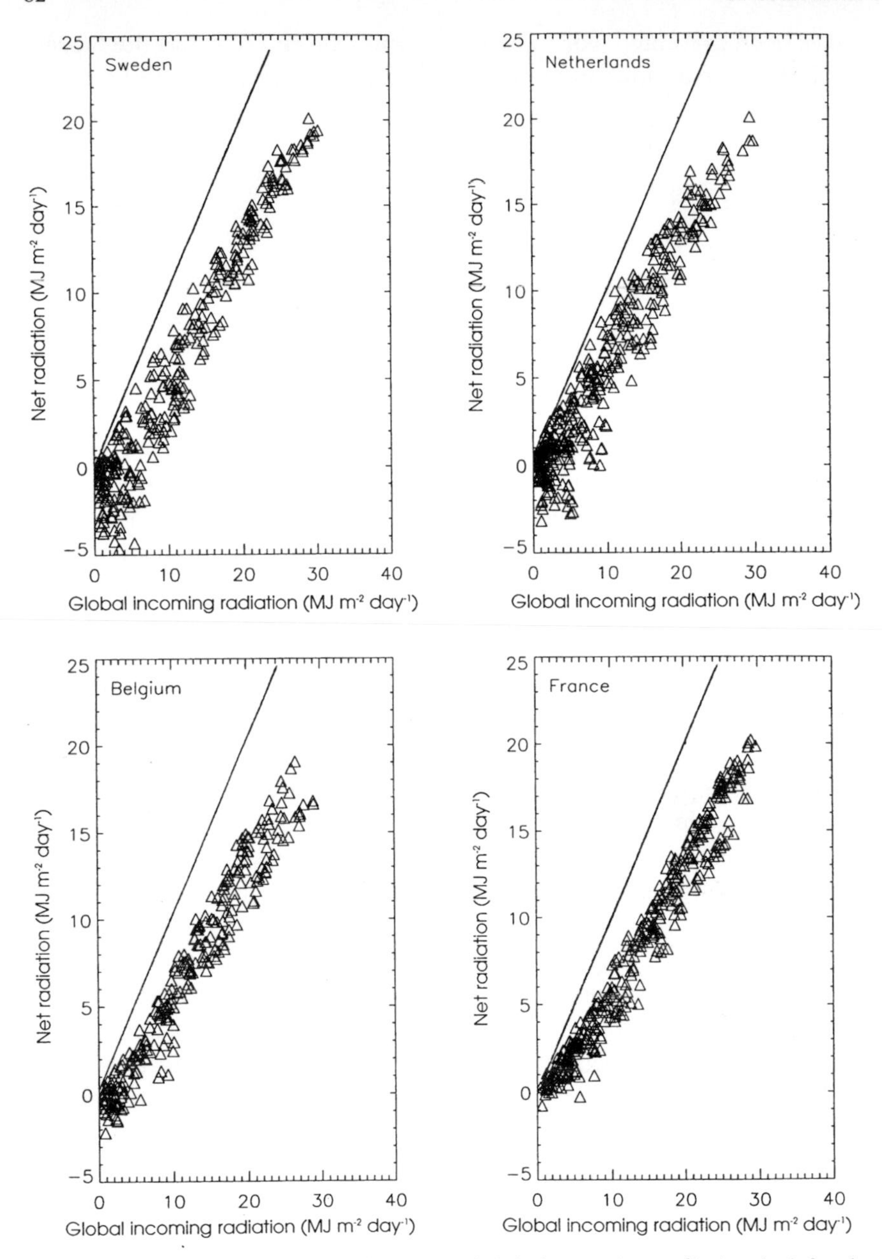

Fig. 5.3. Net radiation (R_n) as a function of global incoming radiation (R_g) for the year 1997 at four of the five sites of the EUROFLUX pine transect. Positive values represent downward fluxes. Both terms are in MJ m^{-2} day^{-1}. Each *point* represents a daily average. The regression equation and correlation coefficient are shown in Table 5.3

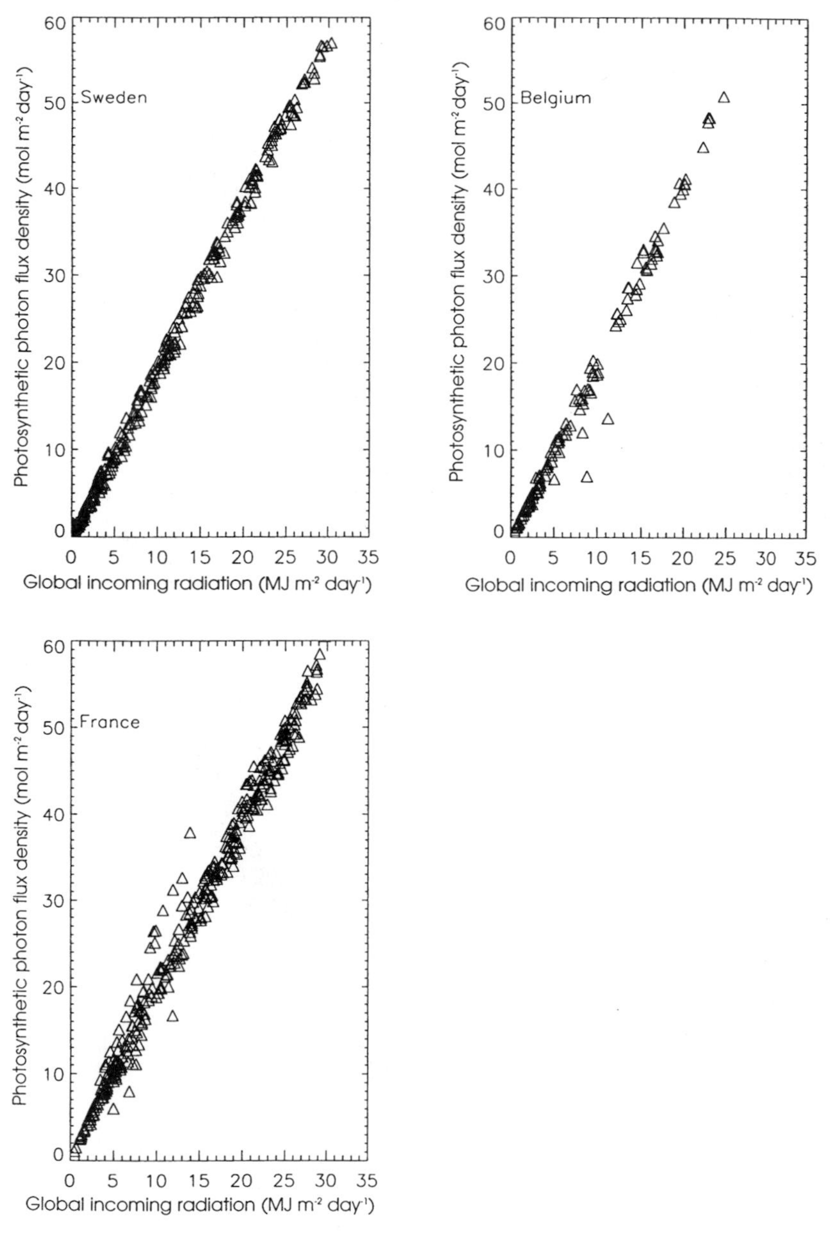

Fig. 5.4. Photosynthetic photon flux density (PPFD, in mol m^{-2} day^{-1}) as a function of global incoming radiation (R_g; in MJ m^{-2} day^{-1}) for 1997 at three of the five sites of the EUROFLUX pine transect. Positive values represent downward fluxes. Each *point* represents a daily average. The regression equation and correlation coefficient are shown in Table 5.4

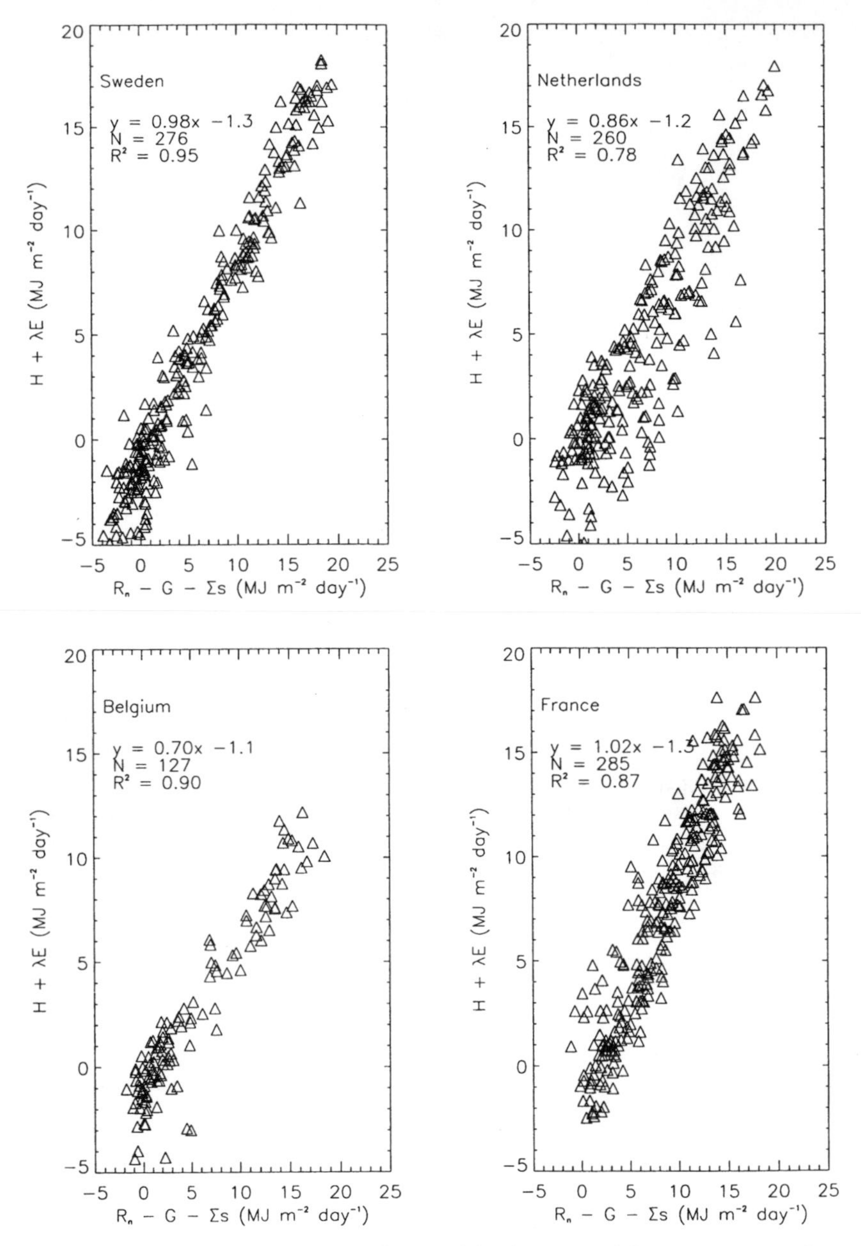

Fig. 5.5. Energy budget closure for four of the five sites of the EUROFLUX pine transect during 1997. Each *point* represents a daily average. All parameters are in MJ m^{-2} day^{-1}. *H* sensible heat flux; λE latent heat flux; R_n net radiation; *G* soil heat flux; Σs sum of latent and sensible heat storage in the canopy air (no vegetation heat storage considered). All fluxes are defined positive-upwards except for R_n

summertime at all five sites. Differences in Bowen ratios among sites were much more pronounced in spring than in summer (Fig. 5.6a,b).

5.3.4 Carbon Dioxide Exchanges

Extrapolation of available results to evaluate the long-term effects of an increase in atmospheric CO_2 concentration on net exchange of CO_2 and water vapor between forests and the atmosphere at the ecosystem and regional scale requires further consideration. The annual results for 2 years of NEE measurements at the fives sites are presented in Figs. 5.7 and 5.8. It is evident from these figures that in Finland and Sweden (and to some extent also in the Netherlands), there are no net CO_2 fluxes nor net sinks in winter. The growing season is defined here as the period during which there is a net carbon fixing (photosynthetic) activity of the forest canopy. In Finland the winter – i.e., the period during which there is no net carbon fixing activity of the forest canopy – may extend from day 280 to day 120. In Sweden the long net inactivity was probably related to the frozen soil. In Finland there can be photosynthesis in April or even in March, but this could not be detected by the eddy covariance measurements since soil respiration is dominating the net flux. In Finland the length of the growing season is not related to the frozen soil, but analysis has shown that the determining factor for the start of photosynthesis or net sink activity is cumulative air temperature. This has been observed from both chamber measurements as well as from eddy covariance measurements. The soil temperature remains very stable until mid-May and does not reveal anything of length of the growing season. In more southern sites, the period of net photosynthetic activity of the forest canopy is very long. At the French site where the soil is never frozen, the winter CO_2 balance was positive, i.e., the absolute value of respiration was larger than the absolute value of photosynthesis, at least in November and December.

A number of observations become clear from Figs. 5.7 and 5.8. First, it seems that the sites in Sweden, France, and the Netherlands had larger day-to-day variability than the sites in Belgium and Finland. Furthermore, the annual trends in net ecosystem exchange (NEE) were very similar everywhere, except for Sweden. From looking at the 25-day running means, it also seems that there was a "default" trend, which can be described as symmetric and somewhat bell-shaped, with a small release of CO_2 in the winter, and larger uptake in the summer. This is seen in Finland and France (both years), and Belgium and the Netherlands (1997 only). The ones that did not fit this pattern (Sweden, and then Belgium and the Netherlands in 1998) showed rather sharp transitions, which were probably associated with some sort of extreme conditions or disturbance. For example, in 1998 around day 230 (in late August and September) both forest sites in Belgium and in the Netherlands showed a sharp transition toward being sources. This coincided neatly with extreme

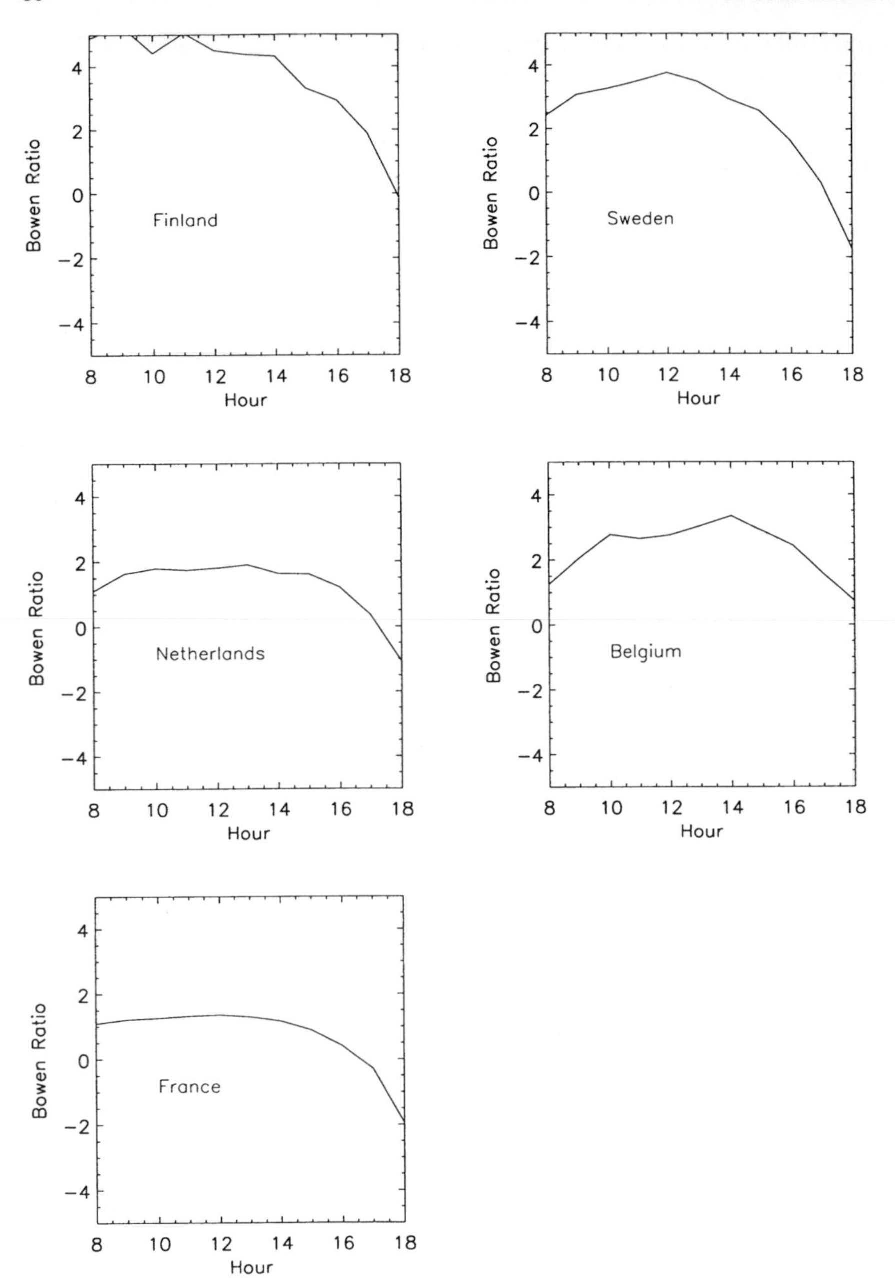

Fig. 5.6. **a** Daytime trend in Bowen ratio (ratio of sensible to latent heat fluxes from the ecosystem) for the five sites of the EUROFLUX pine transect. These data represent monthly ensembles for each hour of day during the month of April 1997. **b** Daytime trend in Bowen ratio (ratio of sensible to latent heat fluxes from the ecosystem) for the five sites of the EUROFLUX pine transect. These data represent monthly ensembles for each hour of day during the month of July 1997

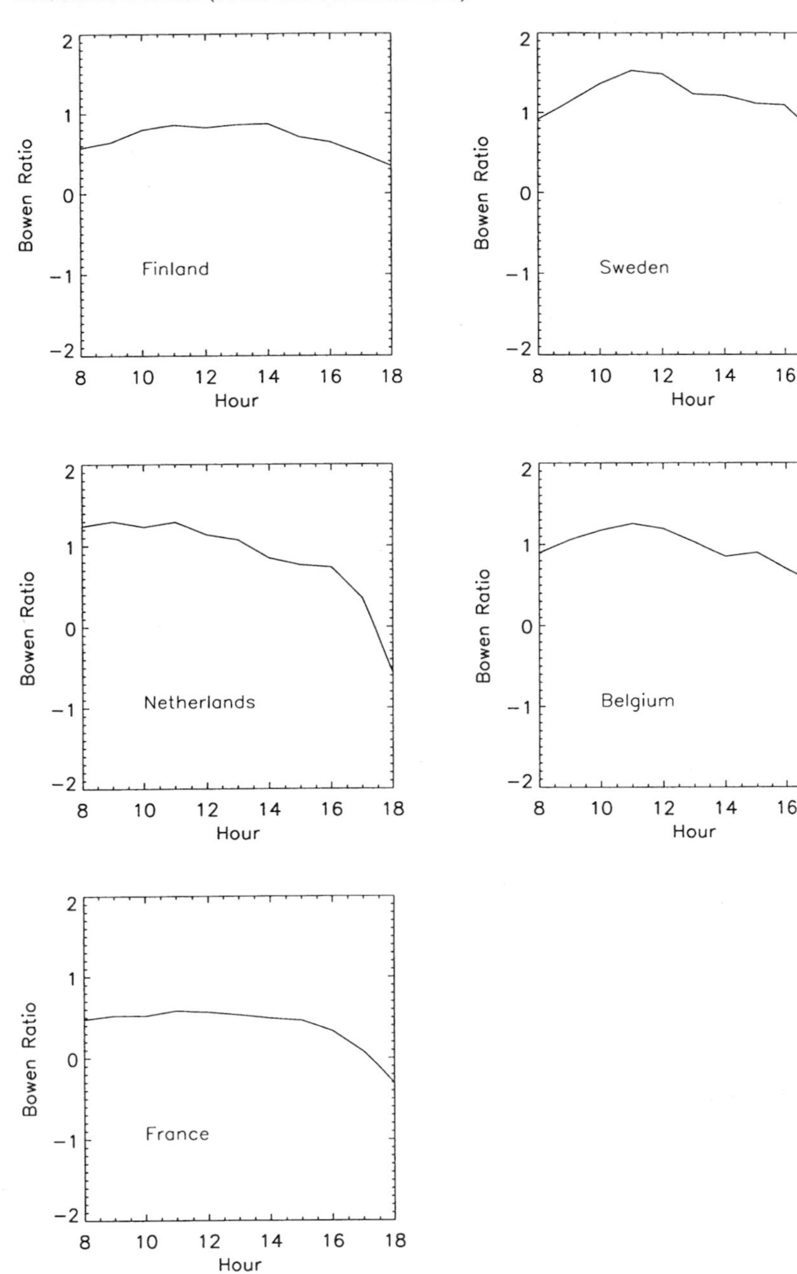
Finland
Sweden
Netherlands
Belgium
France
Bowen Ratio
Hour
2
1
0
−1
−2
8
10
12
14
16
18

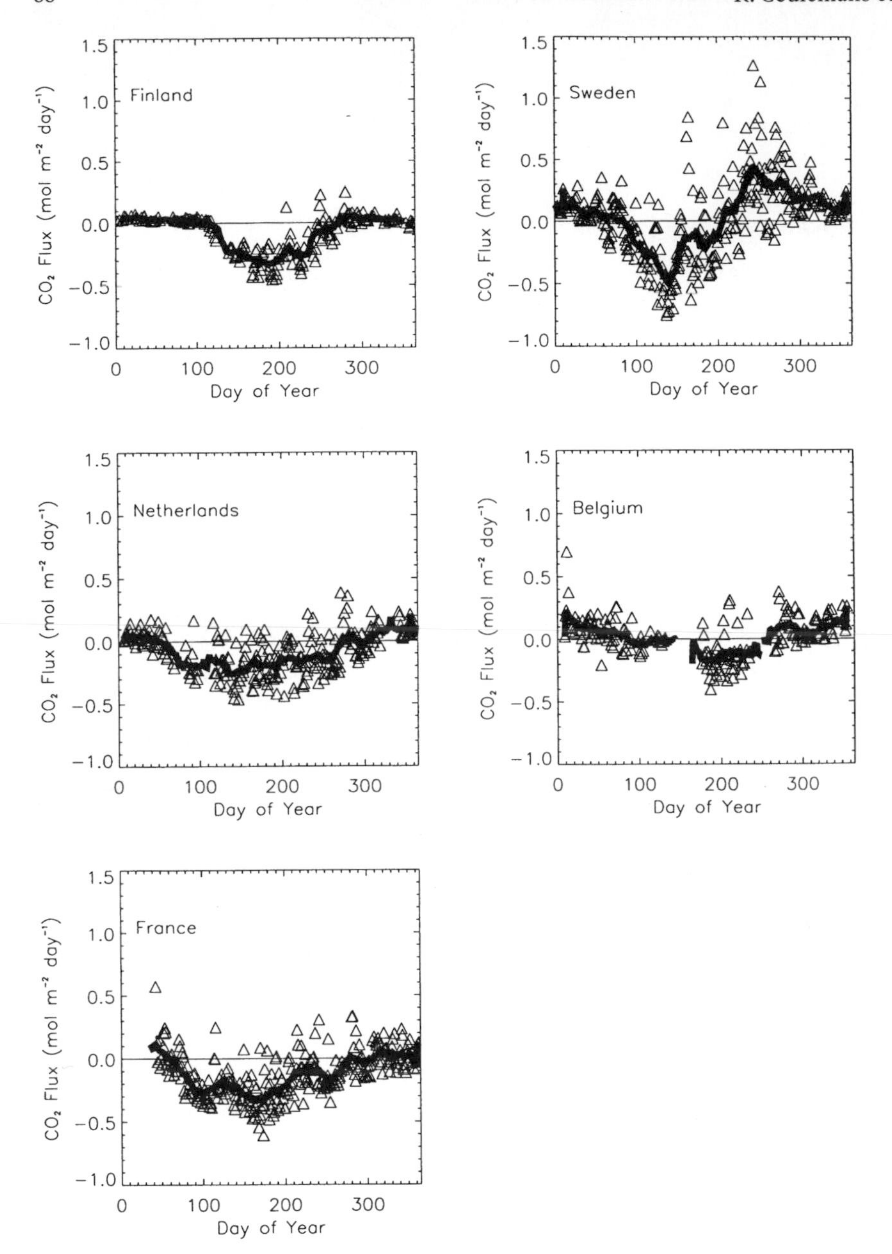

Fig. 5.7. Annual evolution of daily mean CO_2 flux (positive fluxes denote release by the ecosystem) during 1997 for the five sites of the EUROFLUX pine transect. The *solid line* represents a 25-day running average

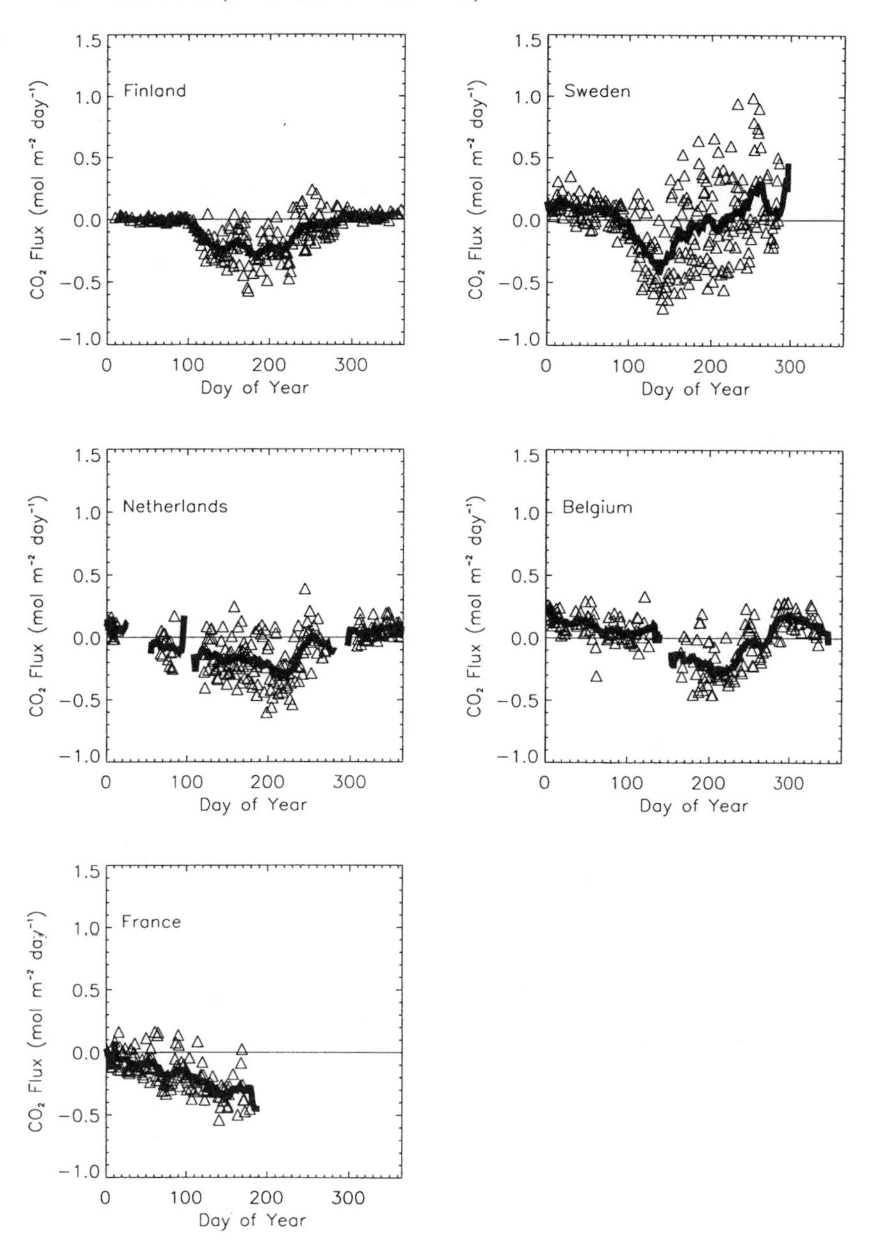

Fig. 5.8. Annual evolution of daily mean CO_2 flux (positive fluxes denote release by the ecosystem) during 1998 for the five sites of the EUROFLUX pine transect. The *solid line* represents a 25-day running average. There are no data for the French site after June

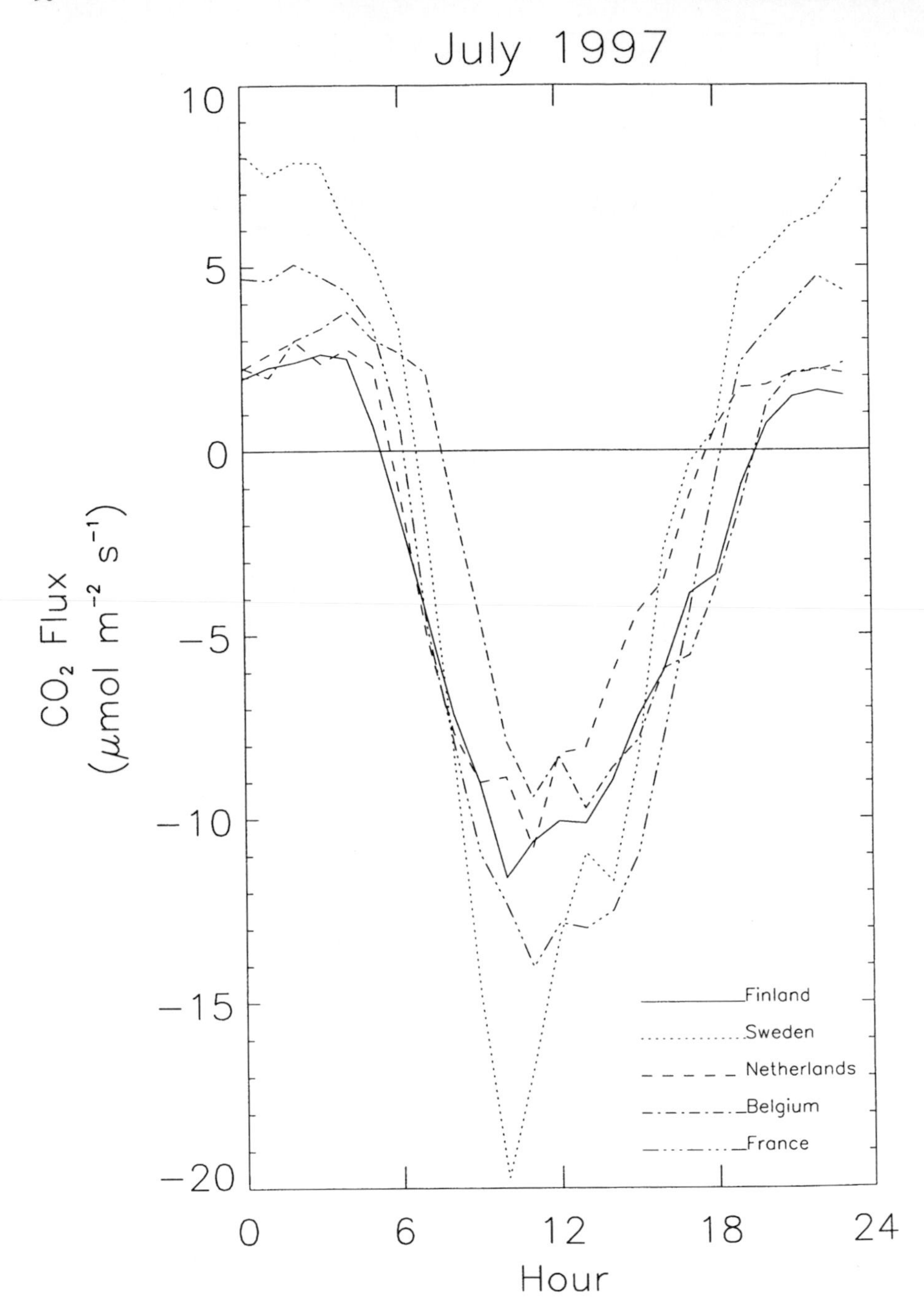

Fig. 5.9. Diurnal trend in CO_2 flux (positive fluxes denote release by the ecosystem) for the five sites of the EUROFLUX pine transect. These data represent monthly ensembles for each and every hour during the months of July 1997 (**a**) and January 1998 (**b**)

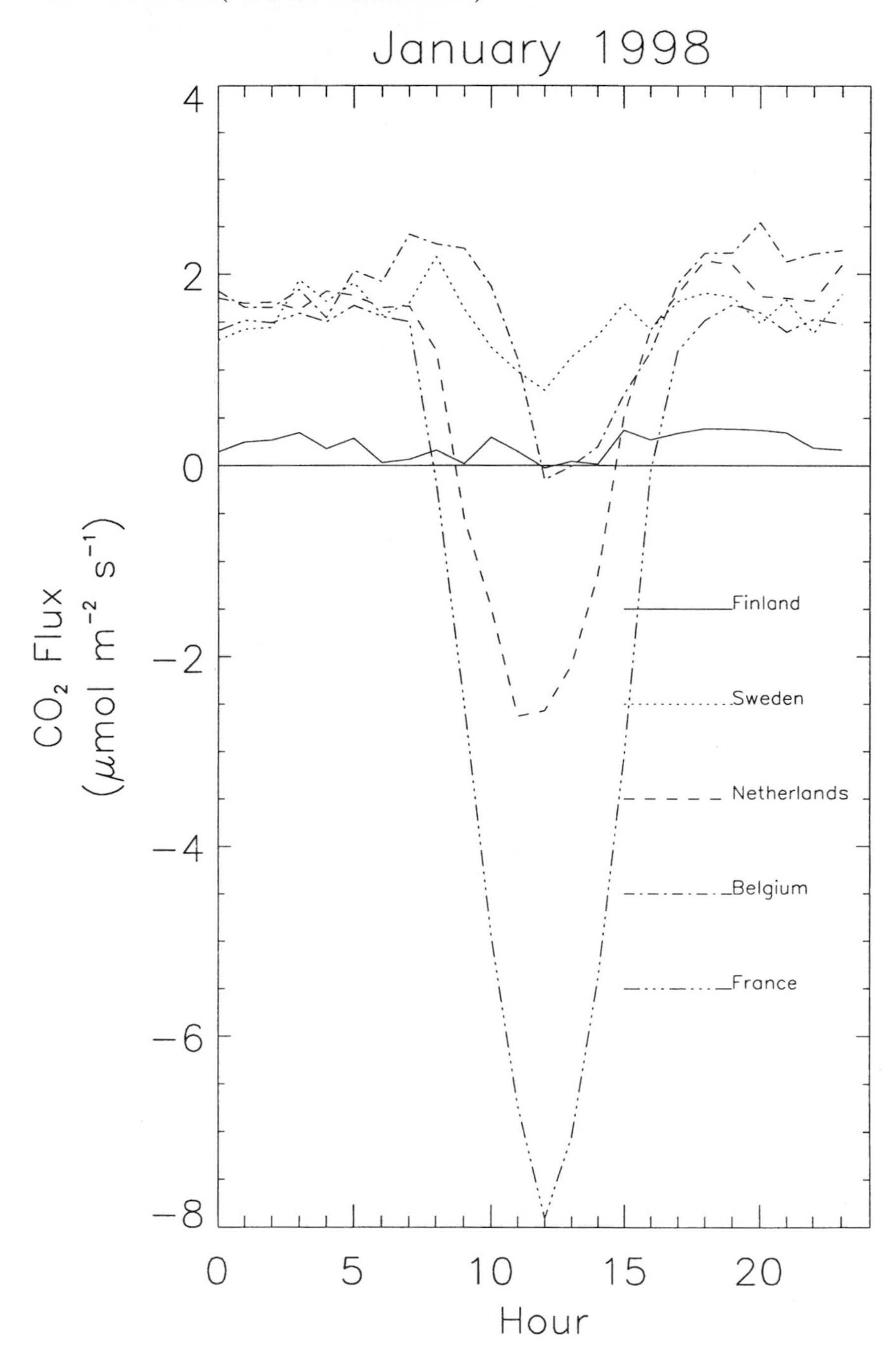
January 1998
4
2
0
−2
−4
−6
−8
CO_2 Flux
(μmol m^{-2} s^{-1})
Finland
Sweden
Netherlands
Belgium
France
0
5
10
15
20
Hour

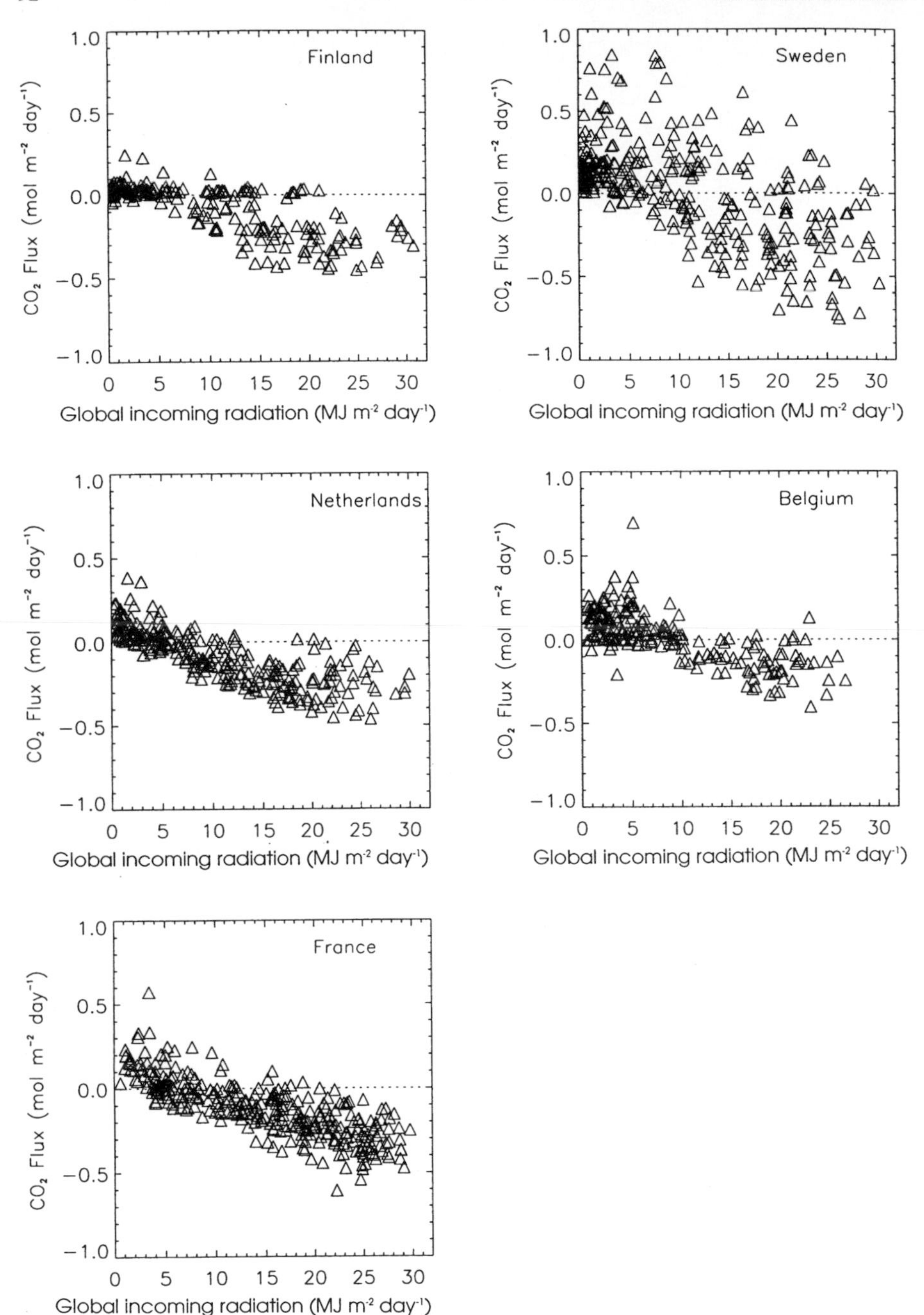

Fig. 5.10. **a** Response of the CO_2 flux (in mol m^{-2} day^{-1}, positive fluxes denote release by the ecosystem) to global radiation (R_g, in MJ m^{-2} day^{-1}) for 1997 at the five sites of the EUROFLUX pine transect. Each *point* represents a daily average. **b** Response of the CO_2 flux (in mol m^{-2} day^{-1}, positive fluxes denote release by the ecosystem) to global radiation (R_g, in MJ m^{-2} day^{-1}) for 1998 at the five sites of the EUROFLUX pine transect. Each *point* represents a daily average

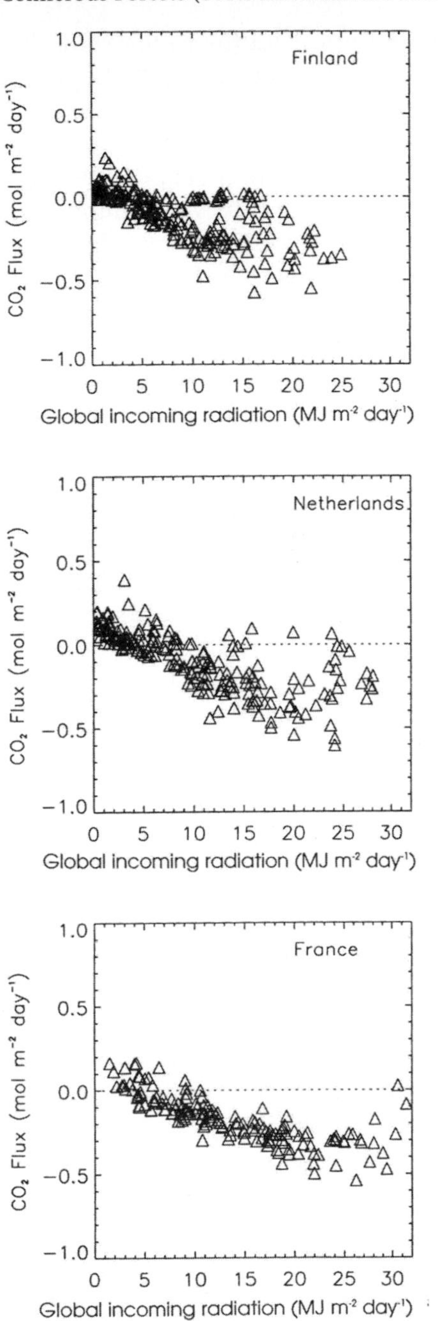
Finland
Netherlands
France
CO_2 Flux (mol m^{-2} day^{-1})
Global incoming radiation (MJ m^{-2} day^{-1})
1.0
0.5
0.0
−0.5
−1.0
0
5
10
15
20
25
30

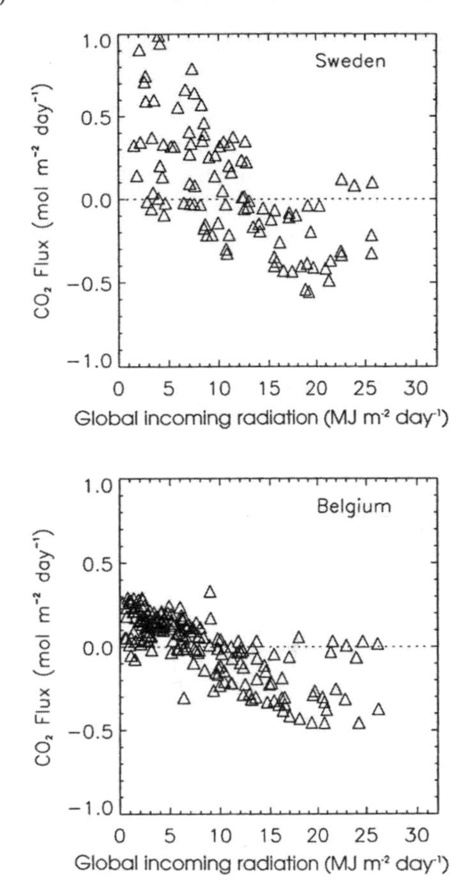
Sweden
Belgium
CO_2 Flux (mol m^{-2} day^{-1})
Global incoming radiation (MJ m^{-2} day^{-1})
1.0
0.5
0.0
−0.5
−1.0
0
5
10
15
20
25
30

precipitation. Sweden was different from Belgium and the Netherlands, but actually rather similar between the two years: the forest rapidly became a strong sink in the spring, but around day 135 something happened and it became a weaker sink in the summer, and finally a strong source in the fall. In considering a plausible explanation, we lack precipitation data for 1997, and the year 1998 showed no extreme drought or flood. There was nothing obvious in the temperature record, except that the high respiration period (days 135–250) went along with the warmest part of the year. One hypothesis is that above a certain soil temperature, the temperature dependence of respiration increases, which results in significant enhancement of ecosystem respiration.

The diurnal cycles of NEE for the five different sites are presented in Fig. 5.9a,b.What strikes one most from Fig. 5.9a,b is the small difference between the five sites in summer, despite the differences in species composition, LAI and tree density. Finland and the two sites in Belgium and the Netherlands were very similar, but the sites in France and Sweden had much larger signals (both photosynthesis and respiration, day and night). The trend in CO_2 flux followed the trend in day length. In July (Fig. 5.9a), Finland had the longest days, and the longest period of uptake, while France had the shortest. The peak in photosynthesis seemed earlier in Sweden and Finland. While all of the data were on European standard time, not solar time, the sites in Finland and Sweden are way east of the other sites, which might possibly account for this. The larger nighttime flux in France could be explained by the higher soil temperature. In January (Fig. 5.9b), France and the Netherlands seemed to photosynthesize, Belgium was not very clearly pronounced in this regard (it was very cold in January 1998), Sweden was somewhat in between, and Finland showed a zero flux both day and night. The responses of NEE to radiation for the five ecosystems are presented in Fig. 5.10a,b.

The relationships between daily NEE and incoming global radiation (Rg) were not very different from site to site (with the exception of Sweden). This means that pines react in all experimental sites more or less the same way to incoming radiation. Differences among the various sites and plots are seen as the number of points with low Rg, which increases for the northern sites. Further, the repartitioning of overcast was very different: the difference in repartitioning of Rg is rather clear from Fig. 5.10a,b. Possible explanations for the differences observed in the fluxes of carbon and water among the five sites are differences in: (1) soil moisture (periods of drought, differences in precipitation, soil moisture holding capacity), (2) soil type, (3) nitrogen deposition and fertilization conditions of the soil, (4) understory species composition (competition with overstory) (Table 5.2).

The large variability in NEE in Sweden (Figs. 5.9 and 5.10) largely depends on the relatively large ecosystem respiration that this site exhibits. As soil respiration is very sensitive to temperature and is highly variable (exponential temperature dependence with a high sensitivity to temperature; Janssens et al. 2001), the variability in NEE can be very large because of the variation in temperature. This very pronounced temperature dependence might not be so clear at the other sites of the pine transect. Another point is that there is of course also a variability in uptake because of the very variable radiation conditions. The daily ecosystem respiration has been calculated and represented using the temperature-dependent function of Lindroth et al. (1998). When the difference between adjacent days is considered, it is clear that the day-to-day variation was typically in the order of ca. 10 g m^{-2} day^{-1} (Grelle and Lindroth 1996; Lindroth et al. 1998). This is of the same order of magnitude as the NEE and, thus, the large day-to-day variability in NEE can most certainly be ascribed to the variation in temperature with corresponding (enhanced) variation in ecosystem respiration. Lindroth et al. (1998) have analyzed nighttime fluxes from the tower in Sweden under good mixing conditions and could only find a very minor dependency on wind direction. Furthermore, the day-to-day variation in Sweden can hardly be an instrumental error because (1) the eddy covariance system in Sweden shows one of the best energy balance closures of all the systems (Fig. 5.5), (2) the CO_2 flux is measured with the same instrument as the water vapor, and (3) all three flux systems on the tower exhibit almost exactly the same pattern of variability during daytime. Finally, interannual variability (in climatic conditions and fluxes) is a predominant factor in the variation observed in NEE among the different sites (Goulden et al. 1996; Baldocchi et al. 1997).

5.4 Conclusions

The climate does vary among the five different sites of the EUROFLUX pine transect as illustrated by air temperature, precipitation, and incoming radiation. However, what is striking from the results presented in this chapter on pine forest sites, is the similarity of the NEE response to similar incoming radiation stimuli. Without any doubt soil respiration is very important (Valentini et al. 2000; Janssens et al. 2001) and might, for instance at the French site, dramatically reduce the NEE. Some of the differences among the sites might also be related to the fact that maximum photosynthetic rate varies a lot along the latitudinal gradient and is coupled with stomatal conductance (Luoma 1997; Valentini et al. 2000). Finally, issues such as management of the site through the years are likely important factors determining NEE. At the Sweden site, for example, past drainage has likely enhanced soil aeration and increased heterotrophic respiration rates and accelerated decomposition of old soil organic matter (Lindroth et al. 1998), resulting in its very high soil and ecosystem respiration rates among the pine forests studies in the transect.

References

Aubinet M, Grelle A, Ibrom A, Rannik U, Moncrieff J, Foken T, Kowalski AS, Martin PH, Berbigier P, Bernhofer C, Clement R, Elbers J, Granier A, Grünwald T, Morgenstern K, Pilegaard K, Rebmann C, Snijders W, Valentini R, Vesala T (1999) Estimates of the annual net carbon and water exchange of forests: the EUROFLUX methodology. Adv Ecol Res 30:114–175

Baldocchi DD, Vogel CA, Hall B (1997) Seasonal variation of carbon dioxide exchange rates above and below a boreal jack pine forest. Agric For Meteorol 83:147–170

Berbigier P, Bonnefond J-M, Mellmann P (2001) CO_2 and water vapour fluxes for 2 years above Euroflux forest site. Agric For Meteorol 108:183–197

Campbell GS (1997) An introduction to environmental biophysics. Springer, Berlin Heidelberg New York, 159 pp

de Pury DGG, Ceulemans R (1997) Scaling-up carbon fluxes from leaves to stands in a patchy coniferous/deciduous forest. In: Mohren GMJ, Kramer K, Sabate S (eds) Impacts of global change on tree physiology and forest ecosystems. Kluwer, Dordrecht, pp 263–272

Gholz HL, Linder S, McMurtrie RE (1994) Environmental constraints on the structure and productivity of pine forest ecosystems: a comparative analysis. Ecol Bull (Copenh) 43:1–198

Gond V, de Pury DGG, Veroustraete F, Ceulemans R (1999) Seasonal variations in leaf area index, leaf chlorophyll and water content; scaling-up to estimate fAPAR and carbon balance in a multilayer, multispecies temperate forest. Tree Physiol 19:673–679

Goulden ML, Munger JW, Fan SM, Daube BC, Wofsy SC (1996) Exchange of carbon dioxide by a deciduous forest: response to interannual climate variability. Science 271:1576–1578

Grelle A, Lindroth A (1996) Eddy-correlation system for long term monitoring of fluxes of heat, water vapour, and CO_2. Global Change Biol 2:297–307

Janssens IA, Lankreijer H, Matteucci G, Kowalski AS, Buchmann N, Epron D, Pilegaard K, Kutsch W, Longdoz B, Grunwald T, Montagnani L, Dore S, Rebmann C, Moors EJ, Grelle A, Rannik U, Morgenstern K, Oltchev S, Clement R, Gudmundsson J, Minerbi S, Berbigier P, Ibrom A, Moncrieff J, Aubinet M, Bernhofer C, Jensen NO, Vesala T, Granier A, Schulze ED, Lindroth A, Dolman AJ, Jarvis PG, Ceulemans R, Valentini R (2001) Productivity overshadows temperature in determining soil and ecosystem respiration across European forests. Global Change Biol 7:269–278

Jarvis PG, Miranda HS, Muetzelfeldt RI (1985) Modelling canopy exchanges of water vapor and carbon dioxide in coniferous forest plantations. In: Hutchinson BA, Hicks BB (eds) The forest-atmosphere interaction. Reidel, New York, pp 521–542

Kowalski AS, Overloop S, Ceulemans R (1999) Eddy fluxes above a Belgian, Campine forest and relationships with predicting variables. In: Ceulemans R, Veroustraete F, Gond V, Van Rensbergen J (eds) Forest ecosystem modelling, upscaling and remote sensing. SPB Academic, The Hague, pp 3–17

Lindroth A, Grelle A, Moren AS (1998) Long-term measurements of boreal forest carbon balance reveal large temperature sensitivity. Global Change Biol 4:443–450

Luoma S (1997) Geographical pattern in photosynthesis light response of *Pinus sylvestris* in Europe. Funct Ecol 11:273–281

Rannik Ü (1998) On the surface layer similarity at a complex forest site. J Geophys Res 103:8685–8697

Richardson DM (1998) Ecology and biogeography of *Pinus*. Cambridge Univ Press, Cambridge, 527 pp

Stenberg P, Kuuluvainen T, Kellomaki S, Grace JC, Jokela EJ, Gholz HL (1994) Crown structure, light interception and productivity of pine trees and stands. Ecol Bull (Copenh) 43:20–34

Valentini R, Matteucci G, Dolman AJ, Schulze ED, Rebmann C, Moors EJ, Granier A, Gross P, Jensen NO, Pilegaard K, Lindroth A, Grelle A, Bernhofer C, Grünwald T, Aubinet M, Ceulemans R, Kowalski AS, Vesala T, Rannik U, Berbigier P, Loustau D, Gudmundsson J, Thorgeirsson H, Ibrom A, Morgenstern K, Clement R, Moncrieff J, Montagnani L, Minerbi S, Jarvis PG (2000) Respiration as the main determinant of carbon balance in European forests. Nature 404:861–865

Vesala T, Aalto P, Altimir N et al (1998) Long-term measurements of atmosphere-surface interactions in boreal forest combining forest ecology, micrometeorology, aerosol physics and atmospheric chemistry. Trends Heat Mass Momentum Transfer 4:17–35

6 Spruce Forests (Norway and Sitka Spruce, Including Douglas Fir): Carbon and Water Fluxes and Balances, Ecological and Ecophysiological Determinants

C. Bernhofer, M. Aubinet, R. Clement, A. Grelle, T. Grünwald, A. Ibrom, P. Jarvis, C. Rebmann, E.-D. Schulze, J.D. Tenhunen

6.1 Introduction

Natural forests with a high percentage of spruce will be found in Europe only in subalpine and alpine regions or in the boreal forests of Scandinavia and Russia. Nevertheless, spruce belongs to the most important European tree species. Due to its favorable architecture and rapid growth, management practices have led to widespread monospecific spruce forests in a latitude band between 45 and 55°N. Only recently have other species such as beech and oak been included in afforestation efforts at former pure spruce sites. So today spruce is still an important forestry tree in central Europe, with a coverage for instance of about 33% of all forested areas in Germany (Bundesministerium für Ernährung, Landwirtschaft und Forsten 1999).

This chapter deals not only with Norway spruce, but with a variety of dense conifers (*Picea abies, Picea sitchensis, Pseudotsuga menziesii*). Two of the Norway spruce sites, the Sitka spruce, and the Douglas fir sites were intensively modeled (Falge et al., this Vol.). They all showed similar features despite their different species composition. This justifies considering all sites along with the four Norway spruce stands that dominate this group of dense conifers within EUROFLUX.

Fluxes of energy, water, and carbon of spruce forests have been investigated by a multitude of methods in the past. The considerable insight that has been gained into characteristics and processes of coniferous stands is demonstrated, e.g., by Jarvis et al. in Monteith's, *Vegetation and the Atmosphere* (Monteith 1976), or in Larcher's textbook, *Ökophysiologie der Pflanzen* (Larcher 1994). Almost continuous flux measurements of water vapor and carbon dioxide by micrometeorological methods had been made and analyzed for a spruce forest near Munich/Bavaria (Ebersberger Forst) by Hager as early as 1972. He reported an annual budget of 26.3 μg ha^{-1} of CO_2 sequestration for 1972 (Hager 1975), demonstrating the large potential sink of carbon in fast growing spruce stands in central Europe.

Ecological Studies, Vol. 163
R. Valentini (Ed.) Fluxes of Carbon, Water and Energy of European Forests

6.2 Material and Methods

The EUROFLUX spruce sites reflect both, the "natural" and the "artificial" abundance of spruce forests in EUROPE (see Table 6.1). They include a Sitka spruce plantation (*Picea sitchensis*) in Scotland (15 years), three managed Norway spruce (*Picea abies*) forests in Germany, and a boreal Norway spruce site in Sweden (30 years). The German sites differ in climate, age (45 to 110 years) and altitude (380 to 780 m a.s.l.). The soil characteristics differ considerably from dystrophic cambisol to sandy till and podzolized brown earth; they reflect at the same time the individual site history ranging from former peat land (Scotland) to areas that were forests since the end of the last ice age (Tharandt, Germany, and Flakaliden, Sweden). None of the sites is "natural", but Flakaliden and Weidenbrunnen/Fichtelgebirge, Germany would also be expected to have a large spruce percentage under non-managed conditions. Tharandt is used as a typical example for a managed spruce forest outside its region of natural abundance, while Flakaliden is used as an example of such a forest within the area of natural abundance of spruce. These two sites contribute the majority of graphs and figures; other sites show similar behavior unless specified.

The methods applied within EUROFLUX are discussed in detail elsewhere in this publication. All measurements conform to the EUROFLUX standards, and all flux software has been tested (Aubinet et al. 2000). Flux measurements of carbon dioxide and water vapor were all done with the eddy covariance technique. Eddy covariance measurements typically have source areas (often called "footprints") that are larger than a typical afforestation plot. Site characteristics such as perfect homogeneity and level terrain are not typical for forested areas in central and western Europe. At the same time, the sites are subject to standard forestry practices. The spruce sites reflect this shortcoming; e.g., at Tharandt there is a regeneration plot (understory of *Fagus sylvatica*) in a small portion of the footprint and several single non-evergreen trees (*Larix decidua, Pinus sylvestris*). We do not expect major effects due to such heterogeneities as the contributing leaf area is still very small.

Table 6.1. Location, climate, and ecological parameters, reference year 1999

Site	Weiden-brunnen (GE1)	Tharandt (GE2)	Flakaliden (SW2)	Aberfeldy (UK1)	Vielsalm (BE1)	Solling (GE ext)
Coordinates	50°09′N 11°52′E	50°58′N 13°34′E	64°07′N 19°27′E	56°37′N 3°48′W	50°18′N 6°00′E	51°46′N 9°35′E
Altitude (m)	780	380	225	340	450	505
Mean annual air temperature (°C)	5.8	7.7	1.0	8.0	7.7	6.6
Sum of annual precipitation (mm)	890	820	570	1400	1000	1045
Species	*Picea abies*	*Picea abies (dominating)*	*Picea abies*	*Picea sitchensis*	*Picea abies, Pinus sylv., Pseudotsuga menziesii*	*Picea abies*
Age (years)	45	113 (108)[a]	30	15	65 (Douglas fir)	110
Canopy height (m)	19	29 (27)	8	6.8	35	29
Tree density (no. ha^{-1})	1000	440 (480)	2100	2500	200	461
Mean tree diameter (cm)		33 (33)			48	37
LAIp	5.5	8 (7.6)	2.4	8	5.3	7
Above ground wood biomass (kg m^{-2})	17.7	22.4 (21.3)			42.6	24.4
Soil type	Brown earth	Podzolized brown earth	Sandy till	Stony podzolized brown earth	Dystrophic cambisol	Sandy podsolized brown earth

[a] Values in parenthesis give turbulent source area weighted values according to Mellmann (pers. commun.); the dominating spruce forest part at Tharandt was founded by seeding in 1887.

6.3 Results

6.3.1 Radiation

Table 6.2 summarizes main radiation characteristics of the dense conifer sites for 1996 through 1998. Albedo values are averages derived from the ratio of the sum of reflected and the sum of incoming solar radiation. Figure 6.1 demonstrates the typical dependence for the Weidenbrunnen site on zenith angles, with a minimum around 8 % and values close to 20 % for times close

Table 6.2. Radiation characteristics of dense conifer stands across Europe

	Albedo α (%)	Solar Radiation R_g ($W\ m^{-2}$)			PPFD Q ($k\ mol\ m^{-2}$)			Net Radiation R_n ($W\ m^{-2}$)		
		1996	1997	1998	1996	1997	1998	1996	1997	1998
Weiden-brunnen	~8		117.3	111.0		8.13	6.99		70.4	61.2
Tharandt	7.6	111.2	125.3	116.2	6.88	8.00	7.08	52.7	65.1	61.6
Flakaliden						7.04	5.60		61.1	52.3
Aberfeldy	9.7		96.9	95.8		5.98	5.25		54.8	48.5
Vielsalm			118	98		6.84	5.50		72	63
Solling F1	6	109	119							

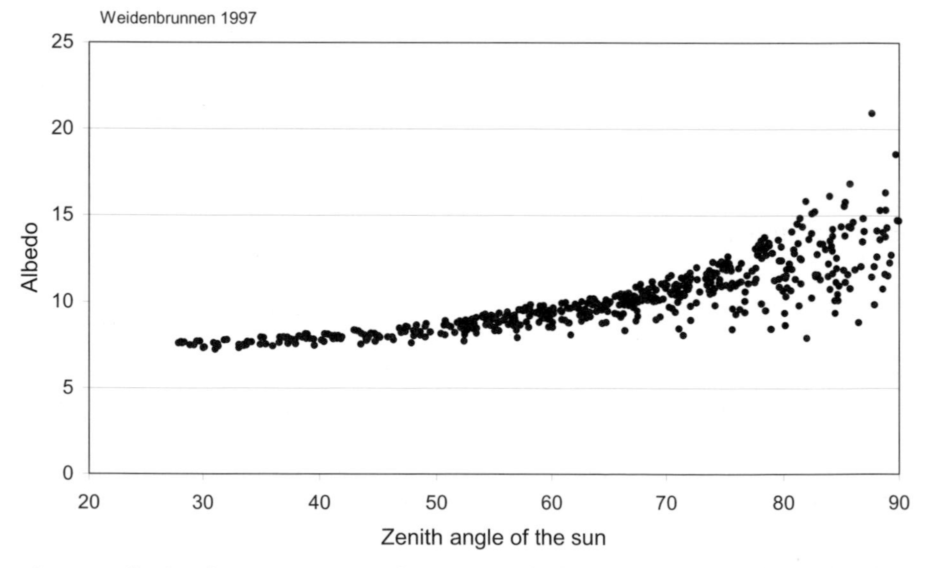

Fig. 6.1. Albedo of a 60 year spruce forest at Weidenbrunnen (Fichtelgebirge) for clear days in 1997

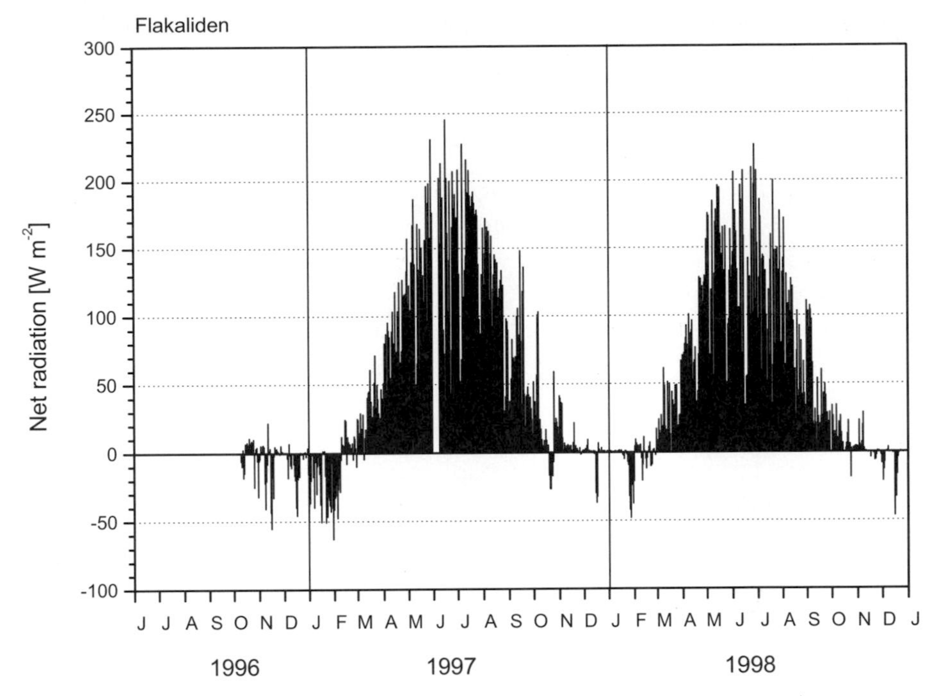

Fig. 6.2. Daily averages of net radiation R_n at the Flakaliden site (1996–1998)

to sunrise or sunset. A considerable difference in radiation characteristics is found among sites and individual years, and no common tendency exists among sites. This indicates a fairly independent weather regime at the sites in Belgium, Germany, Sweden, and Scotland – despite possible complex teleconnections due to the same governing pressure patterns.

A representative ensemble plot for 1998 is given in Fig. 6.10 (upper panel) for January and July. Note the change in absolute value and day length alike that combines to cause extremely differing radiation regimes in winter for high and middle latitudes. In contrast, summer radiation totals are similar, allowing similar photosynthetic activity.

Figure 6.2 shows the net radiation R_n measured at the Flakaliden site in northern Sweden represented by daily averages. The day-to-day variations of R_n are relatively large. During winter time, this may partly be caused by occasional rime and snow on the sensor which could not be maintained regularly due to difficult access.

6.3.2 Momentum

Wind profile or surface parameters derived from direct measurements of u_* by eddy covariance and wind profile measurements are given in Table 6.3. The

Table 6.3. Surface parameters and wind speed of dense conifer stands across Europe. Values in parenthesis represent measurement height and surface parameters as a fraction of canopy height, respectively

	Height h (z_m) (m)	Roughness length z_0 (m)	Displacement height d (m)	Aerodynamic resistance r_{aM} (s m^{-1})	Wind speed u (m s^{-1}) 1996	1997	1998
Weidenbrunnen	19 (32)	2.5 (0.13)	10.5 (0.55)	6.8		2.9	3.2
Tharandt	26.5 (42)	2.25 (0.08)	20.5 (0.77)	11.8	3.2	3.3	3.6
Flakaliden	8 (14)				3.5	3.4	3.2
Aberfeldy	6.8 (13)	0.6 (0.09)	3.4 (0.50)	16.0		2.8	2.9
Vielsalm	35 (41)	3 (0.09)	22 (0.63)				
Solling F1	29 (39)	2.6[a] (0.09)	22.7[a] (0.78)	10	2.8	3.2	

[a] Laubach et al. (1994)

parameters were computed utilizing the following relationships for friction velocity u_*, roughness length z_0, displacement height d, and aerodynamic resistance r_{aM} (for definitions and an illustrative figure see also Monteith and Unsworth 1990):

$$u_* = \sqrt[4]{\overline{u'w'}^2 + \overline{v'w'}^2} \tag{6.1}$$

$$r_{aM} = \frac{u_m}{u_*^2} \tag{6.2}$$

$$z_0 = \frac{(z_m - d)}{e^{\frac{u_m k}{u_*}}} \tag{6.3}$$

where u_m is horizontal wind speed at level of measurement z_m and k is the von Karman constant (0.41).

Surface Parameters. Estimates of z_0 and d are either based on profile measurements of at least three levels well above the canopy or from single-level eddy covariance measurements (see, e.g., Rotach 1994). The latter methods are still under discussion (Martano 2000). Here, the roughness length is computed under neutral conditions with a first guess approach of zero plane displacement d=0.67 h (Oke 1987). Later an iterative procedure is used to optimize z_0 and d by minimizing the differences between modeled and measured wind profiles according to the logarithmic wind profile equation:

$$u_m = \frac{u*}{k} \ln \frac{z_m - d}{z_0} \tag{6.4}$$

The procedure requires wind profile measurements and had been applied for Aberfeldy. Figure 6.3 shows the frequency distribution of roughness length z_0 and displacement height d. This statistical investigation based on half-hourly values without stability correction demonstrates the applicability of the approach. For zero plane displacement, about 71 % lie within the relatively narrow window of 2.5 to 4.0 m, and for roughness length about 62 % between 0.25 and 1.0 m. Theoretically, all analyses of z_0 have to be confined to a neutral stratification defined here according to the stability function $(z-0.67\ h)/L$ within –0.1 and 0.1, where L is the Monin-Obukhov length:

$$L = -\frac{u*^3}{k(g/\theta)\overline{w'\theta'}} \tag{6.5}$$

Doing this yields z_0=0.6 m and d=3.4 m, which is well within the range given above.

For Tharandt, four different methods to derive surface parameters d and z_0 were tested (Queck, pers. commun.). They comprise the standard profile

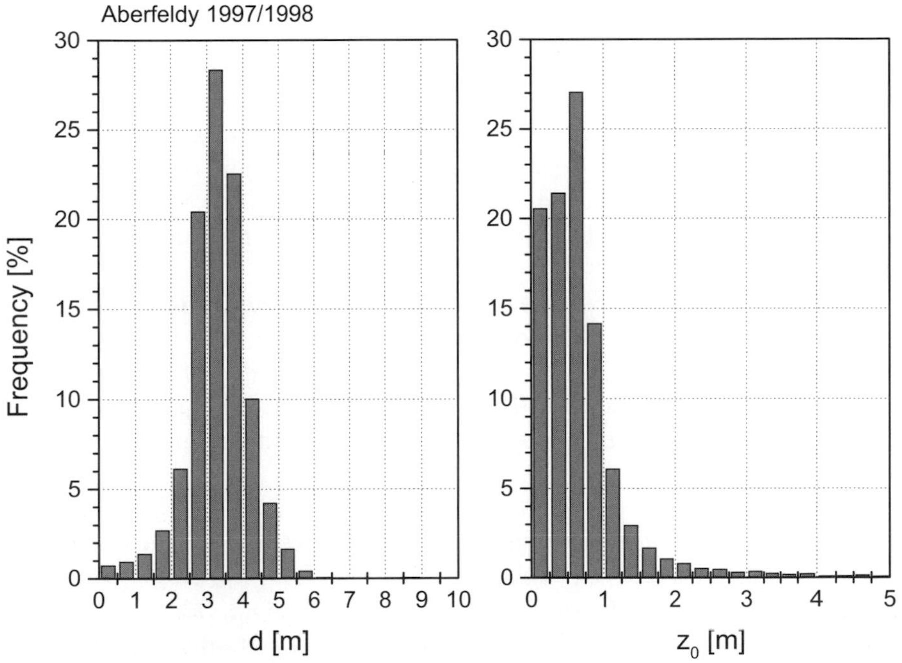

Fig. 6.3. Frequency distribution of half-hourly values for zero plane displacement d and roughness length z_0 at Aberfeldy (1997–1998)

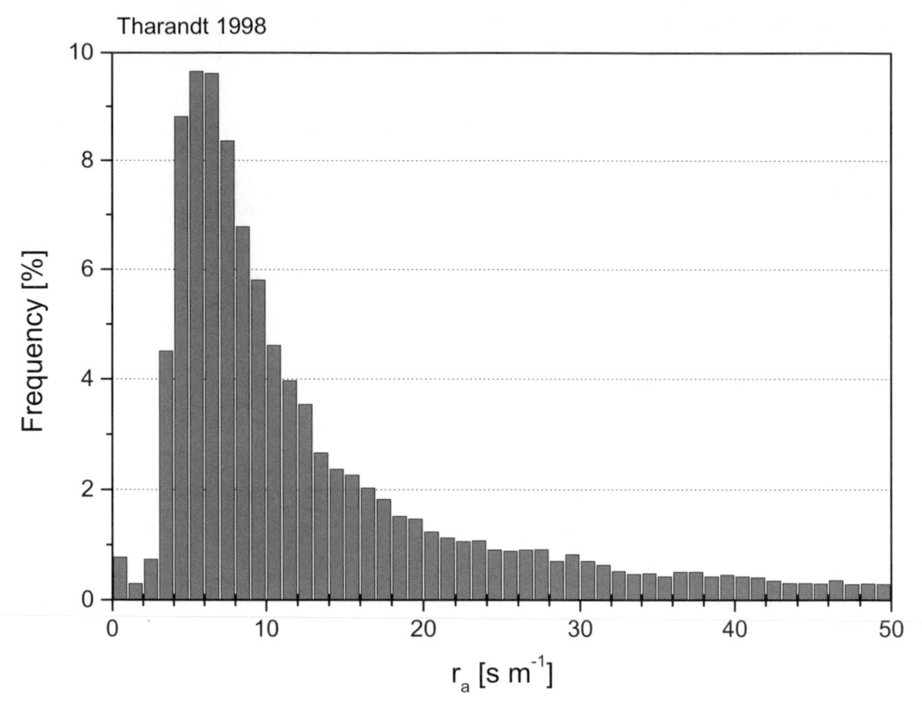

Fig. 6.4. Frequency distribution of half-hourly values of the aerodynamic resistance at Tharandt (1998)

method (with a limited campaign data set), two pure eddy covariance methods (Rotach 1994; de Bruin and Verhoef 1997), and a mixed profile–eddy covariance method. The results show a large variability and only two were in some agreement with standard assumptions (like d=0.67 h). Finally, an average based on de Bruin and Verhoef (1997) and the complete data set of 1999 was used.

This exercise demonstrate the still unsolved parameterization problem (at least for sites with some heterogeneity). At the same time the relation to canopy height h (see Table 6.3) for d (0.54 to 0.76 h) and z_0 (0.05 to 0.13 h) is close enough to conservative estimates to recommend using z_0 ~0.1 h and d ~0.67 h for rough inelastic canopies like the dense conifers under investigation.

Aerodynamic Resistance. Aerodynamic resistances were computed from measured wind speed and friction velocity rather than from rearranging Eq. (6.4). However, doing this would result in a different value for r_{aM} (12.4 s m^{-1} instead of 16 s m^{-1} in case of Aberfeldy) that holds for neutral conditions only. Which one is more appropriate depends on the application, i.e., expected similarity in vertical source distribution of momentum and heat flux, and on stability.

Note also that "secondary flux variables" such as the resistances or, later, the decoupling coefficient, have to be recomputed from mean values (e.g., mean r_{aM} from mean wind speed and mean friction velocity) to arrive at consistent data sets. These values can differ substantially from other statistical characteristics, like the mode of r_{aM} shown in Fig. 6.4, which is around 6 s m^{-1}. This is about half of the mean value of 11.8 s m^{-1} given in Table 6.3.

6.3.2.1 Heat Storage

Heat storage occurs as temperature changes in the soil, air, and biomass beneath a reference level (sensible heat), as water vapor content changes predominantly in the air (latent heat), and as carbon content of the biomass changes (metabolic heat). In general, soil heat flux can be an important contribution to the energy balance for bare or sparsely vegetated ground, as can sensible heat storage in air and biomass for tall vegetation such as forests (Oke 1987). This intermediate storage of heat is especially important when the closure of the energy balance is used as a final check on the consistency of turbulent heat fluxes with a time resolution of 30 min or so.

The energy balance of a homogeneous vast plant canopy on flat terrain can be written as:

$$R_n - G - J = H + L.E \tag{6.6}$$

where R_n is net radiation, G is soil heat flux, J is heat flux due to storage in canopy air and biomass (canopy heat flux), H is sensible heat flux and $L.E$ is latent heat flux. The left side of Eq. (6) is often called the "available energy" (AE), indicating that this amount is available for partitioning between the two turbulent fluxes. J is the sum of several storage fluxes:

$$J = J_H + J_E + J_{veg} + J_c \tag{6.7}$$

where J_H and J_E stand for sensible heat storage flux in the air and latent heat storage flux in the air, respectively, J_{veg} for sensible flux due to biomass temperature changes, and J_c for the heat stored due to photosynthetic activity or released due to respiration.

Please note that the sign convention used leads typically to positive signs during daytime and negative signs during night. Also that advection by lateral fluxes is excluded (due to the assumption of a homogeneous site). It is evident that no real site will be ideally homogeneous, always leading to some advection.

6.3.2.2 Soil Heat Flux

Soil heat flux is either determined by heat flux discs, computed from temperature changes, or derived by a combination of both methods. Heat flux discs are put horizontally in flat ground at variable depths, each giving the heat flux density through the disc. The minimum installation is at one level close to the soil surface (typically 1 cm below). Results depend on calibration of heat flux discs and on the disturbance of the soil due to the installation. The missing flux due to storage above the discs has to be added during data processing. The computation from soil temperature changes relies on temperature measurements at several soil depths:

$$G = -\int_0^z \varrho_s c_s \frac{\delta T_s}{\delta t} dz = \int_0^z C_s \frac{\delta T_s}{\delta t} dz \tag{6.8}$$

where ϱ_s and c_s are soil density and soil specific heat, respectively, and C_s is heat capacity. T_s is soil temperature measured at various depths z: heat flux discs are used at levels of 0.2 to 0.5 m where temperature changes are small and thermometers are used above where an ample temperature signal exists. However, under a forest cover with almost complete crown closure, soil heat flux has proven to be small and different results of various methods have been found not to effect the energy balance considerably (e.g., Vogt et al. 1995).

6.3.2.3 Canopy Heat Flux

All terms on the right side of Eq. (6.7) can be evaluated according to the following set of equations:

$$J_H = \int_0^{z_m} \varrho_a(T_a) c_p \frac{\delta T_a}{\delta t} d z \cong \overline{\varrho_a} c_p z_m \frac{\Delta \overline{T_a}}{\Delta t} \tag{6.9}$$

$$J_E = \int_0^{z_m} L(T_a) \frac{\delta \varrho_v}{\delta t} dz \cong \overline{L} z_m \frac{\Delta \overline{\varrho_v}}{\Delta t} \tag{6.10}$$

$$J_{veg} = \int_0^{z_0} \varrho_{veg} c_{veg} \frac{\delta T_{veg}}{\delta t} dz \cong m_{veg} c_{veg} \frac{\Delta \overline{T_{veg}}}{\Delta t} \tag{6.11}$$

$$J_C = -\mu \overline{w' \varrho_C'} \tag{6.12}$$

Heat stored in the canopy air between soil and reference level, J_H, is computed from air temperature changes over time t, where $\rho_a(T_a)$ (kg m^{-3}) is air density, T_a (°C) air temperature and c_p (J kg^{-1} K^{-1}) is specific heat capacity of air at constant pressure. Latent heat stored in the canopy air, J_E, is computed from vapor density changes, where ρ_v is vapor density and $L(T_a)$ (J kg^{-1}) latent heat of vaporization.

Heat stored in the aboveground biomass (including litter), J_{veg}, is computed from biomass temperature, T_{veg}, canopy density, ρ_{veg}, and canopy-specific heat capacity, c_{veg}. Heat stored or released due to biotic activity, J_c, can be evaluated from carbon dioxide flux density (Laubach 1996), where $\overline{w'\rho_c'}$ is flux density (kg m^{-2} s^{-1}) from eddy covariance measurements of CO_2 density fluctuations, ρ_c', and vertical wind fluctuations, w', and μ is the specific energy of conversion due to photosynthesis (10.88 10^6 J kg^{-1}). CO_2 storage in the canopy air is included in the flux at this point.

Methods used at the sites differ somewhat but all included at least one level (reference height) of temperature and vapor density changes, heat flux discs and the continuous measurements of carbon fluxes. Often biomass temperature is replaced by air or soil temperature depending on individual experiences and typical thermal inertia at the site. Considerable effort had been made at Tharandt to estimate biomass temperature (installation of eight independent thermometers) as well as the amount of biomass at Tharandt and Weidenbrunnen (Mellmann 1998; Mund et al. 2002). Results show the importance of biomass heat storage at tall forest sites and is demonstrated in Table 6.4 for July 1998. However, storage fluxes reduce to about ±8 W m^{-2} (Tharandt) for daily and nullify for yearly averages (computed for 1998).

Table 6.4. Mean maximum storage fluxes at dense conifer stands across Europe taken from ensemble means of July 1998 (all in W m^{-2})

	Soil heat flux	Flux due to sensible heat storage in the air	Flux due to latent heat storage in the air	Flux due to heat storage in the biomass	Heat flux due to carbon release
Weidenbrunnen	11.6	4.6	2.3	12.1	6.4
Tharandt	10.7	14.3	7.6	10.8	5.4
Flakaliden	3.2				
Aberfeldy	5.4	4.2	0.5	2.3	7.4
Vielsalm	11	14	8	13	
Solling F1	18	18	8	24	2.5

6.3.3 Energy Fluxes

Energy fluxes are measured by the eddy covariance method at each site, allowing direct comparison to the radiative fluxes (evaluating the energy balance closure) and computation of a flux Bowen ratio or the decoupling coefficient Ω (see, e.g., Monteith and Unsworth 1990). As an indicator for the portion of energy fluxes explained by the eddy covariance measurements, the energy balance closure for half-hourly values is given in Fig. 6.5 (plotting the sum of turbulent fluxes *H* and *L.E* against available energy *AE*). An agreement of about 90 % was found. The scatter of up to ±150 W m^{-2} (50 %) is attributed primarily to different footprints of radiative and turbulent fluxes (i.e., advection) and storage; the small offset of about 6 W m^{-2} is attributed to calibration insufficiencies. On a daily basis scatter reduces to ±50 W m^{-2} (25 %), demonstrating at least sufficient internal consistency at this site. Energy balance closure at the dense conifer sites was usually better than 80 % of available energy (Aubinet et al. 2000).

Table 6.5 lists several energy balance characteristics of the investigated sites. Available energy reflects climatic conditions (primarily latitude); energy partitioning also represents site characteristics (primarily water availability). Evapotranspiration amounts between 26 % of available energy (Flakaliden) and 61 % of available energy (Tharandt). The evaporative fraction of R_n is high in winter and can be quite low in a dry summer month (refer to Fig. 6.10,

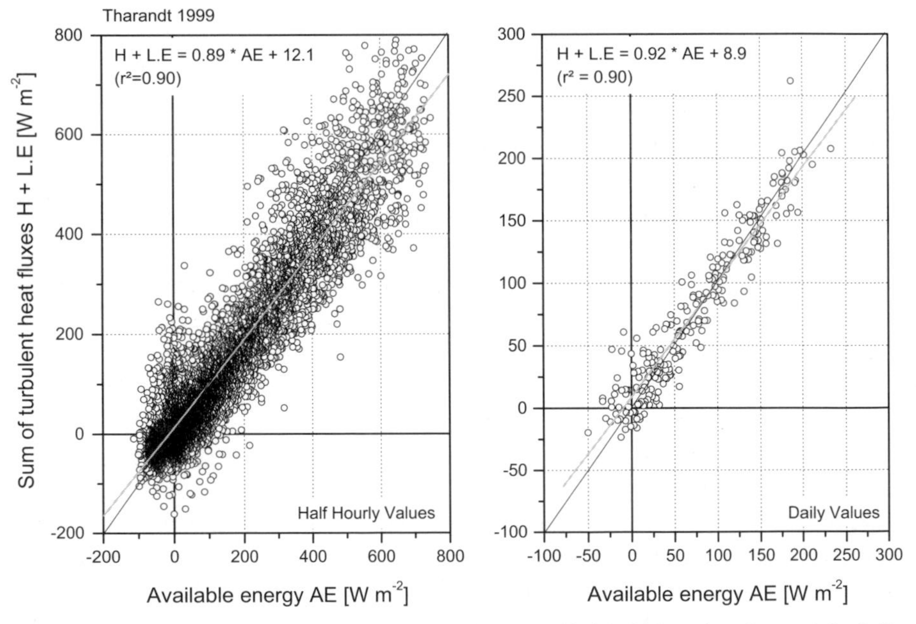

Fig. 6.5. Energy balance closure at Tharandt for 1999 (*left* half-hourly values; *right* daily values)

Table 6.5. Energy fluxes, evaporative fraction, Bowen ratio, canopy resistance, and decoupling coefficient of dense conifer stands across Europe

	Available energy AE (W m^{-2})			Sensible heat flux H (W m^{-2})			Latent heat flux L.E (W m^{-2})		
	1996	1997	1998	1996	1997	1998	1996	1997	1998
Weiden-brunnen		77.2[a]	61.4[b]		13.4[a]	−13.8[b]		27.3[a]	25.3[b]
Tharandt		64.2	60.6		21.1	16.5		37.0	37.2
Flakaliden		56.5	55.6	(7.5)	14.2	6.7	(25.3)	23.3	14.4
Aberfeldy		54.8	48.5		6.1	5.1		16.4	14.1

	Evaporative fraction L.E/AE			Bowen ratio H/L.E			Canopy resistance (s m^{-1})	Decoupling coefficient Ω
	1996	1997	1998	1996	1997	1998	1997/1998	1997/1998
Weiden brunnen		0.35	0.41		0.49	(−0.55)	105	0.19
Tharandt		0.58	0.61		0.57	0.44	83	0.15
Flakaliden		0.41	0.26	(0.30)	0.61	0.47		
Aberfeldy		0.30	0.29		0.37	0.36		

[a] Data for $\lambda.E$ only available for May until December, H and AE are given for the equivalent period.
[b] Data for $\lambda.E$ only available for February until December, H and AE are given for the equivalent period.

middle panel). The decoupling coefficient is generally small (typical for forests), it is expected that a decreasing trend from Aberfeldy (canopy height 6 m) to Tharandt (canopy height 26.5 m) exists.

Figure 6.6 shows ensemble means of canopy and aerodynamic resistances for the summer of 1998. To avoid effects of interception only half-hourly values with canopy resistances above 50 s m^{-1} were included in the analysis. Both resistances exhibit a clear diurnal pattern. Aerodynamic resistances reflect the convective development of the atmospheric boundary layer with higher wind speeds around noon and in the afternoon. Note that typical daytime r_a is around 3 s m^{-1} and typical nighttime values around 15 s m^{-1}. Canopy resistances have their minimum around 80 s m^{-1} with a slight increase in the afternoon to about 120 s m^{-1}.

Figure 6.7 (upper panel) shows the mean daily sensible heat flux at Flakaliden. The negative fluxes during winter are almost of the same magnitude as the positive summer values, and even the duration of periods dominated by positive and negative fluxes, respectively, are fairly similar. Consequently, the overall sensible heat flux is relatively small on an annual basis.

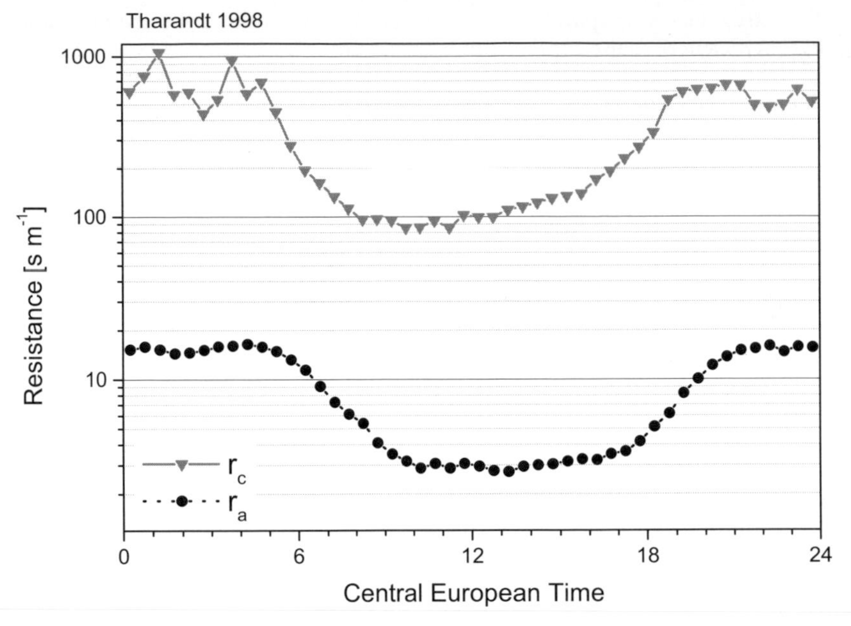

Fig. 6.6. Ensemble plot, i.e., mean diurnal course of canopy resistance and aerodynamic resistance at Tharandt for periods with a dry canopy (May–September 1998)

Additionally, there is a significant soil heat flux at this site because of the fairly open stand structure (data not shown here). The lower panel of Fig. 6.7 shows the mean daily latent heat flux and evaporation at Flakaliden. The fluxes are generally smaller than the sensible heat flux, i.e., the Bowen ration β is slightly larger than unity on a daily basis. During wintertime, there are periods of significant negative water vapor fluxes at Flakaliden. However, these periods contain a large fraction of modeled data with the associated uncertainties that are particularly large during wintertime since the response of fluxes on climatic variables is poor.

An intercomparison of evapotranspiration (ET; mm) is given in Fig. 6.8 for monthly totals of 1997. Maximum ET is found between June and August with sizeable differences between sites. Yearly totals of ET differ considerably – from 291 (Flakaliden) to 482 mm (Tharandt). The ET fraction of precipitation (evaporative fraction) increases from the oceanic climate of Scotland (22 % of the mean annual precipitation of 1400 mm) to the drier climate of central Europe (Tharandt, about 60 % of the mean annual precipitation of 820 mm). These ET values need still additional validation and a data gap filling strategy of their own to reduce the obvious uncertainty involved. Especially, possible errors of the eddy covariance technique during rain might introduce a bias between wetter and drier sites (affecting latent heat fluxes much more than carbon fluxes).

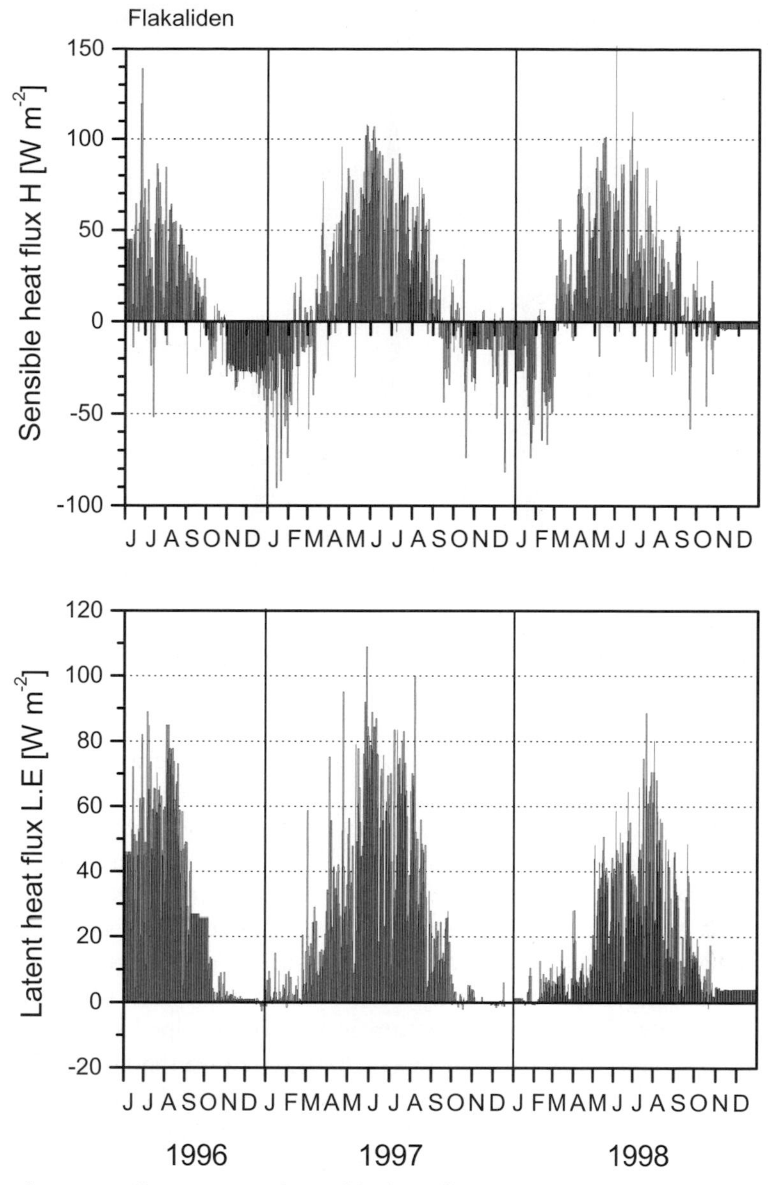

Fig. 6.7. Daily averages of sensible heat flux *H* (*upper panel*) and latent heat flux *LE* (*lower panel*) at Flakaliden (1996–1998)

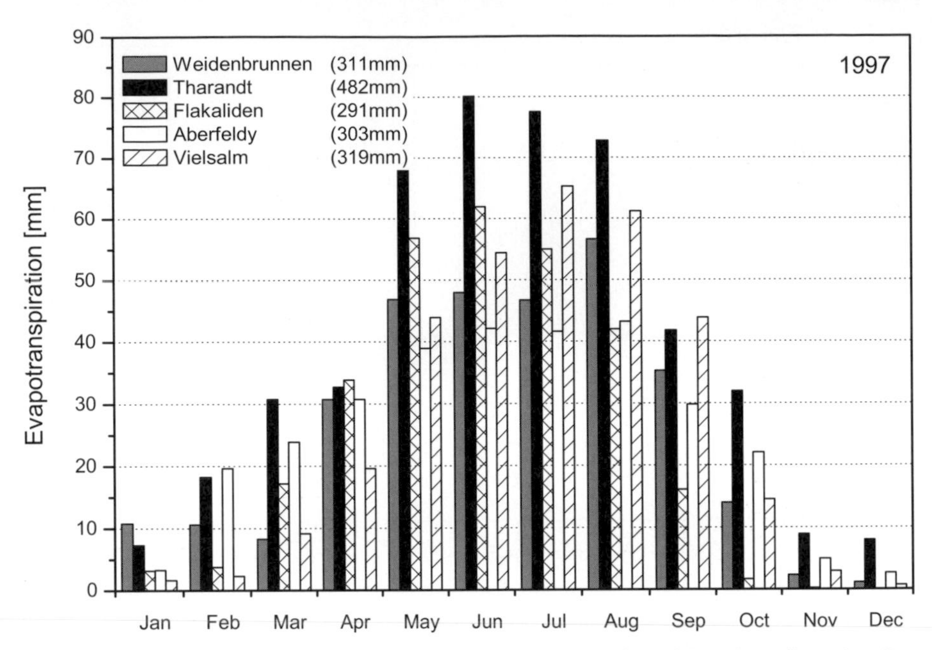

Fig. 6.8. Monthly sums of the evapotranspiration (derived from latent heat fluxes) at five sites (1997; data used with individual gap filling)

6.3.4 Carbon Fluxes

The carbon net ecosystem exchange (NEE) of all sites is based on the maximum available measured data and includes data gap filling (for an in-depth analysis with strategies used and problems involved, see Falge et al. 2001). For data gap filling nonlinear regression analysis is applied (Grünwald and Bernhofer 1998). Daytime analysis uses light saturation curves grouped according to temperature (example curves shown in Fig. 6.9). Based on these data Eq. (6.13) is parameterized for saturation net ecosystem exchange, NEE_{sat}, light use efficiency, a', and daytime respiration, R_d.

$$NEE_d = -\frac{a' \cdot PPFD \cdot NEE_{sat}}{NEE_{sat} + a' \cdot PPFD} - R_d \tag{6.13}$$

with measured daytime net ecosystem exchange NEE_d and photosynthetic photon flux density PPFD. Nighttime fluxes are modeled by a exponential temperature curve (Arrhenius function) to measured fluxes under conditions of sufficient turbulence (nighttime flux correction). That is to say, at night, only half-hourly fluxes where $u^* > 0.3$ m s^{-1} (or a similar turbulence criterion) are used. The presented NEE values are estimated after the implementation of

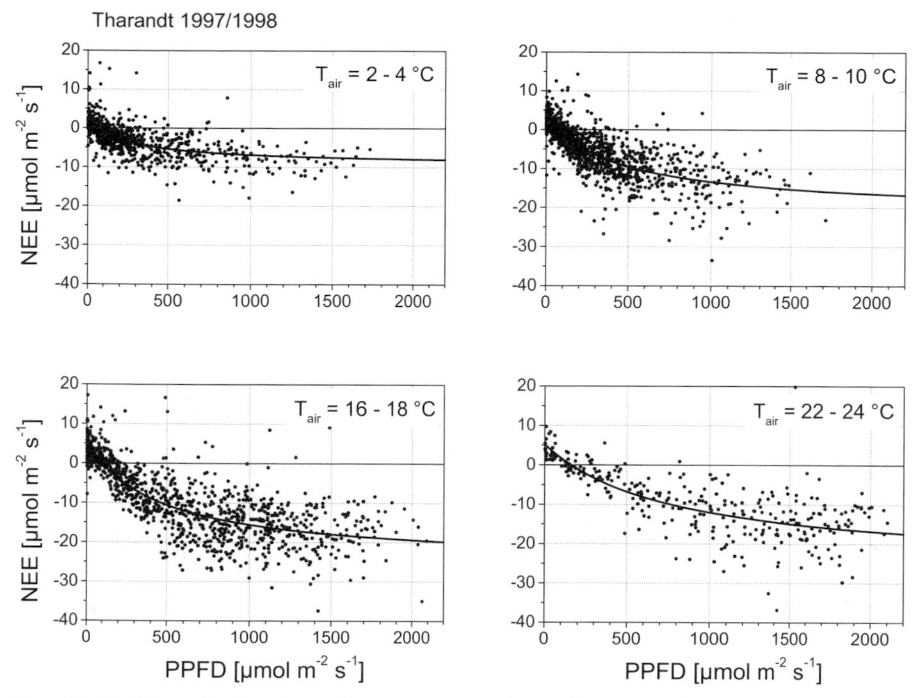

Fig. 6.9. Half-hourly daytime values of NEE in dependence on photosynthetic photon flux density PPFD in four typical 2-K air temperature classes (examples) and respective non-linear light saturation curves (Tharandt, 1998–1999)

data gap filling procedures as well as nighttime flux correction. Total respiration in Table 6.6 is calculated from the sum of measured nighttime fluxes and modeled daytime respiration. Daytime respiration can be modeled alternatively from air temperature-dependent light saturation curves by Eq. (6.13) or from nighttime fluxes. Differences between the two methods due to daytime inhibition of needle respiration were too small to be detected. The nighttime respiration values are the sum of all nighttime fluxes (measured and modeled if necessary). For Tharandt, three different soil moisture classes are modeled separately to include the observed reduction in respiration with decreasing soil moisture. Total ecosystem Respiration (TER) and gross primary production (GPP) in Table 6.6 may include systematic bias because the respiration estimated from flux measurements depends strongly on parameterization of daytime and nighttime respiration and on the definition of the u^* criterion for the nighttime flux correction.

Figures 6.10 and 6.11 summarize the seasonality of the radiation and turbulent regime for the Tharandt site. Figure 6.10 shows mean diurnal courses (ensemble plots) for January and July. The radiation (upper panel) is the main driving force for the turbulent fluxes of heat (middle panel) and carbon dioxide (lower panel). This results in typical diurnal and annual courses of the

Table 6.6. Carbon fluxes for the Tharandt spruce site: net ecosystem exchange (NEE), total ecosystem respiration (TER) derived from soil temperature and moisture; and gross primary production (GPP)

	NEE ($g\ C\ m^{-2}\ year^{-1}$)			TER ($g\ C\ m^{-2}\ year^{-1}$)			GPP ($g\ C\ m^{-2}\ year^{-1}$)		
	1996	1997	1998	1996	1997	1998	1996	1997	1998
Tharandt	(–533)	–614	–588	(1071)	1213	1202	(–1604)	–1827	–1790

ambient CO_2 concentration as indicated in Fig. 6.10. Mean daily amplitudes are around 15 ppm in summer, typical yearly amplitudes of monthly means are about 20 ppm (both values given for Tharandt; about 20 m above zero plane displacement). Figure 6.11 shows daily values of GPP and TER, as well as the respective cumulated daily sums and their differences (NEE) at Tharandt for 1998. The part of the measured carbon dioxide fluxes that could be directly used was 80 %, only 20 % had to be replaced by the above procedure. The resulting yearly NEE is –588 g C m^{-2}. Note the time shift between GPP (maximum around solstice caused by maximum PPFD) and TER (maximum in July/August caused by maximum temperatures). This time shift leads to the dominating influence of spring and early summer months for NEE as demonstrated by the cumulative curves in Figs. 6.11–6.13.

In Fig. 6.12 the daily sums and the cumulative total of CO_2 fluxes of the Swedish site Flakaliden are shown. It shows a clear pattern of summertime uptake and wintertime release of CO_2. The site gained more than 1 kg CO_2 m^{-2} during the observation period. The interannual variations are very large: the sink strength was reduced by more than two thirds between 1997 and 1998. This was mainly caused by a large number of days with a net loss of carbon during the summer of 1998 that was characterized by low solar radiation.

The variations of wintertime fluxes are large, especially during the first winter. This is partly caused by data gaps due to instrument failure that had to be filled by modeling associated with the uncertainties discussed above. During the following winters, the fraction of gaps could be reduced by installation of a heating device.

At least the growing season of 1997 shows a remarkable resemblance between three out of five dense conifer stands in a similar latitude band (similar radiation and temperature, at least during summer) with the exception of Weidenbrunnen (Fig. 6.13). Reasons for the latter are not yet completely clear, but can be either an unusual soil history (and high organic matter with a large portion of soil respiration) or related to the poor condition (and low production) of the spruce stand in the immediate vicinity of the flux tower. Vielsalm and Flakaliden showed about two thirds and one third, respectively, of the sink strength observed at the two similarly carbon accumulating sites, Aber-

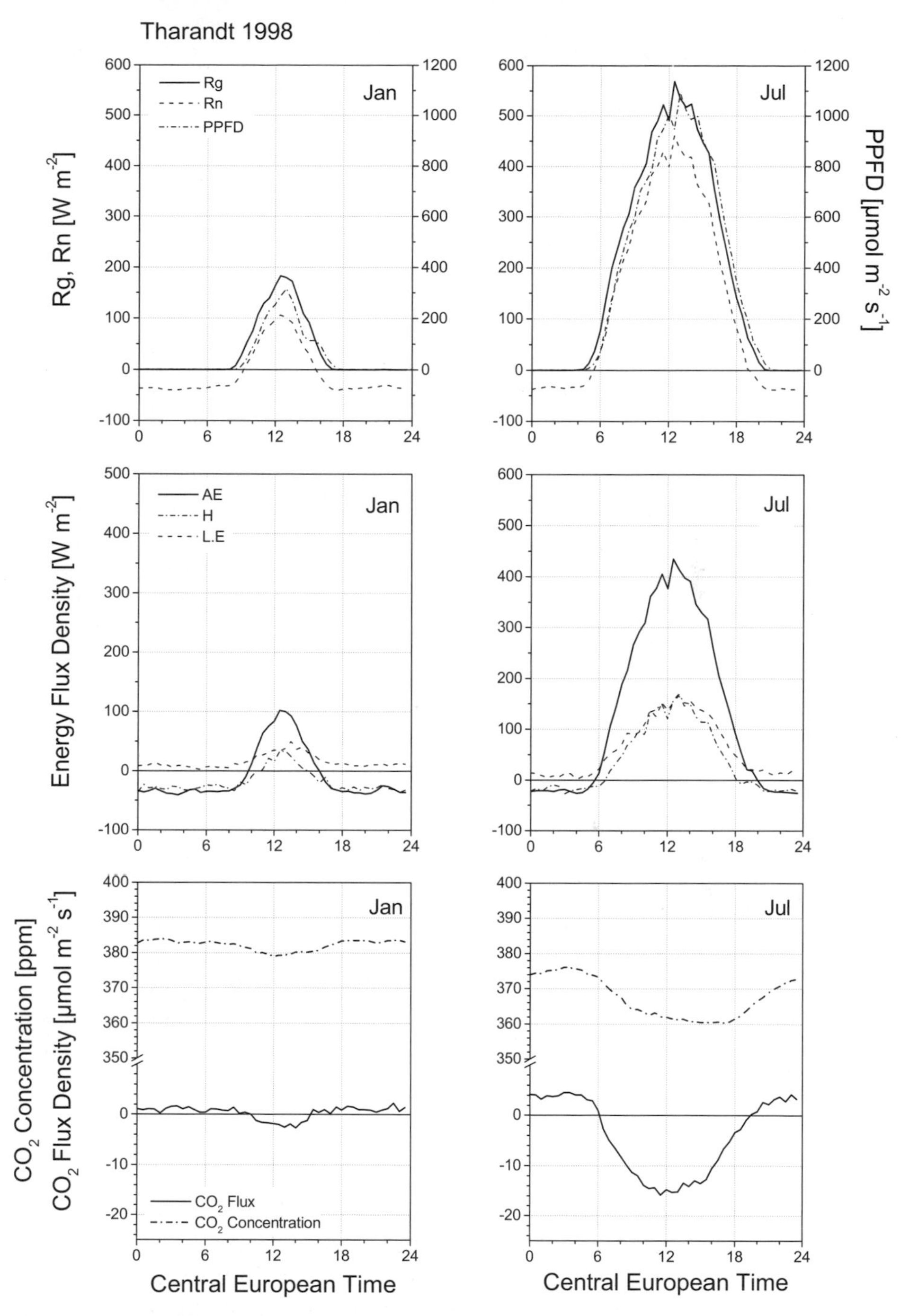

Fig. 6.10. Ensemble plot, i.e., mean diurnal course for an example winter and summer month at Tharandt (1998): solar radiation, net radiation and photosynthetic photon flux density (PPFD) (*upper panel*); available energy, sensible and latent heat flux (*middle panel*); CO_2 concentration and CO_2 flux density (*lower panel*)

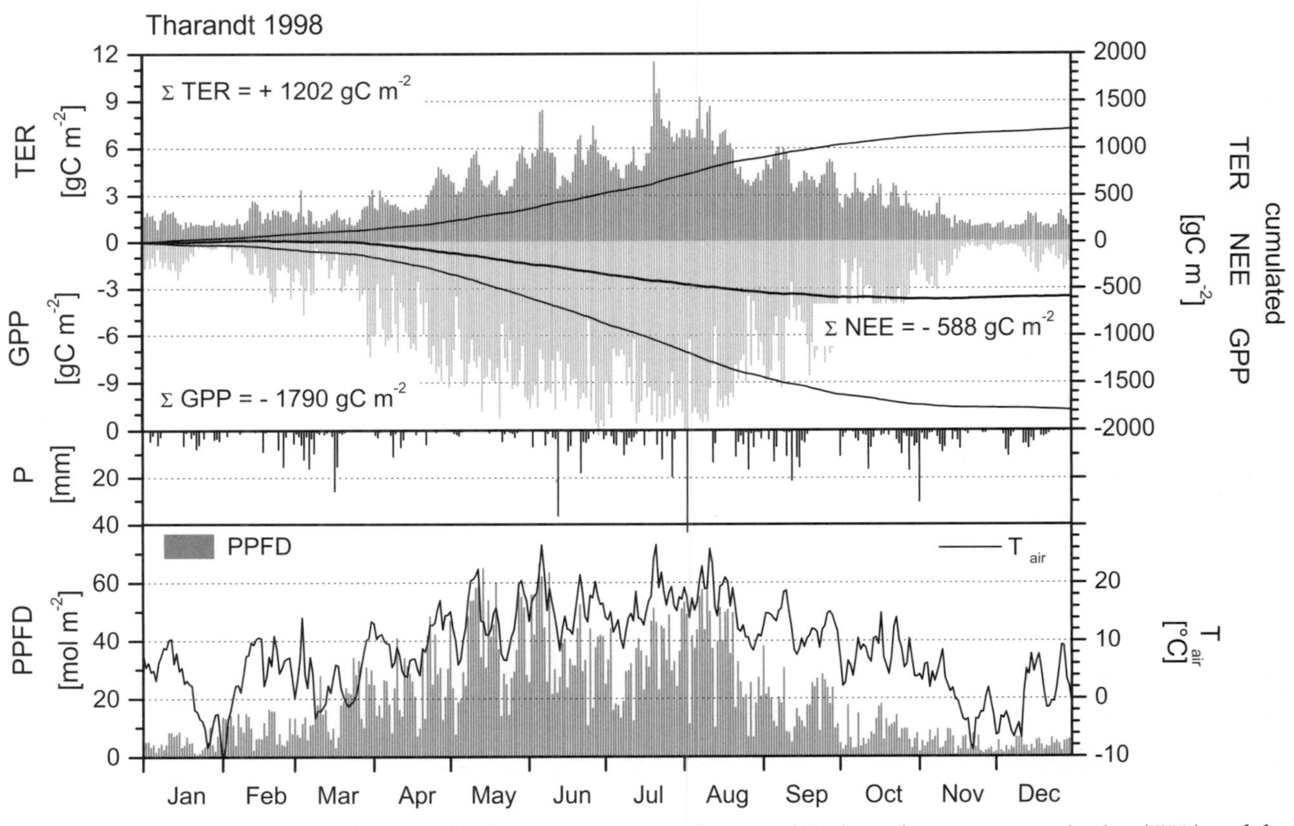

Fig. 6.11. Daily sums of measured and modeled gross primary productivity (GPP), total ecosystem respiration (TER), and the net ecosystem exchange (NEE) as well as precipitation (P), air temperature (T_a) and photosynthetic photon flux density (PPFD) on a daily basis at the Tharandt site 1998

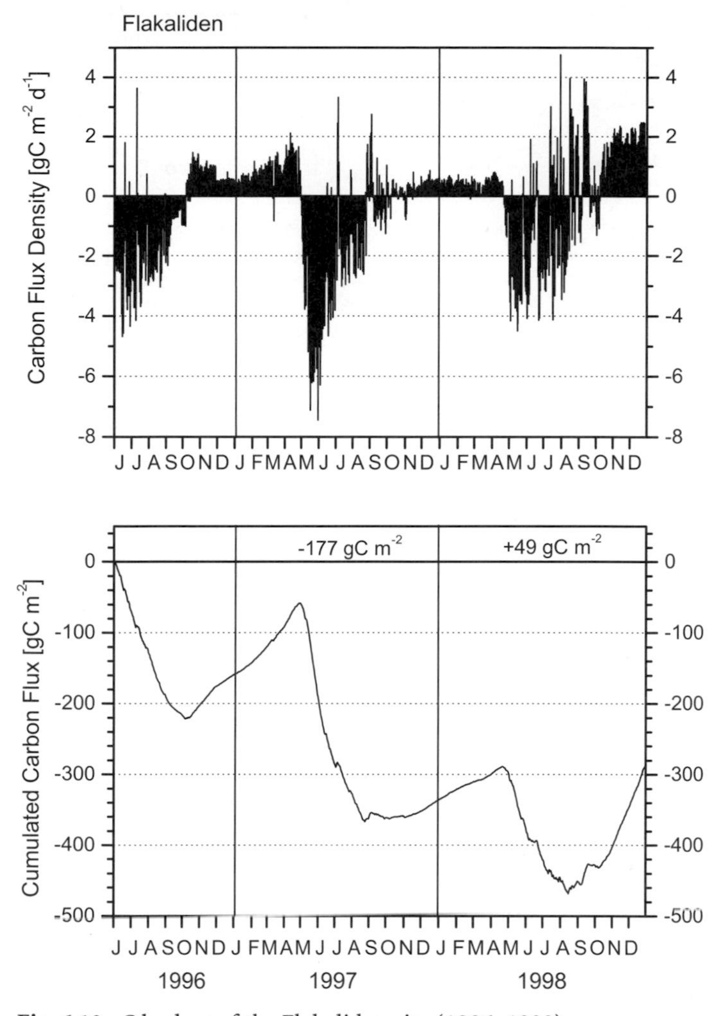

Fig. 6.12. C budget of the Flakaliden site (1996–1998)

feldy and Tharandt. This is attributed more to climate differences than to differences in species composition (see modeled values for Douglas fir in Falge et al., Chap. 8, this Vol.). It is interesting to note that the maximum monthly carbon uptake is at least 100 g C m^{-2} for all dense conifer sites except Weidenbrunnen with little variation between the sites. The yearly total depends on the number of months with similarly favorable growing conditions. This points towards a similar sink potential of spruce forests in Europe that is modified by climate and soil conditions.

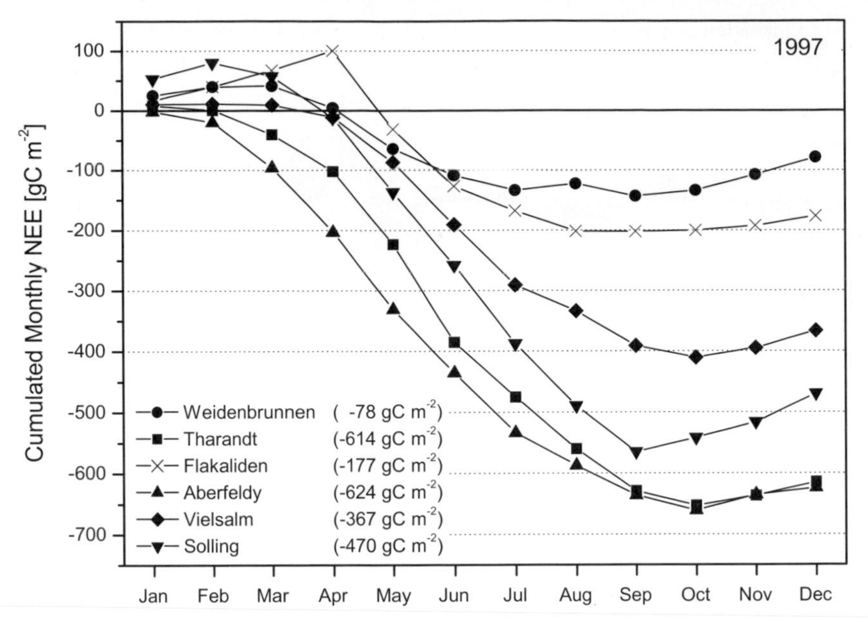

Fig. 6.13. Cumulated monthly sums of the carbon net ecosystem exchange (NEE) at five sites (1997); data with identical gap filling by non-linear regression according to Falge et al. (2001), except for Solling

6.4 Summary and Conclusion

NEE and ET of the five investigated spruce sites along with the Douglas fir stand in Belgium demonstrate remarkable similarities and differences alike. The calculation of statistically correct yearly NEE requires the filling of inevitable data gaps within measurements of the carbon dioxide flux. Furthermore, it is necessary to correct the systematic underestimation of the nighttime carbon dioxide flux (with predominant respiration) during phases of low turbulence.

The main results, governing variables, limiting factors, and open questions are listed below:

- Climatic conditions: the temperature range is similar (5.8 to 8.0 °C) except for Flakaliden (1.0 °C); minimum precipitation is observed in Flakaliden, maximum precipitation in Aberfeldy; the yearly radiation total depends mainly on latitude, though the differences for the summer months tend to be small (see Table 6.1).
- Evapotranspiration: differences observed reflect available water, available energy, coupling, and the need for additional quality checks and data gap filling for evapotranspiration.
- Net ecosystem exchange of carbon: two of the six investigated dense coniferous sites (Aberfeldy, Tharandt) show high net sinks, despite their differ-

ent climate. Two other sites still have sizable sinks (Solling F1, Vielsalm). Considerable interannual variability is demonstrated for Flakaliden, reaching about 40 % of the sink strength of the central and Western European stands for 1997. One site (Weidenbrunnen) is seen to have been only a small sink in 1997 and 1998. A maximum monthly carbon uptake of at least 100 g C m^{-2} is observed in spring for five out of six dense coniferous forest sites.

Reasons for these differences are not yet completely clear. Some issues deserve further attention:

- Measurement technique (e.g., incorporation of advection; see Lee 1998).
- Refinement of measurement screening and filtering, including data gap filling (see Falge et al. 2001).
- Addressing other environmental gradients such as the altitudinal gradient, phenology (even in an evergreen forest), and atmospheric nitrogen input.
- Effects of drought and/or frost.
- Respiration (soil, biomass) and site history (amount of organic matter and nutrients available).

There are various possible reasons for the observed differences in carbon sequestration of the six sites investigated. We believe that climate, stand age, leaf area, and soil conditions, including nutrients, are all important. Climatic conditions in spring and summer are thought to be the main reason for similarities and differences between sites and years alike; soil and stand conditions explain additional site differences. It is therefore not possible to attribute the differences to a single cause. But if one compares the two examples for "natural" and "artificial" abundance of spruce, Flakaliden (sink 1997: 177 g C m^{-2} $year^{-1}$) and Tharandt (sink 1997: 614 g C m^{-2} $year^{-1}$), it is interesting to see that the high productivity of the Tharandt site explains the attraction to spruce in the past. With their ability to utilize radiation input even in winter, spruce forests in Central Europe (where they do not represent the potential natural vegetation) have a sink strength similar to that of productive deciduous forests (Valentini et al. 2000). However, the differences in interannual variability need to be confirmed by longer observation periods.

Acknowledgements. The work presented would not have been possible without the help of numerous Ph.D. students and technicians who form the backbone of all the teams. Only a few have been included in the list of authors. They deserve our special thanks.

References

Aubinet M, Grelle A, Ibrom A, Rannik Ü, Moncrieff J, Foken T, Kowalski AS, Martin PH, Berbigier P, Bernhofer C, Clement R, Elbers J, Granier A, Grünwald T, Morgenstern K, Pilegaard K, Rebmann C, Snijders W, Valentini R, Vesala T (2000) Estimates of the annual net carbon and water exchange of European forests: the EUROFLUX methodology. Adv Ecol Res 30:113–175

Bernhofer C, Feigenwinter C, Grünwald T, Vogt R (2002) Spectral correction of water and carbon flux for EC measurements at the Anchor Station Tharandt. Tharandter Klimaprotokolle (in press)

Bundesministerium für Ernährung; Landwirtschaft und Forsten (1999) Bericht über den Zustand des Waldes – Ergebnisse des forstlichen Umweltmonitoring. http://www.bml.de/wald_forst/waldzustandsbericht_1999/index.htm

De Bruin HAR, Verhoef A (1997) A new method to determine zero-plane displacement. Bound Layer Meteor 82:159–164

Falge E, Baldocchi D, Olson RJ, Anthoni P, Aubinet M, Bernhofer C, Burba G, Ceulemans R, Clement R, Dolman H, Granier A, Gross P, Grünwald T, Hollinger D, Jensen N-O, Katul G, Keronen P, Kowalski A, Ta Lai C, Law T, Meyers BE, Moncrieff J, Moors E, Munger JW, Pilegaard K, Rannik Ü, Rebmann C, Suyker A, Tenhunen J, Tu K, Verma S, Vesala T, Wilson K, Wofsy S (2001) Gap filling strategies for defensible annual sums of net ecosystem exchange. Agric For Meteorol 107:43–69

Grünwald T, Bernhofer C (1998) Data gap filling with regression modelling. Forest ecosystem modelling, upscaling and remote sensing. SPB Academic, The Hague, pp 61–67

Hager H (1975) Kohlendioxyd-Konzentrationen, -Flüsse und -Bilanzen. Münchner Universitätsschriften, Fachbereich Physik, Wissenschaftliche Mitteilung 26, University of Munich, Meteorologisches Institut

Jarvis PG, James GB, Landsberg JJ (1976) Coniferous forests. In: Monteith JL (ed) (1976) Vegetation and the atmosphere, vol 2. Academic Press, London, pp 171–240

Larcher W (1994) Ökophysiologie der Pflanzen, 5th edn. Ulmer, Stuttgart

Laubach J (1996) Charakterisierung des turbulenten Austausches von Wärme, Wasserdampf und Kohlendioxid über niedriger Vegetation anhand von Eddy-Korrelations-Messungen. Wissenschaftliche Mitteilungen aus dem Institut für Meteorologie der Universität Leipzig und dem Institut für Troposphärenforschung e.V. Leipzig, vol 3, 139 pp

Laubach J, Raschendorfer M, Kreilein H, Gravenhorst G (1994) Determination of heat and water vapour fluxes above a spruce forest by eddy correlation. Agric For Meteorol 71:373–401

Lee X (1998) On micrometeorological observations of surface-air exchange over tall vegetation. Agric Forest Meteor 91:39–49

Lloyd CR, Culf AD, Dolman AJ, Gash JH (1991) Estimates of sensible heat from observations of temperature fluctuations. Bound Layer Meteor 57:311–322

Martano P (2000) Estimation of surface roughness length and displacement height from single-level sonic anemometer data. J Appl Meteor 39:708–715

Mellmann P (1998) Die Bedeutung der Speicherterme bei zeitlich hochauflösender Verdunstungsbestimmung am Beispiel der Ankerstation Tharandter Wald. Diplomarbeit, Institut für Hydrologie und Meteorologie, TU Dresden

Monteith JL (ed) (1976) Vegetation and the atmosphere, vol 2. Case studies, Academic Press, London, 439 pp

Monteith JL, Unsworth M (1990) Principles of environmental physics. Arnold, London, 291 pp

Mund M, Kummetz E, Hein M, Bauer GA, Schulze E-D (2002) Growth and carbon stocks of a spruce forest chronosequence in central Europe. For Ecol Manage (submitted)
Oke TR (1987) Boundary layer climates, 2nd edn. Methuen, London, 435 pp
Rotach MW (1994) Determination of the zero plane displacement in an urban environment. Bound Layer Meteor 67:187–193
Valentini R, Matteucci G, Dolman AJ, Schulze E-D, Rebmann C, Moors EJ, Granier A, Gross P, Jensen NO, Pilegaard K, Lindroth A, Grelle A, Bernhofer C, Grünwald T, Aubinet M, Ceulemans R, Kowalski AS, Vesala T, Rannik Ü, Berbigier P, Lousteau D, Gudmundsson J, Thorgeirsson H, Ibrom A, Morgenstern K, Clement R, Moncrieff J, Montagnani L, Minerbi S, Jarvis PG (2000) Respiration as the main determinant of carbon balance in European forests. Nature 404:861–865
Vogt R (1995) Theorie, Technik und Analyse der experimentellen Flußbestimmung am Beispiel des Hartheimer Kiefernwaldes. Stratus 3, Abteilung für Meteorologie und Klimaökologie, MCR-Lab, Geogr Institut, University of Basel

7 Evergreen Mediterranean Forests. Carbon and Water Fluxes, Balances, Ecological and Ecophysiological Determinants

G. TIRONE, S. DORE, G. MATTEUCCI, S. GRECO, R. VALENTINI

7.1 Introduction

The area covered by Mediterranean sclerophyllous forests and woodlands has been reported to amount to 8.5 Gha, with biomass of 50 Gt and a total productivity of 6 Gt (Whittaker and Likens 1975). Mediterranean-type ecosystems occur in five different regions of the world, all between a latitude of 30° and 40°. In the Northern Hemisphere they occur around the Mediterranean Sea and the coastline of California (USA), while in the Southern Hemisphere they occur in the Cape region of South Africa, the southwestern part of Australia, and the Chilean coastline (Mooney et al. 1974; Miller 1981a; Hobbs et al. 1995).

Mediterranean ecosystems have been acknowledged for their ecological, hydrological, and landscape-protective importance. Furthermore, they have been subjected to a historical long-term human influence, with both positive and negative effects (Hobbs et al. 1995). The ecosystems of the Mediterranean basin have been defined as "total human ecosystems", due to the long and strong relationships between man and ecosystem in those regions. Mediterranean vegetation, due to the resilience acquired through its evolution, often has an efficient soil protection role, and provides space for recreation. Nevertheless, in some areas, these ecosystems are threatened by excessive human pressure, causing negative impacts through overgrazing and forest fires (Quezel 1979; Naveh 1987).

The Mediterranean climate is characterized by warm and dry summers and mild, relatively humid winters. Autumn–winter rain is typical in Mediterranean regions, but the amount and distribution are very variable, with total rainfall ranging from 250–300 mm $year^{-1}$ in central Chile and southern California to 2,000 mm $year^{-1}$ in some areas of South Africa (Hobbs et al. 1995).

These peculiar conditions impose a number of stresses to the ecosystem and have caused the development of similar vegetation structures, dominated

Ecological Studies, Vol. 163
R. Valentini (Ed.) Fluxes of Carbon, Water and Energy of European Forests

by sclerophyllous shrubs, through the processes of convergent evolution (Gigon 1979). However, depending on the ecosystem type selected for comparison in each region, it is possible to find both strong similarities and dissimilarities, and simple comparisons of vegetation types are unlikely to give a true overview of the convergence patterns in Mediterranean ecosystems (Hobbs et al. 1995).

In the last 20–30 years, research efforts have been dedicated to the study of dynamics and functionality of Mediterranean ecosystems, trying to link the climatic and structural analogies to common ecophysiological and dynamic traits (Miller 1981b; Kruger et al. 1983; Tenhunen et al. 1987; Davis and Richardson 1995; Kalin Arroyo et al. 1995). Most of the emphasis of these comparisons has been dedicated to shrublands, while fewer studies have involved grassland and forests (Romane and Terradas 1992; Hobbs et al. 1995). Furthermore, the comparison cannot be made only at a species or structural level, but the studies should encompass everything up to the ecosystem level, with the aim to understand the overall dynamics of the vegetation communities (Kummerow et al. 1981).

Mediterranean ecosystems are characterized by the occurrence of various stresses, particularly drought and high temperature. In this respect, according to various climate scenarios, the current trends of global climatic change should have a relevant impact in Mediterranean areas, particularly through the potential increase in the evaporative demand by the atmosphere and the changed distribution of rainfall. A more thorough understanding of the impact of global change on Mediterranean ecosystems will also be crucial for their management (Moreno and Oechel 1995).

The number of forest sites in which canopy flux measurements by the eddy covariance technique are currently performed has increased to more than 100. Nevertheless, even though one of the first study of eddy flux measurements over a forest stand was performed in a Mediterranean *maquis* (Valentini et al. 1991b), currently very few flux towers are located in Mediterranean ecosystems, most of them included in the MEDEFLU EU project (Miglietta and Peressotti 1999). In the EUROFLUX network, only one site is located in an evergreen broad-leaved Mediterranean ecosystem, the *Quercus ilex* L. forest of Castelporziano.

This site was selected for study of the long-term response of an old coppice stand that has been undergoing conversion to a high-stand management system after a fire that occurred in the 1950s. The history of the site is representative of other *Q. ilex* forests in Italy and its climate is typically Mediterranean, even if not extreme. Furthermore, the studied site is also representative of the typical vegetation of coastal plains that was once present in large areas of Italy.

In a recent National Forest Inventory, pure *Quercus ilex* L. forests, mainly coppice, have been quantified in 150,000 ha, while Mediterranean *maquis*, dominated by this species, covers nearly 200,000 ha. Most of these stands are located along the coast and on hilly terrain, in areas of great environmental value. In this respect, the focus on biomass production is changing, as other forest functions, such as carbon sequestration and soil protection, increase in importance.

This study aimed not only to quantify the ecosystem carbon balance, but also to gain a deeper understanding of the functionality of Mediterranean evergreen oak forests, and a more complete and modern knowledge of the ecophysiological mechanisms underlying ecosystem productivity in these forests.

7.2 Materials and Methods

7.2.1 Site Description

The experimental site was a 54-year-old coppice under conversion to high forest, inside the Presidential Reserve of Castelporziano (41°45′N, 12°22′E), 25 km W of Rome. The natural reserve covers about 4800 ha, of which 85.6 % are forests. The climate is Mediterranean type humid–subhumid with average yearly rainfall of 780.8 mm and an air temperature of 15.6 °C (Bruno et al. 1977).

The stand was classified as an *Asplenio–Quercetum ilicis* association (Lucchese and Pignatti 1990). The forest structure is characterized by two layers: a dominant layer of *Q. ilex* L., 12–15 m high, and a shrub layer, 2–4 m high, with *Phillyrea latifolia* L., *Pistacia lentiscus* L., *Erica arborea* L., *Cistus salvifolius* L., *Cistus incanus* L., *Cytisus scoparius* L., and some vines (*Clematis flammula* L., *Hedera helix* L., *Rubia peregrina* L., *Smilax aspera* L.). Large trees more than 100 years old, and up to 17 m high and 170 cm in diameter, represent only 1 % of the stand.

The herbaceous layer is represented by *Brachipodium sylvaticum* (Hudson) Beauv., *Cyclamen repandum* Sibth et Sm. *Carex distachya* Desf., *Asperula laevigata* L., and *Alliaria petiolatia* Bieb (Gratani and Crescenti 1994). The current *Q. ilex* forest represents a secondary succession after a fire destroyed the area in 1944. In 1985 the stand was converted from coppice to high forest.

The morphology of the reserve is mostly flat with altitudes ranging between 0 and 85 m a.s.l. The soil profile of the experimental site is type A–C, with a fairly deep, well draining A horizon, rich in humus and calcium carbonate. The soil is sandy (75.2 % coarse sand, 11.2 % fine sand, 3.7 % coarse silt, 2.2 % fine silt, and 6.7 % clay). The soil organic matter is 3.4 % between 0–20 cm and 1.1 % between 20–40 cm. Organic carbon is 1.9 % in the organic horizon with a C/N ratio of 66, and is 0.65 % with a C/N ratio of 100 in other mineral horizons (Benedetti et al. 1995).

Table 7.1. Site and stand characteristics of the *Quercus ilex* forest of Castelporziano

Site characteristics	
Altitude (m)	2.8
Climate	Mediterranean
Mean yearly temperature (°C)	15.6
Mean maximum temperature (°C)	25.8
Mean minimum temperature (°C)	6.3
Mean rainfall 1997–1998(mm)	781
Prevailing wind direction	NE–SW
Soil	Sandy, calcareous
Soil depth (cm)	40–60
pH	5.8
Vegetation	*Asplenio-Quercetum ilicis*
Age (years)	54
Mean diameter (cm)	16
Mean height (m)	12.5
LAI (m^2 m^{-2})	3.2–3.8
Aboveground biomass (t ha^{-1})	103.7
Wood increment (m^3 ha^{-1})	4.5

The climate is characterized by a dry season extending for 4 months, from the beginning of May to the end of August (Greco 1998). Water deficit during the dry season is 406 mm, while excess rainfall in others seasons is 410 mm. The scattered, low summer rains are offset by high night humidity, which over a period of 10–15 days can contribute 0.2 – 0.8 mm of precipitation.

During the dry seasons in 1996 and 1997, the predawn water potential (PWP) and the midday water potential (MWP) showed values of –2.8 MPa at predawn and –3.6 MPa at midday. A summary of the site characteristics is presented in Table 7.1.

7.2.2 Dendrometric Survey

In order to quantify the stand biomass and productivity, three 400-m^2 test areas were selected to represent the different structures and varying composition of the *Quercus ilex* forest. Height, stem diameter at breast height, height at crown insertion, crown projection, sprout number on every stump, and position of every stump on the *x–y* axis were measured. Tree distribution and biomass were calculated as an average of the three sample areas. Eight *Q. ilex* trees and six shrubs were harvested and used as model trees for biomass determination. During the years 1997–1998 phenological measurements were carried out to establish growth times for branches, twigs, and stems (Fioravanti 1999). The growth of new twigs were recorded on three *Q. ilex* trees at three canopy heights. For every height four twigs, one for each direction (N, S, W, E) were measured. The same measurements were carried out for *Phillyrea latifolia*, *Erica arborea*, and *Pistacia lentiscus*. Stem growth was determined with 12 dendrometric bands (Dial-Dendro, Relaskop-Technit Vertriebsges, Salzburg, Austria) applied to trees of four different diametrical classes – 12.5, 17.5, 22.5, and 27.5 cm – and measurements were taken every 15 days. Litter production was calculated using eight litter traps, each with an area of 0.50 m^2, installed 1.50 m from ground level.

7.2.3 Eddy Covariance Measurements

The eddy covariance system was installed in 1996 on a 14-m scaffolding tower following the protocols of the EUROFLUX network (Grelle and Lindroth 1996; Moncrieff et al. 1997; Aubinet et al. 2000). The system consisted of a three-dimensional sonic anemometer (model R2 Gill Instruments, Lymington, Hamps., UK) positioned 19 m from the ground, and a fast-response infrared gas analyzer (IRGA: model LI – 6262, LI-COR). An external pump (ASF 7012; Thomas Industries, Sheboygan, WI, USA) installed downstream from the analyzer, provided a flow rate of 9 l min^{-1}. There were two Teflon filters (Acro 50 PTFE 1 μm, Gelman Ann Arbor, MI, USA), one at the beginning of the air tube and another at the input of the analyzer. The gas analyzer was calibrated every 2 weeks. The data were processed with the software EFIS (Eddy Fluxes Integrated Software) designed by our group in cooperation with ACSI Informatica (Rome). Measurements were taken continuously from June 1996 to March 1999. Missing data were about 5 % (20 days) during 1997 and 10 % (39 days) during 1998.

7.2.4 Meteorological and Micrometeorological Measurements

The air temperature profile was measured at five heights: above the canopy at 16.5 and 14.6 m, inside the canopy at 11.6 m, below the canopy at 9.0 m, and inside the shrub layer at 2.4 m. Relative humidity (HP 100A, Rotronic, Bassersdorf, CH), photosynthetic active radiation (LI-190, LI-COR), diffuse photosynthetic active radiation (SK215, Skye, Llandrindod Wells, UK), rain (Skye), net radiation (Campbell Q-6, Logan, UT, USA) were measured at 14.4 m. Wind speed was measured at 12.5 and 9.0 m. Soil heat flux was measured with two soil heat flux plates (REBS, Seattle, WA, USA) inserted at 0.05 and 0.30 m. Soil temperature was also measured at 0.05 and 0.30 m. Sensor outputs were averaged and recorded using a data logger [CR-10X, with a channel extension AM-416 (Campbell)] on a half-hourly basis.

The footprint for the flux (70 % of total) had an average value of 505 m (±12 m) varying between 467 m in January to 535 m in April. The distance where there was the maximum contribution to the total carbon flux was at 136 m (±7), varying from 177 m in July to 113 m in April (Dore 1999). Because our area was 48 ha and had a rectangular shape (600 × 800 m), it was possible to establish that fluxes were representative of the experimental area.

7.2.5 Ancillary Measurements

Soil moisture was measured using the TDR technique (Trime IMKO GmbH, Ettlingen, Germany) every 15 days at six different points near the tower at three different depths: 0–20, 30–50, and 70–90 cm.

Soil respiration was measured every 15 days with a portable gas exchange system (EGM-1, PPSystem, Hitchin, Herts., UK), with a 10-cm-diameter cuvette, on 28 points (Dore et al. 1998).

7.3 Biomass and Productivity Dynamics of the *Quercus ilex* Forest Stand

The vegetation study confirmed the botanical composition of a typical *Q. ilex* stand: *Quercus ilex*, *Pistacia lentiscus*, *Phillyrea latifolia*, *Erica arborea*, and *Fraxinus ornus*. There were 2647 stumps ha^{-1}: 1232 *Q. ilex*, 1407 evergreen shrubs, and 8 deciduous trees. The total sprouts number was 9241 ha^{-1}, starting from the 2.5-cm-diameter class (Tables 7.2, 7.3).

The dominant layer was composed almost entirely of *Q. ilex,* with a large number of trees in the 17.5-cm-diameter class, while the dominated layer was comprised of evergreen shrubs lower than 3 m. The shrubby layer had slow growth due to *Q. ilex* competition and the impact of ungulates, which overwhelmed the carrying capacity of the forest. This problem was highlighted by damage to lower sprouts and by the occurrence of a limited number of plant seedlings.

Table 7.2. Species distribution of the *Q. ilex* forest of Castelporziano

Species	Stumps (ha^{-1})	(%)	Sprouts (ha^{-1})	Sprouts ($stump^{-1}$)
Erica Arborea	766	28.9	5,175	6.8
Fraxinus ornus	8	0.3	8	1
Pistacia lentiscus	58	2.2	125	2.1
Phyllirea latifolia	583	22.1	1,275	2.2
Quercus ilex	1,232	46.5	2,658	2.2
Total	2,647	100	9,241	3.5

Table 7.3. Dendrometric parameters of the *Q. ilex* forest of Castelporziano

Species	Basal area (m^2 ha^{-1})	(%)	Crown projection (m^2 ha^{-1})	(%)	Av. height (m)
Erica Arborea	0.21	1	3,043	14.1	2.9
Fraxinus ornus	0.61	2.8	338	1.6	12.2
Pistacia lentiscus	0.03	0.1	109	0.5	1.4
Phyllirea latifolia	0.92	4.2	1,765	8.1	2.6
Quercus ilex	20.20	91.9	16,391	75.7	11.9
TOTAL	21.97		21,646		

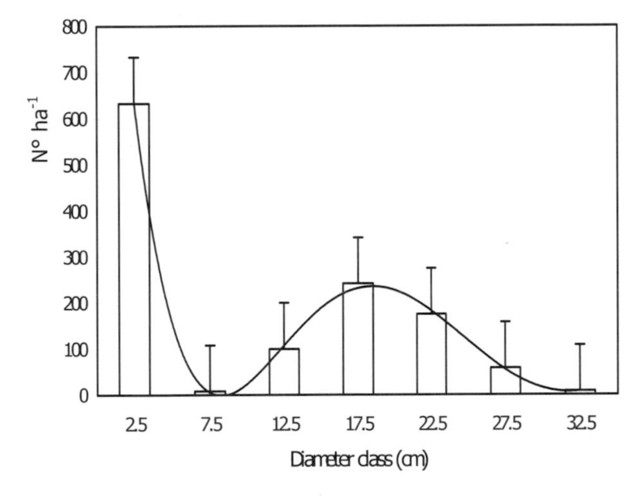

Fig. 7.1. Distribution in diametric classes of the *Quercus ilex* forest in Castelporziano

Q. ilex comprised 75.72 % of the total forest crown projection area and made a full cover, only interrupted over old charcoal deposits. Among the evergreen shrubs, *Phyllirea latifolia* had the highest values of basal and crown projection area, with a tendency towards a morphological arboreal layer. Excluding the 2.5-cm class, which were sprouts from stumps left after conversion to high forest in 1985, the diametrical distribution of *Q. ilex* trees was normal and typical of an even-age forest (Fig. 7.1).

The total biomass was calculated as the sum of aboveground dry matter: branches, leaves, twigs, stems, and belowground dry matter (fine roots <2 mm and medium roots between 2 and 5 mm). The leaf biomass was 7.86 t ha^{-1}, similar to other Mediterranean *Q. ilex* forest (Susmel et al. 1970) such as Orgosolo 7.4 t ha^{-1} (Sardinia, Italy) or another site in Castelporziano (7.8 t ha^{-1}, Bruno et al. 1977). Of the total leaf biomass, 7.19 t ha^{-1} was from *Q. ilex*, particularly the 27.5- and 17.5-cm-diameter classes. A further 0.67 t ha^{-1} was from other evergreen species (58 % *Phillyrea latifolia*). Branch biomass was 15.1 t ha^{-1}, all of which was *Q. ilex*, since in the shrub the branches could not be distinguished from the stems, and all the woody biomass was consequently counted as stem. Branch biomass was concentrated in the 22.5-, 27.5-, and 32.5-cm-diameter classes. Twig biomass was 2.7 t ha^{-1} of which 2.13 t ha^{-1} was *Q. ilex*. For shrubby species the value was 0.63 t ha^{-1} with *Erica arborea* contributing 60.80 % of the shrub total. Stem biomass (bark and wood) was 77.8 t ha^{-1}, with 74.6 t ha^{-1} from *Q. ilex*, almost 96 % of the total, while shrub biomass was only 3.27 t ha^{-1}. Fine root biomass (<2 mm) was estimated for three different layers: 0–15, 15–30, and 30–45 cm, using seasonal coring and ingrowth bags (Muratore 1998). The biomass in the 0–15 cm level was between 2.6 and 6.7 t ha^{-1}, varying with season and subject to quick turnover. Root biomass in the 15–30 cm level was between 1.5 and 5.1 t ha^{-1}. In the 30–45 cm level the root biomass decreased to 0.5–1.8 t ha^{-1}. The biomass of medium roots (diameter between 2 and 5 mm) had an allocation pattern similar to the fine roots with a total biomass of 3.6 t ha^{-1} (Muratore 1998). The total stand biomass was 113.22 t ha^{-1} of which 103.67 t ha^{-1} was in the aboveground component, a lower aboveground figure than obtained in a previous study (Bruno et al. 1977). This biomass total corresponded to 51.79 t ha^{-1} of carbon storage of which 47.52 t ha^{-1} was *Q. ilex* and 4.27 t ha^{-1} evergreen shrubs.

Total stand productivity was 16.33 t ha^{-1} year^{-1}, corresponding to 8.89 t ha^{-1} year^{-1} of carbon. In Table 7.4, the different components of forest productivity are shown. Stems and branches accounted for 32 % of the total productivity, or 2.85 t C ha^{-1} year^{-1}. Twigs accounted for 6.55 % of the total, or 0.59 t C ha^{-1} year^{-1} (Fioravanti 1999). Leaf production was 2.74 t ha^{-1} year^{-1}, or 16.77 % of the total, corresponding to an annual carbon storage of 1.51 t ha^{-1} year^{-1}. For fine roots, the productivity was higher in the 0–15 cm layer, 2.84 t ha^{-1} year^{-1}. In the 15–30 cm layer it was 1.99 t ha^{-1} year^{-1}, and in the third level (30–45 cm) was 1.12 t ha^{-1} year^{-1} . Total fine root production in the first 45 cm was 5.96 t ha^{-1} year^{-1}.

The average aboveground litter productivity, calculated during years 1997, 1998, and 1999 was 3.82 or 1.81 t C ha^{-1} year^{-1}. Litter (Table 7.5) was composed mainly of leaves (50.85 %) and to a lesser extent of acorns (36.42 %) and

Table 7.4. Allocation of biomass and carbon productivity

Components	Dry biomass (t ha^{-1} year^{-1})	Carbon (t C ha^{-1} year^{-1})
Stems and branches	5.18	2.85
Twigs	1.07	0.59
Leaves	2.74	1.51
Fine roots (<2 mm)	5.95	3.25
Acorns	1.39	0.69
Total	16.33	8.89

Table 7.5. Annual production and distribution of aboveground litter

Litter components	Biomass (t ha^{-1})	(%)
Leaves	1.94	50.9
Acorns	1.39	36.4
Branches	0.49	12.7
TOTAL	3.82	

branches (12.72 %). Leaf litter decomposition was shown to take about 2 years (data not shown).

The analysis of diametrical and longitudinal growth of twigs showed that, for *Q. ilex*, growth was predominantly in the upper part of the canopy. Longitudinal growth of twigs in *Q. ilex* occurred between mid-March and the end of June, while diametrical growth occurred continuously from March to November. (Figs. 7.2, 7.3). The other shrub species exhibited the same growth patterns, indicating that this is a general feature of this Mediterranean ecosystem.

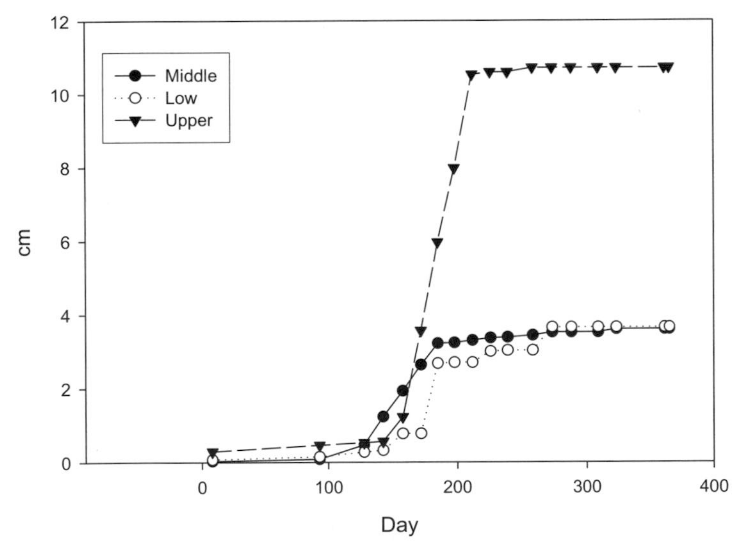

Fig. 7.2. Longitudinal growth of *Q. ilex* twigs

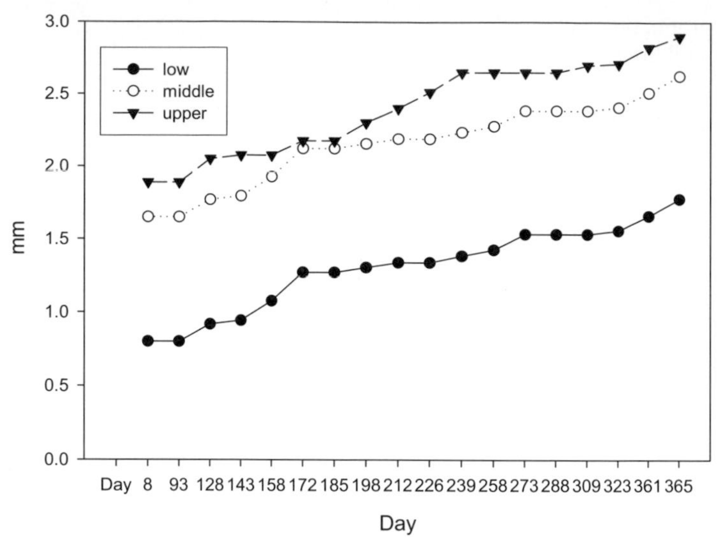

Fig. 7.3. Diametric growth of *Q. ilex* twigs

7.4 Soil and canopy Carbon Fluxes

7.4.1 Daily Carbon Fluxes

Eddy correlation measurements started in October 1996 . During 1997 and 1998, the data coverage was 95 and 89 % of the time, respectively. Four days' data, one day for each season, in 1997 (15 Jan., 14 April, 12 Aug., and 12 Nov.) and 1998 (20 Jan., 13 April, 13 Aug., and 12 Nov.) show the diurnal trend of CO_2 fluxes (Fig. 7.4). The selected days were characterized by clear skies, with wind speed, air temperature, and net radiation that were similar to the monthly average (Tirone et al. 2000). Instantaneous NEE ranged from a maximum of −18 μmol CO_2 m^{-2} s^{-1} in June, to a minimum of −6.0 μmol CO_2 m^2 s^{-1} in January. However, the differences between the seasons were not so marked, indicating favorable conditions for ecosystem carbon assimilation throughout the year. During August, there was a midday reduction in NEE, caused by stomata closure. However, unlike other Mediterranean ecosystems, this stand is typically able to access the shallow ground water table and did not often exhibit signs of water limitation. Although the CO_2 exchange rates were similar among seasons, day length had an important effect on the total net ecosystem exchange, with a maximum of 500 daylight hours in July, and a minimum of 320 daylight hours in January. The day length determined the size of the daily carbon sink more than did the maximum NEE values.

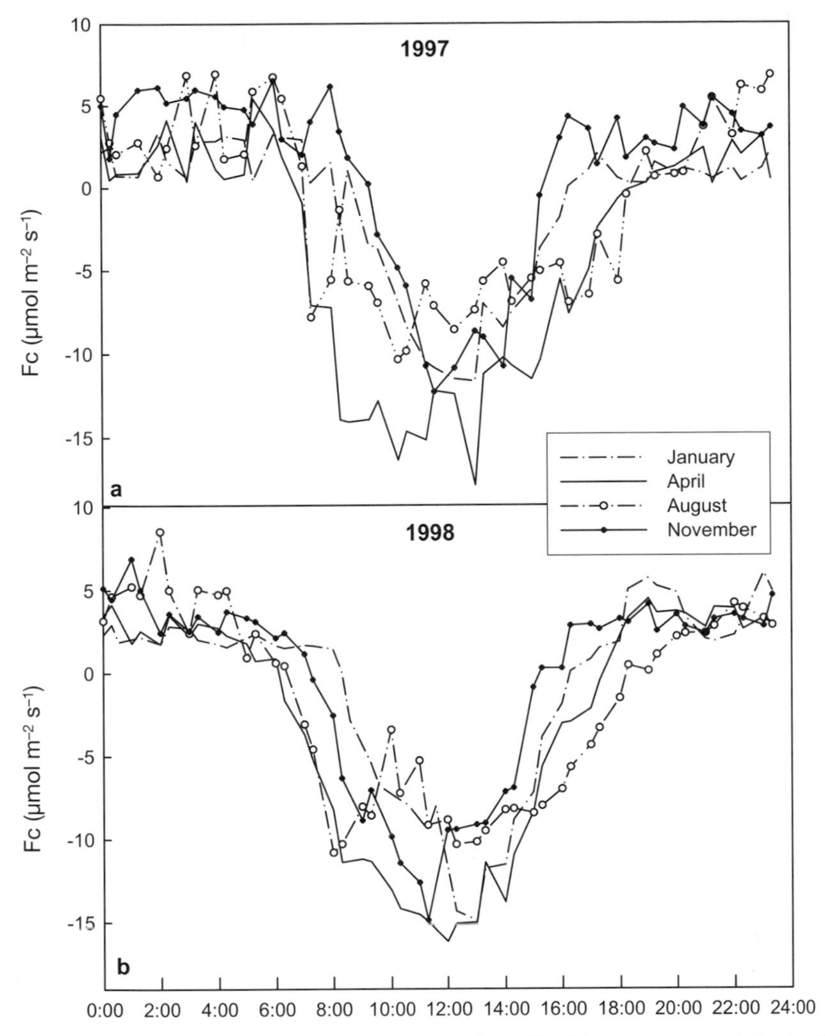

Fig. 7.4. Seasonal typical daily trends of carbon dioxide exchanges of the *Q. ilex* forest for the year **a** 1997 and **b** 1998

7.4.2 Seasonal Carbon Fluxes

The monthly and yearly trends of CO_2 fluxes and seasonal climatic data are presented in Table 7.6. Summer 1998 was warmer and more humid than that of 1997, so the monthly carbon assimilation values in July and August were higher in 1998 than in the comparable period in 1997: -92.55 and −77.01 g C m^{-2} in 1998 versus −70.87 and −46.98 g C m^{-2} in 1997. Monthly carbon assimilation was reduced in May, both in 1997 and 1998 (Tirone et al. 2000; Fig. 7.5). The maximum CO_2 assimilation was between March and July

Table 7.6. Monthly values of cumulated carbon fluxes (Fc cum.), average air temperature (Air temp.), average soil water content (SWC) for the years 1997–1998

Month	Fc cum. (g C m^{-2}) 1997	Fc cum. (g C m^{-2}) 1998	Air temp. (°C) 1997	Air temp. (°C) 1998	SWC (vol%) 1997	SWC (vol%) 1998
January	–21.9	–32.7	8.7	6.9	17.2	15.6
February	–41.6	–46.5	8.6	8.4	15.3	16.6
March	–75.6	–69.4	9.8	9.1	10.2	15.3
April	–80.4	–66.1	10.5	12.4	9.8	16.2
May	–44.2	–24.0	15.7	16.6	11.8	14.3
June	–62.9	–62.3	18.7	19.7	7.3	10.2
July	–70.9	–92.5	20.2	22.9	4.3	7.0
August	–46.9	–77.0	21.0	23.5	5.2	5.0
September	–45.9	–44.7	19.1	19.2	5.9	9.8
October	–29.6	–45.1	14.8	15.8	14.7	13.8
November	–19.9	–22.6	12.4	10.3	20.1	10.5
December	–16.7	–14.7	10.2	9.1	19.2	17.3
Total	–546.7	–597.7				

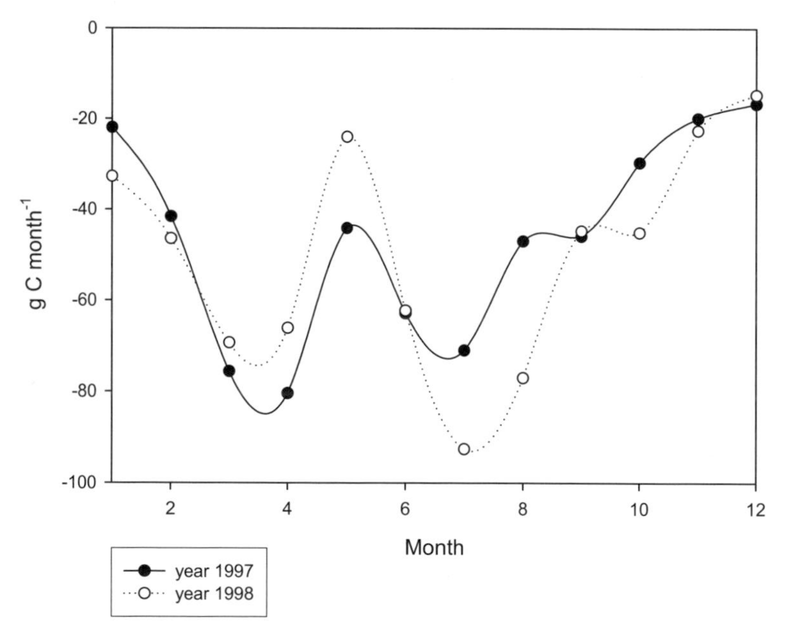

Fig. 7.5. Annual trend of carbon dioxide exchanges for the *Q. ilex* forest

with a considerably reduced assimilation in May in both years. This reduction was caused by increases in nighttime and daytime respiration. Ecosystem respiration in May was 177 g C m^{-2} $month^{-1}$ compared with a yearly average of 82 g C m^{-2} $month^{-1}$ (Dore 1999). In particular, soil respiration was highest between the end of April and the end of May when soil temperature and moisture were optimal (Dore et al. 1997). The observed increase in soil respiration during the spring was observed systematically in 1996, 1997, and 1998 (Dore 1999; Fig. 7.6a,b). Soil respiration was lowest in winter and summer because of temperature and water limitation and highest in spring and autumn. Similarly

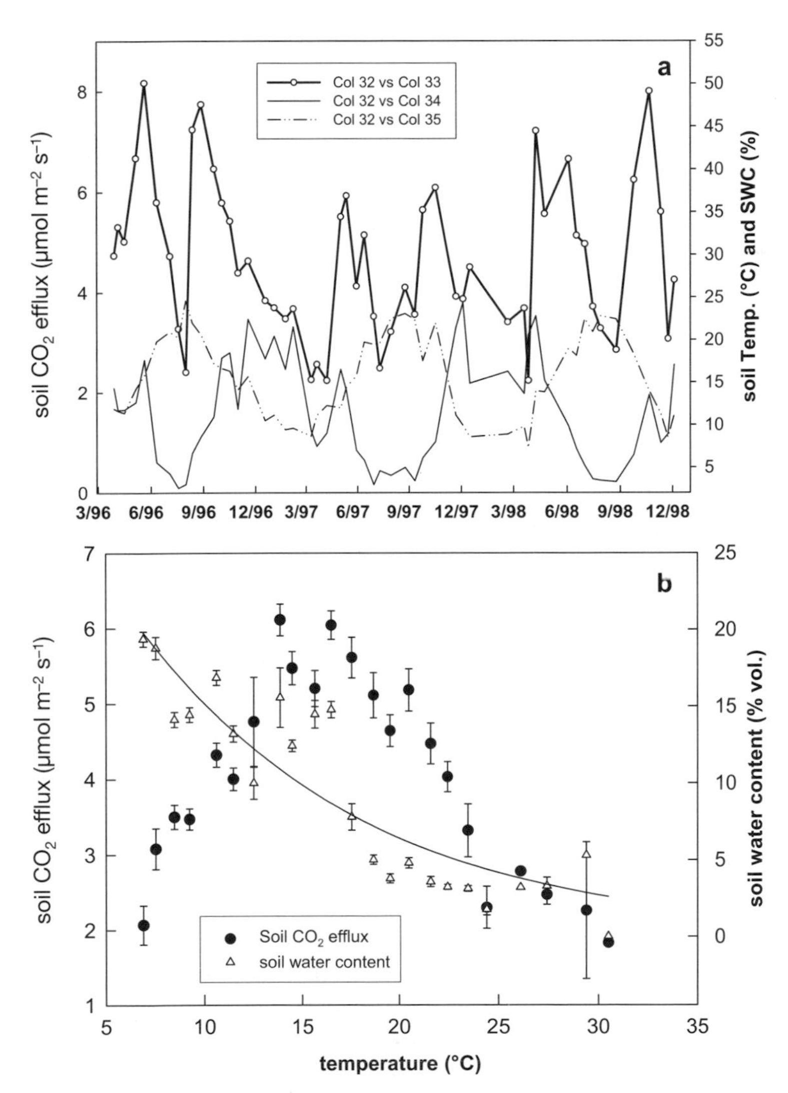

Fig. 7.6. Soil respiration. **a** Annual trend of soil respiration between 1996 and 1998; **b** effects of soil temperature and soil water content on soil CO_2 efflux

high night respiration rates in autumn (5 μmol m^{-2} s^{-1}) were observed in another study on a Mediterranean *macchia (maquis)* (Valentini et al. 1991b).

Monthly and yearly totals of carbon exchange were comparable in the two measurement years, being 546.7 and 597.7 g C m^{-2} $year^{-1}$ in 1997 and 1998, respectively (Tirone et al. 2000). The difference of 51 g C m^{-2} $year^{-1}$ was due to a wetter summer in 1998. In the 2 years, the number of days when the ecosystem was a source of carbon was similar (49 in 1997 and 45 in 1998) as well as was their distribution. During these days, the average carbon lost was 0.46 g C m^{-2} day^{-1} in 1997 and 0.73 g C m^{-2} day^{-1} in 1998. For the remaining days, the average carbon assimilation was 1.807 g C m^{-2} day^{-1} in 1997 (316 days) and 1.981 g C m^{-2} day^{-1} in 1998 (320 days).

For these 2 years this *Q. ilex* forest was a significant carbon sink, although it was lower than recorded for a beech forest in Italy (Matteucci et al. 1999). The evergreen *Q. ilex* forest maintains the capacity to assimilate carbon throughout the year. This forest did not achieve stasis in either winter or summer. Carbon assimilation continued in winter whenever light and temperature were not limiting, or in summer when water was still available. The net exchange was primarily dependent on the number of daylight hours.

7.5 Energy Partition at Canopy Level

The components of the energy balance sensible heat (H), latent heat (LE), net radiation (Rn), and soil heat flux (G) showed a distinct seasonal trend. To compare different trends of energy partitioning we used five average representative days for every season. Every point was the average of 30 min for the 5 days (Fig. 7.7). In general, excluding October, the energy dissipated via sensible heat loss was greater than that used for evapotranspiration. The LE flux was greater in October, although input energy (Rn) was smaller in this month than in May and August. Water availability, after autumn rainfalls, combined with sustained high energy input, was responsible for the significant transpiration.

Soil heat flux (G) had a limited effect on the energy balance of the *Q. ilex* forest, and was only characterized by a change in flux direction from winter (positive during the nighttime and negative during the daytime) to summer (negative during the nighttime and positive during the daytime). Soil during nighttime was colder than the air during winter, and warmer in the summer. The annual trend of the Bowen ratio exhibited a bimodal distribution, with peaks in April–May and August (Fig. 7.8). However, while in August the energy was dissipated via sensible heat, due to water limitation and high radiation, the higher Bowen ratios of the period April–May are less clearly explained. It is possible that the change in season from winter, with limited energy inputs, to spring, with increased solar radiation, precedes the ability of

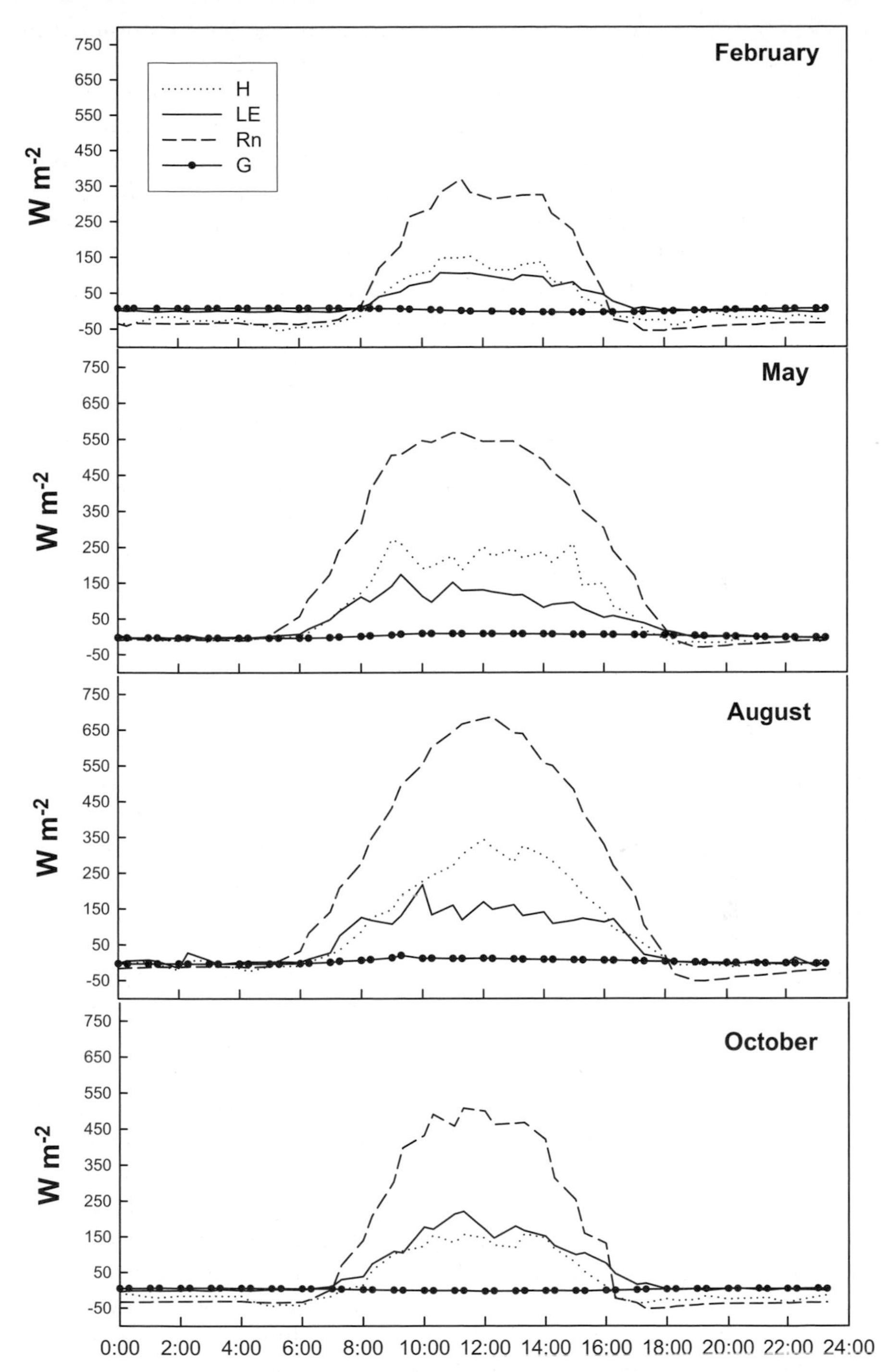

Fig. 7.7. Seasonal variation of the energy components: sensible heat (*H*), latent heat (*LE*), net radiation (*Rn*), and soil heat flux (*G*). The data represent a 5-day average

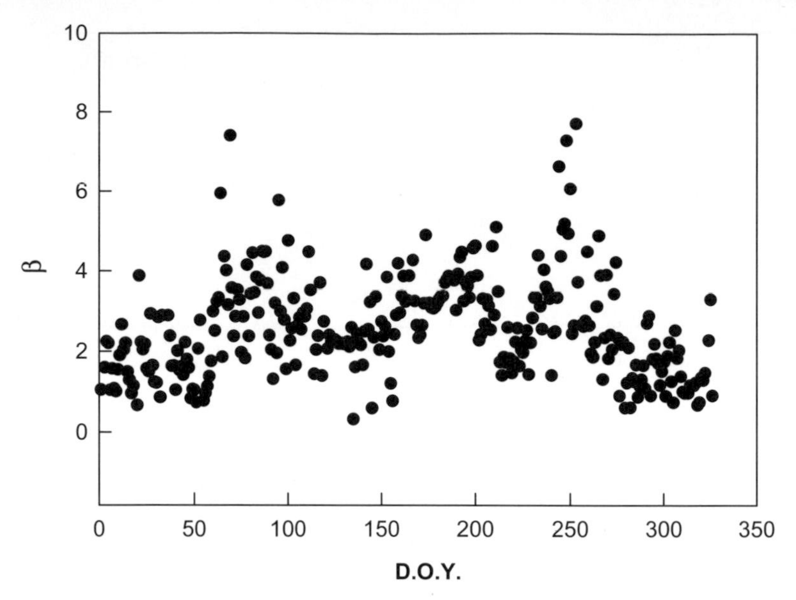

Fig. 7.8. Trend of the Bowen ratio in 1998

the vegetation to operate at a rate able to dissipate the increased energy. To some extent this passage between winter and spring, which is well documented for boreal forests as a physiological "switch," can also be a feature of Mediterranean vegetation. H and LE had a cumulative trend, increasing from January to June, in which latter month water availability began to decrease. At this point LE remained steady (about 85 kW m^{-2} day^{-1}) until August. H reached the maximum value (about 151 kW m^{-2} day^{-1}) in August and afterwards began to decrease.

7.6 Canopy Ecophysiology

The gross primary ecosystem production (GPP), was determined as the sum of net ecosystem production (NEP) measured by eddy covariance, and ecosystem respiration (RE). Daytime RE was calculated by assuming that the relationship between night respiration and air temperature was maintained during the daytime.

Thus GPP has been calculated as:

GPP=NEP+RE

GPP of the *Q. ilex* forest in Castelporziano changed during the seasons, responding to environmental and physiological factors. The comparison of

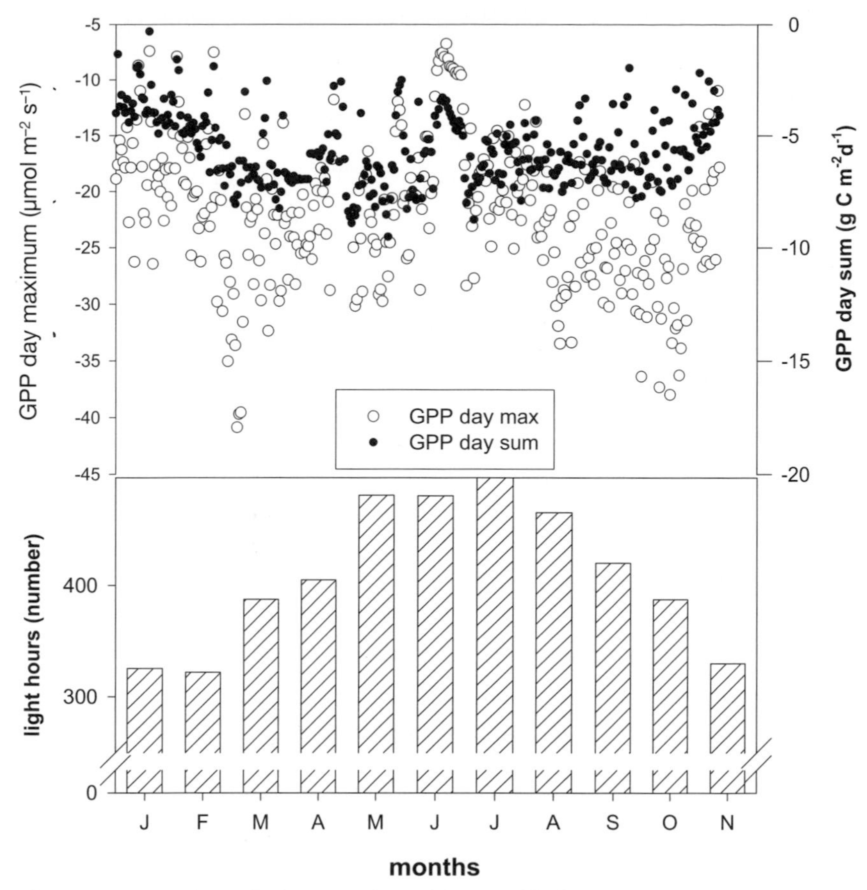

Fig. 7.9. Seasonal trend of GPP and availability of light

the daily maximum values and the daily sum of GPP is shown in Fig 7.9. The maximum daily ecosystem assimilation was highest in spring and autumn (March and October, with values higher than 35 μmol m^{-2} s^{-1}), and lowest in winter and summer. Total daily assimilation was highest in the summer (9.5 g C m^{-2} day^{-1}), except during the driest period, when both maximum GPP and daily GPP were greatly reduced. Total GPP was sensitive to both summer drought and to low winter temperatures, however, carbon uptake was maintained throughout the winter. During summer, water availability often limited GPP, but longer days compensated for this effect. In winter, when light and temperature were favorable, GPP was high, but these conditions were not common. Short days and low temperature resulted in low daily totals of GPP. These seasonal changes were driven by phenological and climatic variables, such as light level and duration, temperature, and water availability. In the following paragraphs some processes, such as the effect of light and canopy conductance on GPP, and temperature on RE, are presented.

Table 7.7. Parameters of the rectangular hyperbola used to fit the monthly light curves. *R* represents the night respiration, *a* the maximum quantum yield, *Amax* the saturation value, when PPFD is 1,600 μmol m^{-2} s^{-1} and r^2 the regression correlation coefficient

Months	*R* (μmol m^{-2} s^{-1})	*a*		*Amax*		r^2	
	NEP	NEP	GPP	NEP	GPP	NEP	GPP
January	3.2	0.11	0.08	−12.4	−17.2	0.98	0.99
February	2.9	0.057	0.054	−18.9	−19.0	0.97	0.99
March	2.3	0.043	0.047	−15.6	−19.7	0.97	0.97
April	4.4	0.044	0.04	−12.4	−18.0	0.96	0.94
May	6.7	0.047	0.064	−13.6	−18.9	0.97	0.99
June	5.9	0.056	0.037	−8.8	−11.8	0.91	0.95
July	3.8	0.034	0.03	−13.4	−12.2	0.96	0.96
August	4.6	0.061	0.049	−12.1	−13.2	0.95	0.94
September	4.8	0.063	0.07	−14.7	−18.0	0.98	0.97
October	5.5	0.071	0.054	−19.4	−26.7	0.96	0.98

7.6.1 Light Response of Canopy Photosynthesis

The relationship between assimilation and photosynthetic photon flux density (PPFD) followed a rectangular hyperbola (Ruimy et al. 1995). The 30-min data of each month were grouped in 100 μmol m^{-2} s^{-1} PPFD interval classes. For each class the average values of PPFD, GPP, and their standard errors were calculated. In Table 7.7 the parameters for each month are presented. The relationship between assimilation (GPP and NEP) and light intensity changed throughout the year (Figs. 7.9 and 7.10). The difference between the two curves represents RE. The apparent quantum yield (a) for GPP, a measure of light use efficiency, was highest in January (0.08), and lowest in July (0.03). Water stress lowered both the initial slope of the curves (*a*), and light saturated photosynthetic capacity (*Amax*). During summer, ecosystem respiration was suppressed by water deficit (Dore 1999), allowing the system to maintain an overall uptake of carbon.

7.6.2 Temperature and Soil Water Response of Ecosystem Respiration

In the absence of water limitation, respiration increased exponentially with air temperature up to an optimum and then decreased linearly (Fig. 7.11). The data points that form the linear decrease are from times of the year with low water availability and high temperatures (open symbols). Analysis of individual months showed that water limitation began in June and lasted all through July and August (Table 7.8). Also, because of the relationship between soil

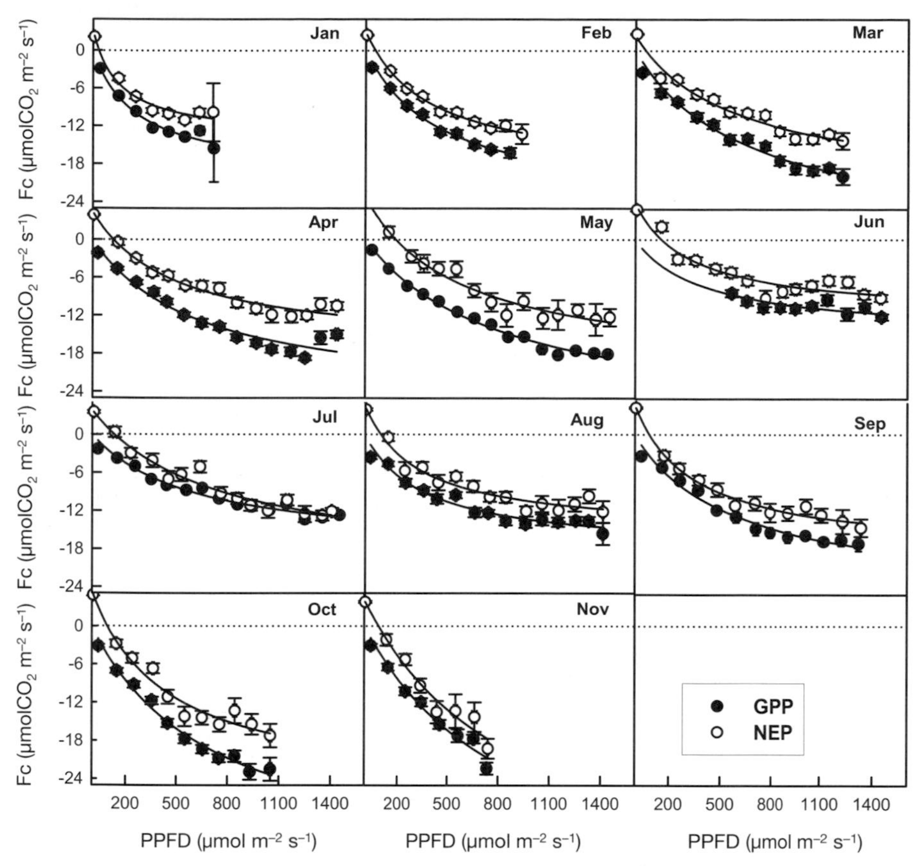

Fig. 7.10. Light response curve and their changes throughout the year

temperature and soil respiration measured by chambers (Fig. 7.6b), there was a clear exponential rise to a soil temperature optimum of 17 °C, followed by a linear decline. Soil respiration was dependent on soil temperature when soil water content was above 10 %. Below this level soil respiration was dependent on soil water content. The Q_{10} (change in respiration per 10 °C change in temperature), calculated for total ecosystem respiration from the nocturnal, non-water stressed, eddy data, was 1.9 and for the soil fluxes measured by chambers, when soil water content was above 10 %, was 1.5. Other studies based on eddy covariance techniques have reported Q_{10} ranging between 1.5–3.7 (Falge et al. 2001). Ecosystem respiration played an important role in the ecosystem carbon balance. The effects of ecosystem respiration on net C exchange changed during the months, being 86 % of GPP in winter and lower in spring and summer (59 % in July). The average NEP/GPP ratio was 0.26 in 1998 (Table 7.9).

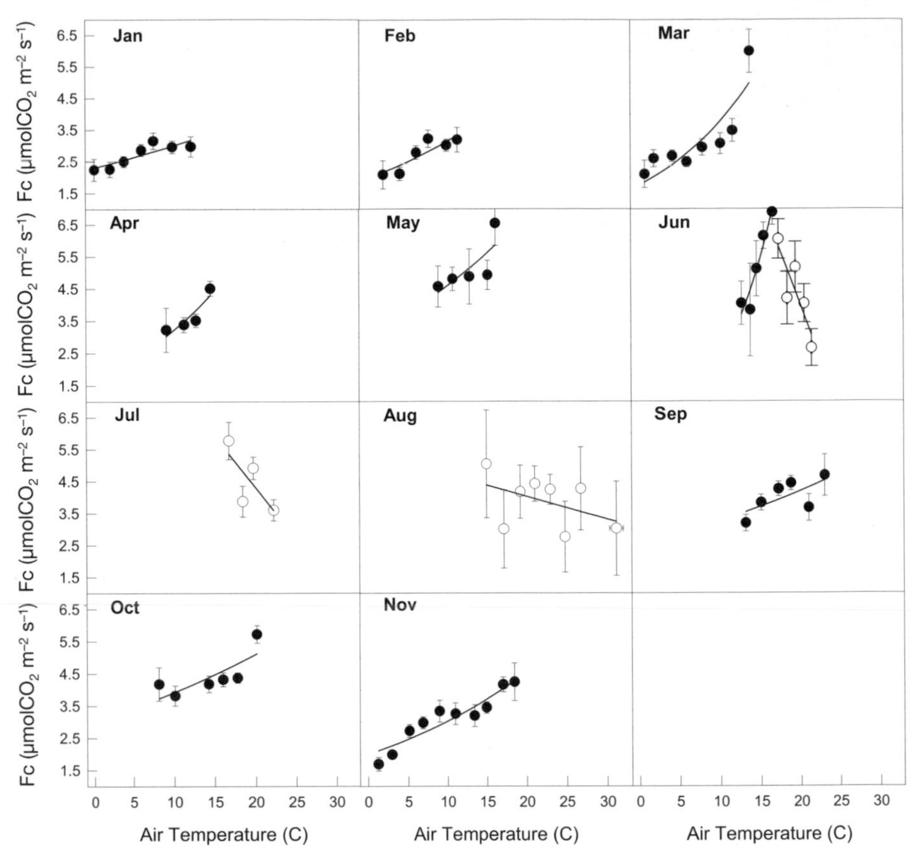

Fig. 7.11. Dependence of ecosystem respiration on temperature and soil moisture. Night carbon exchange measured by eddy covariance, selected by appropriate u*, were grouped for single months in 2° temperature classes. Average values and standard errors shown. Data were fitted with exponential regression curves $RE=a^{bT}$ (*closed circles*) or a linear equation $RE=a + bT$ (*open circles*), according to the time of the year. The parameters and r^2 are reported in Table 7.8

7.6.3 Water Exchanges

Surface conductance (G_s) was calculated from the Penman-Monteith equation. The surface conductance of the *Q. ilex* stand was dependent on VPD (Fig. 7.12a), decreasing when VPD increased (G_s was <1 mm s^{-1} for average daily VPD=2.5 kPa). This forest ecosystem was generally coupled to the atmosphere, following Jarvis and MacNaughton (1986) (omega factor around 0.1) during all the seasons, and strongly coupled to the atmosphere during the month of August (omega factor <0.05) (Dore 1999).

GPP decreased with decreasing Gs (Fig. 7.12b), and was correlated with LE when LE was less than 4 MJ m^{-2} day^{-1}(Fig. 7.12 c). At higher values of LE, indicative of good water conditions, GPP was constant.

Table 7.8. Relationship between air temperature and night ecosystem respiration. The values were fitted with exponential regression curves RE=$a^{b\,T}$ or the linear equation RE=a + bT, according to the time of the year. The parameters and r^2 are reported

months	A Exponential	b Exponential	a Linear	b Linear	r^2
January	2.30	0.026			0.73
February	1.97	0.046			0.77
March	1.75	0.076			0.74
April	1.68	0.064			0.81
May	3.08	0.0397			0.57
June	0.47	0.0413	17.26	−0.66	0.91/0.73
July			10.8	−0.32	0.55
August			4.59	−0.04	0.21
September	2.55	0.024			0.46
October	3.01	0.026			0.59
November	1.99	0.041			0.87

Table 7.9. Monthly variation of ecosystem functional characteristics, 1998

month	Light use efficiency [g C(mol fot^{-1})]	RE night/RE	NEP/GPP	RE/GPP	RE/NEP
January	0.05	0.65	0.14	0.86	6.27
February	0.09	0.55	0.30	0.70	2.33
March	0.10	0.46	0.38	0.62	1.62
April	0.14	0.47	0.35	0.65	1.85
May	0.03	0.49	0.17	0.83	4.75
June	0.04	0.68	0.32	0.68	2.14
July	0.06	0.52	0.41	0.59	1.41
August	0.05	0.53	0.31	0.69	2.27
September	0.04	0.57	0.19	0.81	4.15
October	0.05	0.61	0.13	0.87	6.43
November	0.13	0.61	0.26	0.74	2.91
December	0.09	0.63	0.20	0.80	3.91
Year mean	0.07	0.56	0.26	0.74	3.34
SD	0.04	0.07	0.10	0.10	1.75
SD/average	0.5	0.1	0.4	0.1	0.5

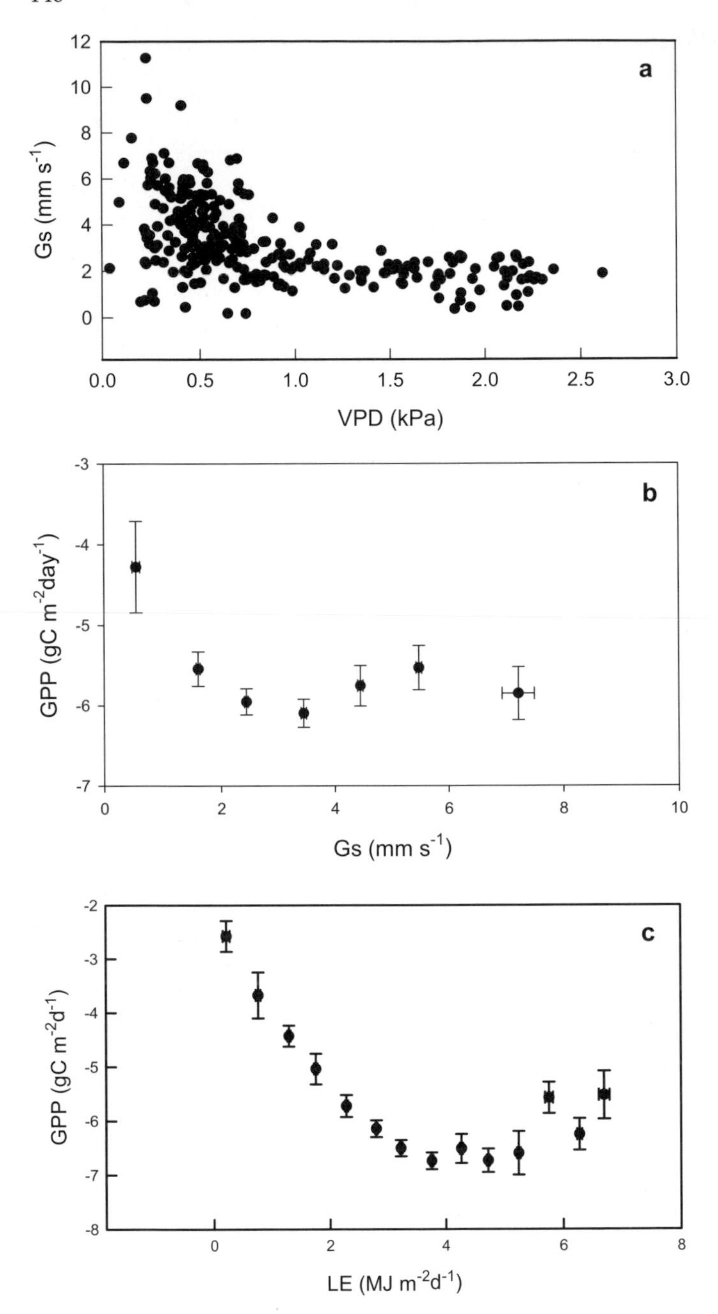

Fig. 7.12. **a** Response of the surface conductance, calculated using the Penman-Monteith equation, to VPD; **b** response of GPP to surface conductance Gs; **c** relationship between GPP and evapotranspiration LE

7.7 Conclusions

The evergreen *Q. ilex* forest showed significant rates of carbon uptake throughout the year due to the mild winter and moderate water limiting conditions in the summer. In particular, during winter significant rates of carbon sequestration were measured when light and temperature were not limiting. The difference in carbon balance between winter and summer was mainly determined by the length of the day. Water stress played a minor role except during a short period in August when carbon exchange rates were greatly reduced. Respiration is an important component of the ecosystem carbon balance, accounting for an average of 70 % of the total GPP. Respiration dynamics in response to climate were also interesting for this Mediterranean ecosystem. Despite high temperatures in summer, soil and ecosystem respiration both showed a significant reduction in rates, leading to the conclusion that soil moisture is the limiting factor for respiration. This particular dynamic results in a great reduction of net carbon uptake in spring, when the higher respiration rates are not yet balanced by a parallel increase of photosynthesis.

Energy dissipation was characterized by a sensible heat flux greater than evapotranspiration for most of the year, particularly in August (when water was limited) and spring (when leaf physiology was probably still limited by temperature).

References

Aubinet M, Grelle A, Ibrom A, Rannik U, Moncrieff J, Foken T, Kowalski AS, Martin PH, Berbigier P, Bernhofer C, Clement R, Elbers J, Granier A, Grunwald T, Morgenstern K, Pilegaard K, Rebmann C, Snijders W, Valenini R, Vesala T (2000) Estimates of the annual net carbon and water exchange of forests: the EUROFLUX methodology. Adv Ecol Res 30:113–175

Benedetti A, Alianello F, Dell'Orco S, Pinzari F, Trinchera A (1995) Impact of different xenobiotic factors on soil fertility (in Italian). II. Seminario Tenuta Presidenziale di Castelporziano, 12 Mar 1995

Bruno F, Gratani L, Manes F (1977) Primi dati sulla biomassa e produttività della lecceta di Castelporziano (Roma): biomassa e produzione di Q. ilex. Ann Bot XXXVVI:109–117

Davis GD, Richardson DM (eds) (1995) Mediterranean-type ecosystems: the function of biodiversity. Ecological studies 109. Springer, Berlin Heidelberg New York

Dore S, Muratore G, Tirone G (1997) Soil CO2 emission, comparison between a beech and a oak forest (in Italian). In: Borghetti M, Scarascia Mugnozza G. (eds) Atti del I convegno della SISEF La ricerca italiana per le foreste e la selvicoltura 4–6 giugno 1997, Legnaro

Dore S, Muratore G, Tirone G (1998) Emissioni di anidride carbonica dal suolo: confronto tra una faggeta e una lecceta. Atti I congresso SISEF: la ricerca italiana per le foreste e la selvicoltura, Legnano (PD), pp 57–62

Dore S (1999) Carbon and energy exchanges in forest ecosystems: comparison of three different study cases. PhD Thesis (in Italian), University of Padova, Italy

Falge E, Baldocchi D, Olson R, Anthoni P, Aubinet M, Berhofer C, Burba. G, Ceulemans R, Clement R. Dolman H, Granier A et al (2001) Gap filling strategies for defensible annual sums of net ecosystem exchange. Agric For Meteorol 107:43–69

Fioravanti R (1999) Produttività ed allocazione della biomassa in una lecceta della Tenuta Presidenziale di Castelporziano. Università degli Studi della Tuscia, Viterbo, Tesi di laurea

Gigon A (1979) CO_2-gas exchange, water relations and convergence of mediterranean shrub-types from California and Chile. Oecol Plant 14:129–150

Gratani L, Crescente MF (1994) Economia nell'uso delle risorse delle specie sempreverdi mediterranee. Atti V colloquio su Approcci metodologici per la definizione dell'ambiente fisico e biologico mediterraneo

Gratani L, Marzi P, Crescente MF (1992) Morphological adaptions of *Q. ilex* leaves in Castelporziano forests. Vegetatio 99/100:83–96

Greco S (1998) Analisi degli scambi gassosi e di energia di una biocenosi a *Q. ilex*. Università degli Studi di Padova, Tesi di dottorato

Grelle A, Lindroth A (1996) Eddy-correlation system for long term monitoring of fluxes of heat, water vapor and CO_2. Global Change Biol 2:297–307

Hobbs RJ, Richardson DM, Davis GW (1995) Mediterranean-type ecosystems: opportunities and constraints for studying the function of biodiversity. In: Richardson DM, Davis GW (eds) Mediterranean-type ecosystems. The function of biodiversity. Ecological studies 109. Springer, Berlin Heidelberg New York, pp 1–42

Jarvis PG, McNaughton KG (1986) Stomatal control of transpiration: scaling up from leaf to region. Adv Ecol Res 15:1–49

Kalin Arroyo MT, Zedler PH, Fox MD (eds) (1995) Ecology and biogeography of Mediterranean ecosystems in Chile, California and Australia. Ecological studies 109. Springer, Berlin Heidelberg New York, 455 pp

Kruger FJ, Michell DT, Jarvis JUM (eds) (1983) Mediterranean-type ecosystems: the role of nutrients. Ecological studies 43. Springer, Berlin Heidelberg New York, 450 pp

Kummerow J, Montenegro G, Krause D (1981) Biomass, phenology and growth. In: Miller PC (ed) Resource use by chaparral and matorral. Ecological studies 39. Springer, Berlin Heidelberg New York, pp 69–96

Lucchese F, Pignatti S, (1990) Overview of the coastal Lazio vegetation (in Italian). Accad Nazionale Lincei 264:5–48

Matteucci G, De Angelis P, Dore S, Masci A, Valentini R, Scarascia Mugnozza G (1999) Il bilancio del carbonio nelle faggete: dall'albero all'ecosistema. Ecologia strutturale e funzionale di faggete italiane. Scarascia Mugnozza, Edagricole, pp 133–181

Miglietta F, Peressotti A (1999) Summer drought reduces carbon fluxes in Mediterranean forest. Global Change Newsl 39:15–16

Miller PC (1981a) Conceptual basis and organization of research. In: Miller PC (ed) Resource use by chaparral and matorral. Ecological studies 39. Springer, Berlin Heidelberg New York, pp 1–15

Miller PC (ed) (1981b) Resource use by chaparral and matorral. Ecological studies 39. Springer, Berlin Heidelberg New York, 400 pp

Moncrieff JB, Massheder JM, de Bruin H, Elbers J, Friborg T, Heusinkveld B, Kabat P, Scott S, Soegaard H, Verhoef A (1997) A system to measure superface fluxes of momentum, sensible heat, water vapor and carbon dioxide. J Hydrol 188/189:589–611

Mooney HA, Parsons DJ, Kummerow J (1974) Plant development in Mediterranean climates. In: Lieth H (ed) Phenology and seasonality modelling. Springer, Berlin Heidelberg New York, pp 255–267

Moreno JM, Oechel WC (eds) (1995) Global change and Mediterranean-type ecosystems. Ecological studies 117. Springer, Berlin Heidelberg New York, 527 pp

Muratore G (1998) Studio dell'emissione di CO_2 dal suolo: il caso della lecceta di Castelporziano. Università degli Studi della Tuscia, Viterbo, Tesi di laurea, pp 60–66

Naveh Z (1987) Landscape ecology, management and conservation of European and Levant Mediterranean uplands. In: Tenhunen JD, Catarino FM, Lange OL, Oechel WC (eds) Plant response to stress. Functional analysis in Mediterranean ecosystems. NATO ASI Series G.15. Springer, Berlin Heidelberg New York, pp 641–657

Quezel P (1979) "Matorrals" mediterraneens et "Chaparrals" californiens. Quelques aspects comparatifs de leur dynamique, de leurs structures et de leur signification ecologique. Ann Sci Orest 36:1–12

Richardson D.M, Davis GW (eds) Mediterranean-type ecosystems. The function of biodiversity. Ecological studies 109. Springer, Berlin Heidelberg New York, 367 pp

Romane F, Terradas J (eds) (1992) *Quercus ilex* L. ecosystems: function, dynamics and management. Kluwer, Dordrecht, 280 pp

Ruimy A, Jarvis PG, Baldocchi DD, Saugier B (1995) CO_2 fluxes over plant canopy and solar radiation: a review. Adv Ecol Res 26:1–64

Susmel L, Viola F, Bassato A (1970) Ecologia della lecceta del Supramonte di Orgosolo. Estratto Annali di Economia Montana delle Venezie, vol X

Tenhunen JD, Catarino FM, Lange OL, Oechel WC (eds) (1987) Plant response to stress. Functional analysis in Mediterranean ecosystems. NATO ASI Series G.15. Springer, Berlin Heidelberg New York, 668 pp

Tirone G, Valentini R, Dore S (2000) Variazioni stagionali e annuali dei flussi di carbonio di un ecosistema forestale mediterraneo della Tenuta Presidenziale di Castelporziano. Atti II° congresso SISEF: applicazioni e prospettive per la ricerca forestale italiana, Bologna, 21–24 Oct 1999, pp 381–385

Valentini R, Scarascia Mugnozza G, De Angelis P, Bimbi R (1991a) Carbon dioxide and water vapour exchanges of a Mediterranean macchia canopy. Plant Cell Environ 14:1–8

Valentini R, Scarascia Mugnozza G, De Angelis P, Bimbi R (1991b) An experimental test of the eddy correlation technique over Mediterranean macchia canopy. Plant Cell Environ 14:987–994

Valentini R, Baldocchi DD, Thenunen JD (1999) Ecological controls on land-surface atmospheric interaction. In: Thenunen JD, Kabat P (eds) Integrating hydrology, ecosystem dynamic and biochemistry in complex landscape. Dahlem Workshop Reports, Chichester, pp 117–146

Whittaker RH, Likens GE (1975) The biosphere and man. In: Lieth H, Whittaker RH (eds) Primary productivity of the biosphere. Springer, Berlin Heidelberg New York, pp 305–328

8 A Model-Based Study of Carbon Fluxes at Ten European Forest Sites

E. Falge, J. Tenhunen, M. Aubinet, C. Bernhofer, R. Clement,
A. Granier, A. Kowalski, E. Moors, K. Pilegaard, Ü. Rannik,
C. Rebmann

8.1 Introduction

During the International Biological Program (IBP), systematic investigations of carbon flow through ecosystems and the structuring of carbon pools within ecosystems were carried out for the first time across the existing spectrum of biome types. Since then, concerted efforts have been made to understand the changes occurring in ecosystem carbon pools and carbon flow during aging and maturation of the system, in response to disturbances, and as symptomatic of ecosystem status or "health" (see review by Reiners 1983, Odum 1985). In other words, thorough knowledge of ecosystem carbon pools and carbon flow for a wide variety of ecosystem types and states is viewed as a key means by which better understanding of ecosystem response, resilience, and sustainable function may be achieved.

The development of robust methods to measure overall net ecosystem CO_2 exchange (NEE) over long-term periods (Valentini et al. 1999), thus obtaining estimates of net ecosystem production (NEP), and the implementation of these methods within a network context (Baldocchi et al. 1996) has provided us with the opportunity to evaluate patterns in NEP and ecosystem respiration (Reco) required to support it along environmental gradients, among ecosystem types, and with respect to chronosequences related to development. Changes in ecosystem status both in terms of normal process variability and in response to stress should be apparent in the balance of GPP (gross primary production) and Reco that make up NEE and NEP. Observations of NEE were carried out during the period 1996–1999 at forest sites of the EUROFLUX network. We examine here the 1997 course of CO_2 exchange at ten locations studied by EUROFLUX, viewing the observed similarities and differences in terms of species physiology, stand structure, and climate effects. Since all locations are northerly sites, soil drought is expected to play

Ecological Studies, Vol. 163
R. Valentini (Ed.) Fluxes of Carbon,
Water and Energy of European Forests

a minor role in determining NEE and we have initially assumed that drought did not occur at these sites during 1997.

We attempt to identify a set of mathematical functions and simulation tools that are useful, appropriate, and adequate to summarize the annual course of NEE, GPP, and Reco from a network of flux stations. An initial summary of these data with models of the type applied here will subsequently lead to the design of methods appropriate in many other applications. Since network databases such as those assembled in EUROFLUX permit us to focus on response along climate gradients and over long periods of time, we can begin to consider:

1. the degree to which we must simplify models for use in spatial assessments,
2. the amount of complexity required to answer questions about global change at different scales, and
3. whether surface exchange "types" may be identified which lead to an effective classification of landscape functional units in this regard (Reynolds and Wu 1999; Tenhunen et al. 1999; Valentini et al. 1999).

This first across-site and model-based examination of the EUROFLUX data base assembled at ten locations has been undertaken to obtain a simple model that describes the results, that is easily understood, and that will facilitate communication on concepts of data interpretation as well as optimal exploitation of the data. The analysis also permits us to search for patterns of response in the data and to examine data consistency.

8.2 Methods

8.2.1 Observations of NEE at the EUROFLUX Sites

The EUROFLUX network includes fifteen measuring stations (Tenhunen et al. 1998, 1999; Aubinet et al. 2000). EUROFLUX encompasses a large range of latitudes (from 41°45′ to 64°14′N), climates (Mediterranean, temperate, arctic) and species (*Quercus ilex*, *Fagus sylvatica*, *Pseudotsuga menziesii*, *Pinus sylvestris*, *Picea abies*, and *Picea sitchensis*). Information from ten of these sites (Table 8.1) was available from the EUROFLUX data bank for 1997 and was used for this across-site modeling effort directed at analyzing long-term forest ecosystem carbon dioxide exchange. The intent to scale-up measurements from the local scale to the European continent, and to compare forest ecosystem function along continental gradients (Tenhunen et al. 1998), impose on the partners within the program the adoption of common methodologies for the measurement of fluxes and for the correction and treatment of the data. These methodologies used in the EUROFLUX network

Table 8.1. Characteristics of the EUROFLUX sites for which the data analysis and modeling was conducted. Measurements at Vielsalm, Belgium were separated into those originating in pure Douglas fir east of the measurement tower and beech mixed with 20 % conifers west of the tower. Stands were otherwise monospecific except at Brasschaat with 17 % cover by *Fagus sylvatica*, and Sorø with ca. 15 % cover of evergreens. For additional information about the stands see Tenhunen et al. (1998). LAI is on a projected leaf area basis. Climate data are for the observations during 1997

Site	Position	Main species	LAI	Mean T_{air}	Mean T_{soil}	Global shortwave radiation	Precipitation
			(m^2 m^{-2})	(°C)	(°C)	(GJ m^{-2} $year^{-1}$)	(mm $year^{-1}$)
Aberfeldy, Scotland	56°37′N 3°48′E	*Picea sitchensis*	8.0	8.2	6.7	2.98	958
Brasschaat, Belgium	51°18′N 4°31′E	*Pinus sylvestris, Fagus sylvatica*	3.0	10.5	10.1	3.72	673
Hesse, France	48°40′N 7°05′E	*Fagus sylvatica*	5.6	10.2	9.6	4.06	925
Hyytiälä, Finland	61°51′N 24°17′E	*Pinus sylvestris*	3.9	4.3	4.9	3.48	540
Sorø, Denmark	55°29′N 11°39′E	*Fagus sylvatica, evergreens*	4.8	8.4	7.5	3.66	527
Loobos, Netherlands	52°10′N 5°45′E	*Pinus sylvestris*	1.7	9.8	9.6	3.57	766
Tharandt, Germany	50°58′N 13°38′E	*Picea abies*	5.0	8.3	7.1	3.95	714
Vielsalm "beech", Belgium	50°18′N 6°00′E	*Fagus sylvatica*	4.2	8.0	8.0	3.76	758
Vielsalm "fir", Belgium		*Pseudotsuga menziesii*					
Weidenbrunnen, Germany	50°09′N 11°52′E	*Picea abies*	5.3	6.2	6.3	3.61	572

including additional topics relating to the standardization of software, quality control of the measured flux data, discussion of the tower footprints, and especially the site-to-site differences in methodology, have been described in detail in Aubinet et al. (2000).

To provide a set of input and validation data for the ecosystem flux model GAS-FLUX, reliable quality-checked data are identified by a flag setting (flag = 0 indicates measurements with high dependability and accuracy as derived under optimal, well-mixed conditions). For the following model development, estimates of ecosystem respiration are derived by analyzing this subset of data. Flux underestimation during night periods was corrected by replacing the measured fluxes by the simulated efflux estimated with a temperature function derived during well-mixed conditions at night (u^* corrected). The threshold value of u^* that was used to distinguish between stable and well-mixed conditions was site-dependent and varied between 0.15 and 0.5 m s^{-1}. Ecosystem respiration functions were derived from these nighttime fluxes as described in the following section. Ignoring any gap filling methods applied by individual research groups, gap filling to obtain long-term NEE estimates reported in this paper for either measured or modeled data (when meteorology is missing) was conducted in exactly the same manner for all sites. For relatively short periods the appropriate canopy NEE was estimated for prevailing light and temperature conditions from canopy responses derived from data grouped into 2° temperature and 100 μmol m^{-2} s^{-1} light (PAR) classes or model results similarly grouped. The light and temperature responses were based on data from immediately preceding or following time periods (for methodology, see Falge et al. 2001). Long data gaps are difficult to fill, and these problems are discussed in Sections 8.3 and 8.4.

8.2.2 Flux Model Description and Parameterization

The across-site EUROFLUX modeling efforts are not intended to replace the development of detailed models that are undertaken by individual research groups at their own sites. Instead, the focus is at the European level and the intention is to synthesize, generalize, and simplify. Given this objective, we are interested in determining how experimental results can be transferred to user communities in models which capture the most important biological responses, but which describe response in simple enough terms that they can be used in spatial applications (i.e., for regional, country, and continental modeling efforts). An effort was made to develop a general model with a minimum in site specific parameterization. The approach used to summarize NEE response for 1997 at ten EUROFLUX sites is illustrated in Fig. 8.1 and described below.

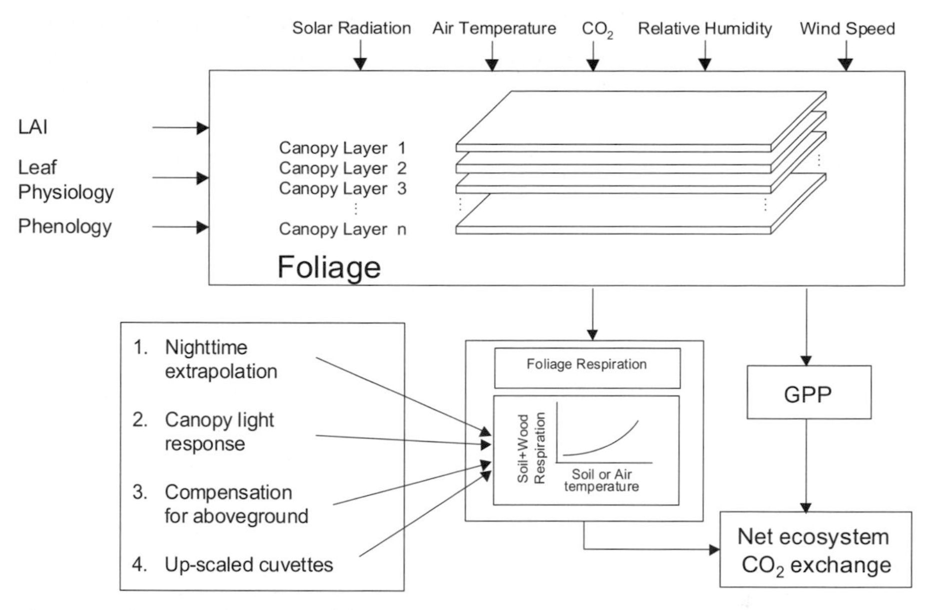

Fig. 8.1. Schematic diagram of the structure of the models applied in analysis of eddy covariance measurements of NEE at ten European forest sites as described in the text. Indicated are the layered canopy model which provides estimates of GPP and foliage respiration, the estimate for respiration of wood and soil which is determined via different methods, and the summation of GPP and Reco to obtain NEE

8.2.3 Foliage Gas Exchange

CO_2 gas exchange of foliage elements arc described with a traditional homogeneous-layered canopy model (GAS-FLUX) that estimates vegetation light interception, canopy microclimate profiles, and leaf or needle gas exchange as regulated via photosynthetic sub-processes and stomatal conductance (cf. Tenhunen et al. 1994; Sala and Tenhunen 1996). The model is similar to others developed for crop and forest stands (Norman 1975, 1980; Ross 1975; Monteith 1976; Baldocchi and Harley 1995; Williams et al. 1996) and utilizes sun and shade light classes for canopy foliage. As indicated in Fig. 8.1, half hour values of global shortwave radiation, air temperature, CO_2 partial pressure, relative humidity, and horizontal wind speed measured directly above the respective forest canopies are input to the model. Model parameterization requires information on leaf area index (LAI; Table 8.1, separated into layers with maximum LAI of 0.5), on leaf physiology (Table 8.2), and on the phenology in the case of deciduous species (which is assumed to influence only the LAI found on a particular date, as shown in Fig. 8.2). LAI was held constant for coniferous stands at the values provided by investigators from each site. In the mixed stands at Vielsalm and Brasschaat (see Table 8.1), the deciduous component of the overall stand was assumed to follow the general phenology used

Table 8.2. Parameters applied to describe leaf physiology at the sites studied. For detailed explanation of the leaf model and use of the parameters see Falge et al. 1996 or Harley and Tenhunen (1991). Output of the model is on a projected surface area basis for beech and on a total surface area basis for dense coniferous canopies and pine. Conversion of rates for dense coniferous canopies to a projected leaf area basis is obtained by multiplying $f(R_d)$, $c(P_{ml})$, $c(Vc_{max})$, gmin, and α by 2.57, and for pine by 2.7

Description	Parameter	Norway Spruce, Sitka Spruce, Douglas Fir	Value
Dark respiration	$f(R_d)$	Original	0.526
		Weidenbrunnen	0.510
		Tharandt	0.636
		Aberfeldy	0.800
		Vielsalm	0.631
	$E_a(R_d)$		63,500
Electron transport capacity	$c(P_{ml})$	Original	14.279
		Weidenbrunnen	13.851
		Tharandt	17.278
		Aberfeldy	21.704
		Vielsalm	17.135
	$\Delta H_a(P_{ml})$		47,170
	$\Delta H_d(P_{ml})$		200,000
	$\Delta S(P_{ml})$		643
Carboxylase capacity	$c(Vc_{max})$	Original	19.69
		Weidenbrunnen	19.099
		Tharandt	23.825
		Aberfeldy	29.929
		Vielsalm	23.628
	$\Delta H_a(Vc_{max})$		75.750
	$\Delta H_d(Vc_{max})$		200,000
	$\Delta S(Vc_{max})$		656
Carboxylase kinetics	$f(K_c)$		299.469
	$E_a(K_c)$		65,000
	$f(K_o)$		159.597
	$E_a(K_o)$		36,000
	$f(\tau)$		2,339.53
	$E_a(\tau)$		–28,990
Light use efficiency	α	Original	0.0150
		Weidenbrunnen	0.0146
		Tharandt	0.0182
		Aberfeldy	0.0228
		Vielsalm	0.0180
Stomatal conductance	gmin		1
	gfac		9.8

Scots Pine	Value	European Beech	Value	Unit
Original	0.716	Original	1.040	(μmol m^{-2} s^{-1})
Loobos	0.716	Hesse	1.040	
Hyytiälä	0.451	Sorø	1.040	
Brasschaat	0.473	Brasschaat	0.686	
		Vielsalm	1.040	
	56,050		58,000	(J mol^{-1})
Original	12.272	Original	35.300	(μmol m^{-2} s^{-1})
Loobos	12.272	Hesse	35.300	
Hyytiälä	7.731	Sorø	35.300	
Brasschaat	8.100	Brasschaat	23.298	
		Vielsalm	35.300	
	44,898		40,000	(J mol^{-1})
	190,000		200,000	(J mol^{-1})
	643		655	(J K^{-1} mol^{-1})
Original	37.074	Original	80.100	(μmol m^{-2} s^{-1})
Loobos	37.074	Hesse	80.100	
Hyytiälä	23.357	Sorø	80.100	
Brasschaat	24.469	Brasschaat	52.866	
		Vielsalm	80.100	
	75,250		69,000	(J mol^{-1})
	230,000		198,000	(J mol^{-1})
	656		660	(J K^{-1} mol^{-1})
	299.469		404	(μmol mol^{-1})
	65,000		59,500	(J mol^{-1})
	159.597		248	(mmol mol^{-1})
	36,000		35,900	(J mol^{-1})
	2,339.53		2,339.53	
	–28,990		–28,990	J mol^{-1}
Original	0.0180	Original	0.0450	(mol CO_2 m^{-2} leaf area)
Loobos	0.0180	Hesse	0.0450	[(mol photons m^{-2} horizontal area)$^{-1}$]
Hyytiälä	0.0113	Sorø	0.0450	
Brasschaat	0.0119	Brasschaat	0.0297	
		Vielsalm	0.0450	
	1		1	(mmol m^{-2} s^{-1})
	12.8		12.0	

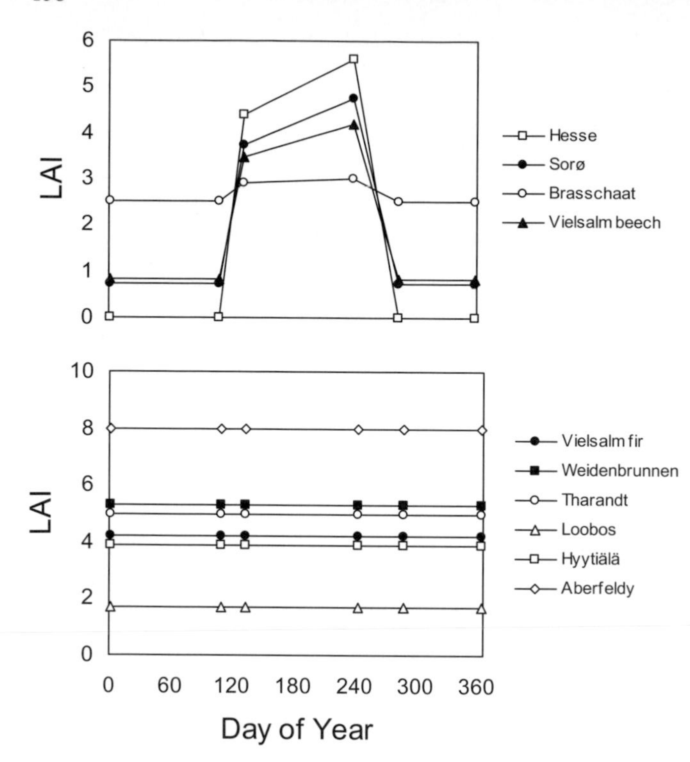

Fig. 8.2. Annual courses for LAI applied in the analysis at each of the ten forest sites

for *Fagus sylvatica* at Hesse and Sorø. The time-dependent change in LAI depicted in Fig. 8.2 is based on monitoring carried out at Hesse, France (Granier et al. 2000).

Leaf physiology was parameterized according to methods described by Harley and Tenhunen (1991) and Falge et al. (1996). Parameters were based on observations of leaf gas exchange of *Picea abies* carried out in the Fichtelgebirge, Germany (Falge et al. 1996), of *Pinus sylvestris* growing at the Hartheim Plantation near Freiburg, Germany (Sturm 1998; Sturm et al. 1998), and of *Fagus sylvatica* at Steigerwald, Germany (Fleck 2001). These baseline gas exchange studies provided a parameter set for each species (indicated as "original" and in boldface type in Table 8.2). However, observed leaf gas exchange capacities (electron transport, carboxylation at 25 °C, if reported) varied considerably among sites, requiring an adjustment of electron transport and carboxylation as indicated in Table 8.2. A similar linear scaling was performed for dark respiration and light harvesting. The shape of the temperature response as determined by the parameters of energy of activation, deactivation, and entropy was assumed not to change. To minimize parameter sets, physiological parameters for Norway spruce were hypothesized to hold for other coniferous species in dense canopies, and the actual parameters used were similarly scaled to the observed needle N content at specific measurement sites.

Stomatal conductance was estimated in proportion to net photosynthesis rate according to the model of Ball et al. (1987; cf. Harley and Tenhunen 1991). Since water stress was assumed not to occur, the critical factor establishing the magnitude of leaf conductance as a function of net photosynthesis rate, i.e., gfac in Table 8.2, remained at the constant values shown.

The aboveground canopy model provides as output an estimate of canopy CO_2 uptake (GPP) and an estimate of foliage respiration required to support CO_2 gain (Fig. 8.1). Potentially, this part of the overall model could additionally provide estimates of woody tissue carbon dioxide exchange, however, the required information to parameterize these fluxes within the canopy model is in general lacking.

8.2.4 Ecosystem Respiration

In order to estimate ecosystem respiration (Reco), estimates for aboveground woody respiration and for soil respiration (root plus microbial) must be added to foliage respiration (Fig. 8.1). A relatively complete set of direct measurements of respiration components was obtained with cuvettes at a few sites (e.g., Hesse, Brasschaat, and Weidenbrunnen), but sampling with cuvettes was in general restricted to soil at 2-week intervals. Thus, estimates of Reco based on chamber measurements are not possible to derive for all sites. However, an estimate for Reco may be obtained from the eddy covariance measurements in two separate ways, which we refer to as (1) the nighttime flux extrapolation method to obtain $Reco_N$ and (2) the canopy light response method to obtain $Reco_L$. $Reco_N$ extrapolates nighttime respiration into the day, $Reco_L$ daytime data into the night. As foliage respiration at a given temperature is reduced in the light (Farquhar and von Caemmer 1982), $Reco_N$ most likely will overestimate, and $Reco_L$ will underestimate ecosystem respiration. In a third approach, if we assume that the aboveground canopy model correctly provides GPP, then Reco may be estimated as the residual between eddy covariance measurements and modeled GPP: (3) the compensation-for-aboveground method to obtain $Reco_C$.

According to the nighttime flux extrapolation method, u*-corrected measurements during night periods were used to obtain parameterizations for an exponential equation [Arrhenius type, see Eq. (8.2) below; Janssens et al., this Vol., their Table 12.1] for calculating $Reco_N$. Since canopy respiration capacity (especially in deciduous species) varies over the course of the year, the data were treated separately for summer (June through September), winter (December through March), spring (April, May), and fall months (October, November).

If we consider a small subset of daytime eddy covariance NEE measurements by sorting the observations into temperature classes of 2 °C, then data collections are obtained similar to that illustrated in Fig. 8.3. The data (sign

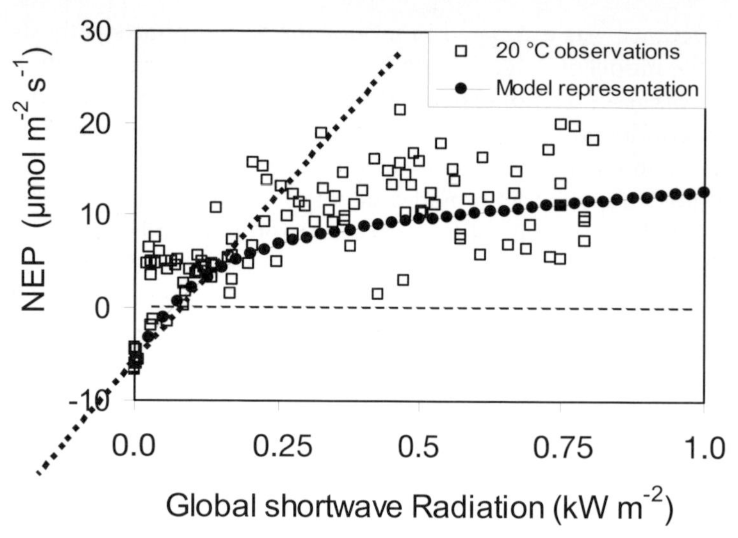

Fig. 8.3. An example of a canopy light response obtained by sorting observations within a limited temperature range from the eddy covariance database. Shown also is a hypothetical representation of the actual canopy behavior and an attempt to linearly extrapolate the initial response to zero light in order to estimate Reco

reversed) may be plotted versus radiation input, since this is the main short-term environmental factor influencing canopy function at relatively constant temperature and atmospheric CO_2. Such light responses were fit to the Michaelis-Menten equation as has been done for other forest stands [Eq. (8.1); Hollinger et al. 1994; Pilegaard et al. 2001].

$$NEP = \frac{\alpha \cdot Q \cdot GPP_{sat}}{GPP_{sat} + \alpha \cdot Q} - Reco_L \tag{8.1}$$

where Q is global shortwave radiation (in kW m^{-2}), α ecosystem light use efficiency, and GPP_{sat} the theoretical value of GPP where light saturation occurs. The extrapolation to zero light provides an estimate of $Reco_L$. From these $Reco_L$ estimates Arrhenius-type temperature dependencies [Eq. (8.2)] were derived for each seasonal period as defined above.

An additional possibility for describing Reco is the compensation method. The site-specific parameterized canopy model provides for each time of daytime observation an estimate of GPP that is usually greater than NEE (recall from Sect. 8.2 that the analysis is based on daytime flag = 0 data). Thus, an estimate for $Reco_C$ is obtained for each daytime half hour that allows for compatibility between NEE measured and GPP calculated: $Reco_C$=GPP+NEE.

$Reco_N$, $Reco_L$, and $Reco_C$ were described as an exponential function of soil temperature at 5 cm depth ($Reco_N$) or air temperature ($Reco_L$ and $Reco_C$) according to the following equation:

$$Reco = R(T_{ref}) \cdot e^{\left(\frac{E_a}{R} \cdot \left(\frac{1}{T_{ref}} - \frac{1}{T_K}\right)\right)} \quad (8.2)$$

where T_{ref} is the reference temperature of 283.16 K, and T_K is air or soil temperature in K. $R(T_{ref})$, the respiration rate at T_{ref}, and E_a, the activation energy in J mol^{-1}, are the fitted parameters. R is the gas constant (8.134 J K^{-1} mol^{-1}).

Combining GPP with one of the Reco estimates ($Reco_N$, $Reco_L$, or $Reco_C$) provides a simulated NEE which may be integrated over time and compared with integrated observations.

8.3 Results

The analysis assumes that a parameterization of the model may be derived to describe average forest gas exchange response in the vicinity of a measurement tower. Due to continual change in the footprint and heterogeneity in forest structure, however, flux measurements are quite variable. Thus, exact criteria that should be used for model fitting and testing are unclear at this time. Nevertheless, we would expect a successful or promising modeling approach for NEE to demonstrate agreement at several time scales: annually in order to quantify vegetation development and growth as well as long-term changes in the overall structuring of ecosystem carbon pools, monthly in agreement with seasonal changes in climate and shifts in phenology, and daily in order to examine linkages between eddy covariance and short-term behavior of physiological processes (stomatal conductance and net photosynthesis regulation).

8.3.1 Model Comparison

Annual NEE estimated from observed data and modeled estimates employing the nighttime flux extrapolation, the light response, and the compensation for aboveground methods are compared in Fig. 8.4. Examined here is overall performance of the three models during daytime, nighttime, and 24-h periods summed over the year. Considering performance during nighttime (middle panel), it must be remembered that data utilized in the parameterization of $Reco_L$ originate under turbulent daytime conditions best suited for the eddy covariance method. However, it extrapolates daytime leaf respiration into night periods without considering the downregulation of daytime to nighttime leaf respiration, and therefore underestimates the measured nighttime fluxes. The compensation method suggests that CO_2 efflux during the night-

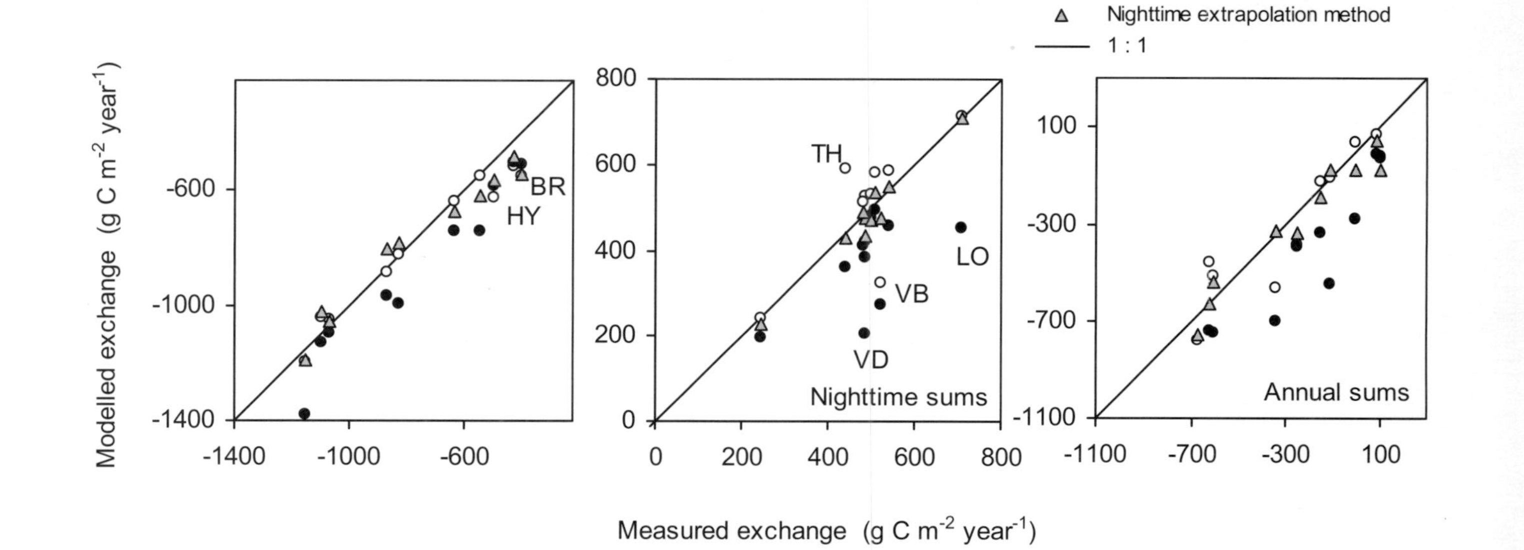

Fig. 8.4. Comparison of the annual integrated carbon dioxide exchange of the ten forest sites during daytime, nighttime, and 24h periods as estimated by measurements, the nighttime extrapolation method, the light response method, and the compensation for aboveground method, all with accompanying gap filling. The response at *BR* Brasschaat, *HY* Hyytiälä, *TH* Tharandt, *LO* Loobos, *VD* and *VB*, Vielsalm Douglas fir and Vielsalm Beech are discussed in the text

time period should be higher than recorded in the database, which is plausible considering uncertainties in eddy diffusivity at night and suspected by numerous researchers (the so-called nighttime flux underestimation, see Lee 1998, Baldocchi et al. 2000 for references). Alternatively, one can consider that the compensation model overestimates CO_2 efflux during nighttime (see below). This could be due to potential errors in GPP estimates carrying over to Reco estimates derived as the difference between modeled GPP and measured NEP, or methodological problems in obtaining the nighttime data. Obviously, the nighttime extrapolation method reproduces nighttime data best: the figure only documents the quality of fitting model parameters and the selected model.

While the light response method might be preferred because it is based only on observed data, CO_2 efflux estimates obtained with this method are less than measured during daytime. This seems unreasonable and it probably results from the inability to accurately determine the light utilization efficiency (initial light response curve) of the canopy. It is also possible that we are not able to recognize the appropriate shape or curvature of the canopy light response from eddy covariance data. Low estimates for ecosystem daytime CO_2 efflux carry over to the nighttime period and enhance the problem of extrapolating downregulated foliage respiration into the night (see above). As a result, both daytime sums and annual sums were systematically more negative than observation, i.e., there is an overestimation of carbon gain caused by a large underestimation of $Reco_L$.

In certain cases identified in Fig. 8.4 and documented in Table 8.3, deviations between estimated and measured nighttime CO_2 efflux seem unreasonably high. So is $Reco_C$ during the nighttime for Tharandt much higher than measured (shown as TH in Fig. 8.4), whereas the model predicted an adequate carbon gain during the daytime period (the period for model fitting), suggesting that other phenomena (for instance leakage or footprint heterogeneity) are influencing the nighttime measurements. Reduction of foliage (based on a potentially incorrect model parameterization for LAI) would reduce the predicted respiration. However, daytime respiration (and assimilation) would be also affected, leading to an underestimation of daytime fluxes. $Reco_L$ underestimates measured nighttime flux especially at Vielsalm (Douglas fir VD, and beech site VB) and Loobos, perhaps due to as yet unexplained advective effects. An additional problem in obtaining good agreement between the model and observations at Brasschaat (shown as BR in Fig. 8.4) relates to a high frequency of data gaps during the course of the year.

The overestimate by the compensation model for daytime NEE at Hyytiälä (shown as HY in Fig. 8.4) is related to winter dormancy. As discussed further below, canopy photosynthetic processes at Hyytiälä were activated 60 days later than predicted by the model and based on the air temperatures recorded. In this case, parameterization for the average conditions leads to a serious overestimation of carbon gain.

Table 8.3. Comparison of integrated gap filled observations and simulation results for net ecosystem CO_2 exchange using nighttime extrapolation modeling method as described in the text. Data are separated for daytime, nighttime, and 24-h total periods and summed over the annual cycle. All results are given in g C m^{-2} and correspond to the values plotted for the nighttime extrapolation method in Fig. 8.4

Site and dominant species	Daytime sums		Nighttime sums		Annual sums	
	Gap-filled observations	Gap-filled model results	Gap-filled observations	Gap-filled model results	Gap-filled observations	Gap-filled model results
Aberfeldy, Scotland						
Picea sitchensis	−1,094	−1,018	488	477	−606	−535
Brasschaat, Belgium						
Pinus sylvestris (mixed stand)	−401	−549	499	472	99	−76
Hesse, France						
Fagus sylvatica	−635	−674	482	487	−153	−192
Hyytiälä, Finland						
Pinus sylvestris	−499	−567	247	228	−253	−338
Loobos, Netherlands						
Pinus sylvestris	−824	−786	708	709	−116	−75
Sorø, Denmark						
Fagus sylvatica (mixed stand)	−546	−624	540	548	−6	−78
Tharandt, Germany						
Picea abies	−1,066	−1,054	443	430	−624	−624
Vielsalm, Belgium						
Fagus sylvatica (mixed stand)	−867	-800	522	476	−345	−330
Vielsalm, Belgium						
Pseudotsuga menziesii	−1,154	−1,190	485	435	−670	−758
Weidenbrunnen, Germany						
Picea abies	−427	−486	511	533	84	46

With respect to seasonal performance, the light response method results in large overestimates in carbon gain, particularly in the summer. This was the case for almost all sites and for all species studied. Thus, the light response method was judged inappropriate as a practical method for summarizing NEE across sites. The performance of the compensation for aboveground was similar to the nighttime flux extrapolation method, however, potential errors in modeled assimilation were compensated by the respiration parameterization, leading to involuntary shifts in the relative contribution of assimilation and respiration rates to the net flux. Thus, further examination of the simulation results is limited to the nighttime flux extrapolation method.

8.3.2 Performance of Nighttime Flux Extrapolation Method at Monthly and Daily Time Scales

To evaluate the performance of the nighttime flux extrapolation method, modeled monthly daytime NEE is shown in Fig. 8.5 in comparison to NEE obtained from daytime measurements. This allows a sort of model testing, because (except for ancillary information such as LAI, leaf physiology) only nighttime data were used for model parameterization. The data are grouped according to the three canopy physiology and structure types included in the study (dense coniferous in Norway and Sitka spruce and Douglas fir, relatively open coniferous as occurs in Scots pine, and broad leaf deciduous as occurs in European beech). As seen in Fig. 8.5, monthly measured and simulated daytime data agree with few exceptions at all times of the year and for all three canopy types, despite the argument that the method might overestimate respiration due to the neglect of downregulation of foliar respiration in daylight. The agreement between daytime measurements and simulation could eventually be a result of a compensation between over-and underestimation, since daytime respiration is calculated from nighttime fluxes over forests – these latter regularly underestimated by the eddy covariance technique (see Lee 1998, Baldocchi et al. 2000). In other words, over- and underestimations compensate, and the model produces *correct* daytime fluxes for the *wrong* reason. Deviations from the one-to-one line are seen for Hyytiälä during spring due to the observed delay in physiological activation after winter, as was mentioned above. Deviations occur at both Weidenbrunnen and Brasschaat due to filling of long gaps during winter (see Fig. 8.7 below). Long gaps leave little chance of estimating correctly from observations, while the model is reasonably reliable if weather data have been recorded throughout.

The annual course of GPP and Reco estimated from the nighttime extrapolation model is shown for all sites in Fig. 8.6. The simulations in the upper panels of the figure indicate the response of the model to the climate conditions measured at each site, while the lower panels indicate response to a "standardized" climate, in this case the meteorological conditions at Hyytiälä,

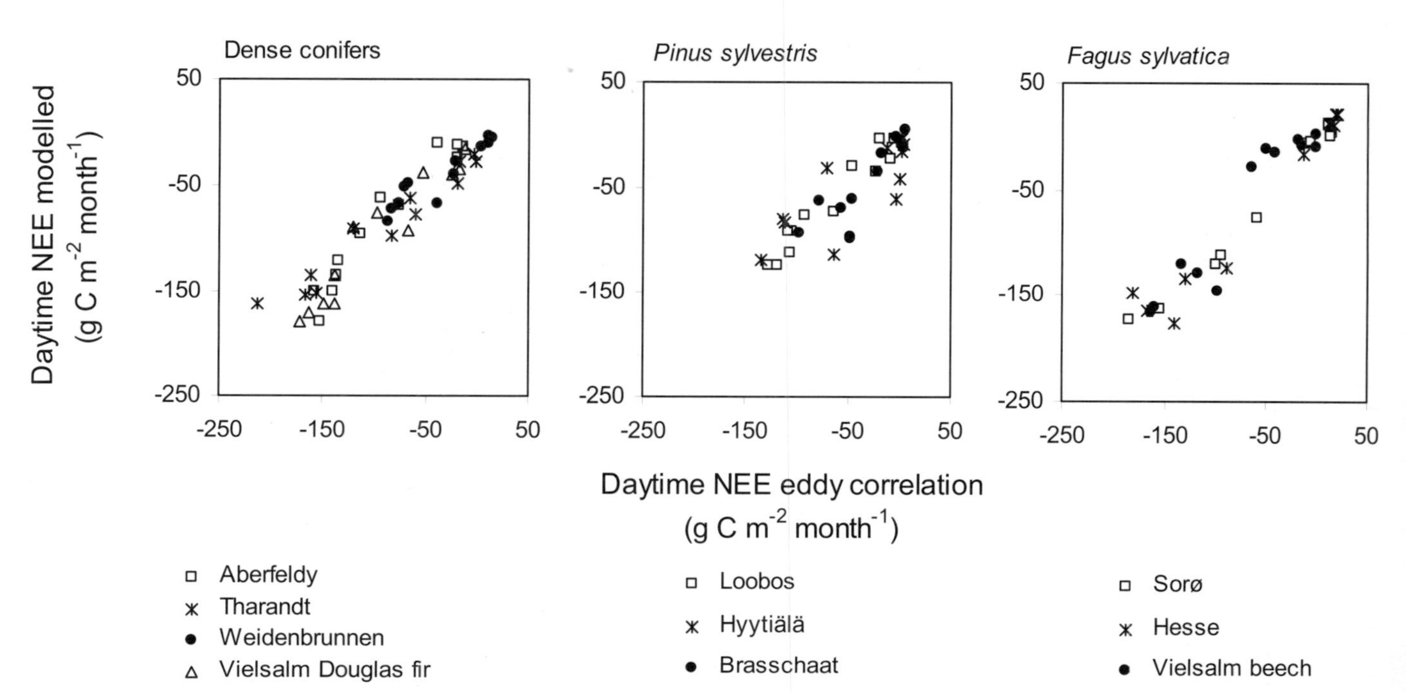

Fig. 8.5. Comparison of the monthly integrated carbon dioxide exchange of the ten forest canopies as estimated by measurements and the nighttime extrapolation model. Canopies are grouped into physiological and structural types. *Symbols* indicate the source of the monthly data

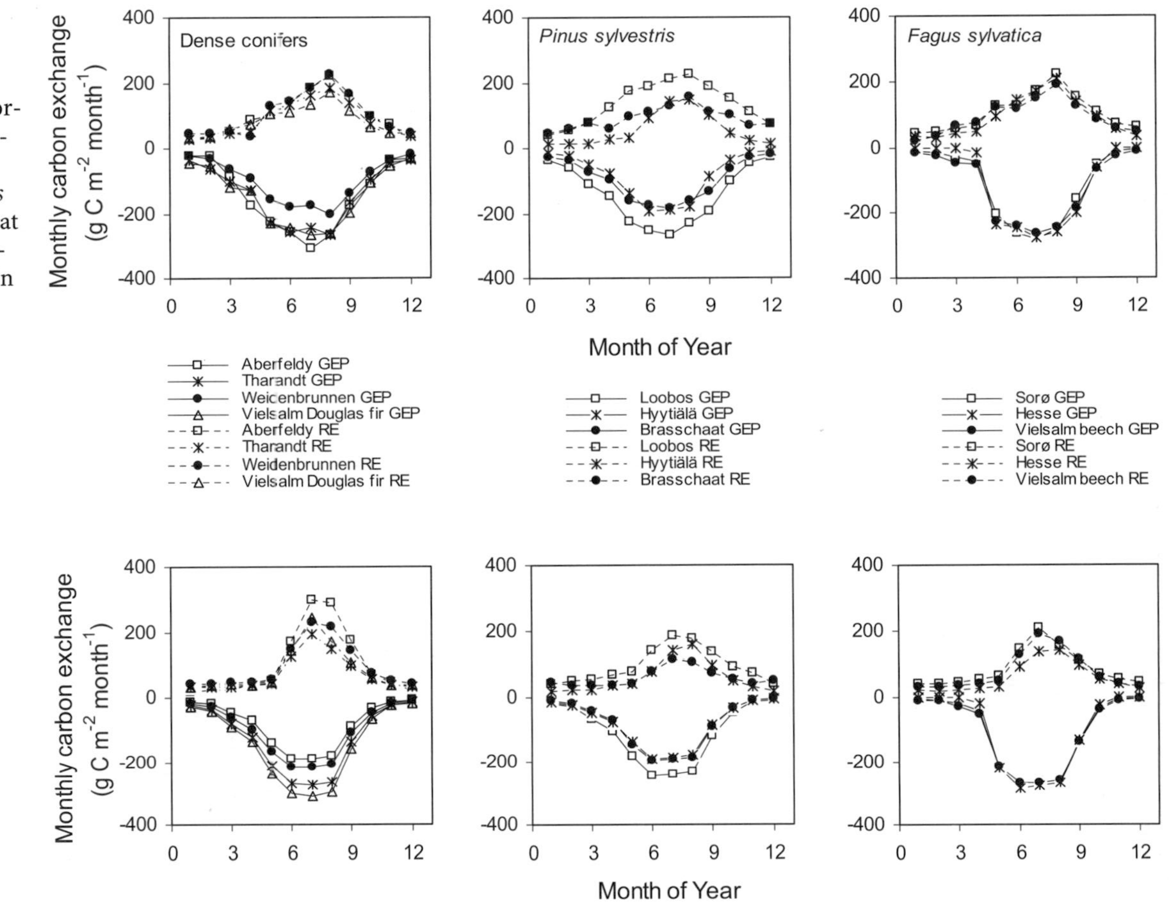

Fig. 8.6. The annual course of monthly GPP and Reco as predicted by the nighttime flux extrapolation method for ten forest stands. *Upper panels* Simulations with climate measured locally at the sites. *Lower panels* Climate measured during 1997 at Hyytiälä, Finland. *Symbols* indicate the specific sites included in each grouping

Finland. In the categories "dense coniferous" and "*Fagus sylvatica*", fluxes simulated with local climate (upper panel) show equal responses to seasonal meteorological changes. In the coniferous categories, fluxes at Loobos and Weidenbrunnen demonstrated exceptions: Loobos showing the highest, Weidenbrunnen the lowest monthly GPP in their category, perhaps due to nutrient availability at those sites. For the other sites, and in each category, differences in Reco are greater than changes observed for GPP. Under Hyytiälä conditions, carbon fixation potentials included in the canopy parameterization are evident. Little difference occurs in GPP within the beech group despite different overall LAI, but GPP in the summer months is greater than in the open coniferous canopies. In dense coniferous sites the combination of LAI, phenology, and physiological characteristics leads to increases in GPP of up to 60 %.

Quantitatively, however, the integrated values for GPP and Reco (Table 8.4) show that small decreases in GPP coupled with small increases in CO_2 efflux can rapidly move the forest ecosystem toward carbon loss. In all of the forest stand simulations, large positive and negative integrals are balanced and the overall net carbon exchange is very sensitive to shifts in the functions used to calculate GPP and Reco as well as to gap filling of either measured or modeled time series data. Simulations for Aberfeldy changed greatly both for GPP and Reco as compared to the other stands when switching to Hyytiälä climate. Fitting of the model parameters over a limited range of conditions introduces in the Aberfeldy case an extreme response to temperature. This is further interesting, since Aberfeldy would be judged as the warmer site based on mean annual temperature. Inappropriate response occurs only for the short period with warm temperatures during summer (means of June, July, and August is 13.3 °C for Aberfeldy, and 16.9 °C for Hyytiälä), but this has a dramatic effect on the overall estimated NEE.

Annual courses for daily modeled and measured NEE are shown for all sites in Fig. 8.7. Measured data are included where gap filling was judged reasonable (at least 24 half-hourly measurements existing on each day upon which to base the gap filling). The scatter of closed and open circles suggests that the model includes appropriate sensitivity to changes in environmental conditions in order to reproduce the measured responses. Systematic differences that require improvement are apparent, e.g., the spring delay in physiological activity at Hyytiälä, whether the assumed leaf phenology is appropriate or whether summer NEE maxima match.

One criterion for examining the model fit is to plot daily sums from the model against those measured, considering only days on which the measurements were complete. Arbitrarily, we hoped to include at least 50 days in the comparison. Although data records from some sites are very complete (e.g., Hesse with 284 complete days), the condition could not be fulfilled for all stands. The comparison for Aberfeldy, Brasschaat, Weidenbrunnen, Vielsalm Beech, and Tharandt required that we use days with up to seven half hour val-

Table 8.4. Simulation results with parameterization via the nighttime extrapolation method for GPP, Reco, and NEE at the ten forest sites described in Table 1. The results include simulations with the climate measured at each site during 1997 and for all sites with the climate occurring at Hyytiälä, Finland during 1997. Fluxes are expressed for all components in g C m^{-2} $year^{-1}$

Simulated under actual climate conditions

Dense conifers	GPP	Reco	NEE	*Pinus sylvestris*	GPP	Reco	NEE	*Fagus sylvatica*	GPP	Reco	NEE
Aberfeldy	–1,721	1,186	–535	Brasschaat	–1,164	1,088	–76	Hesse	–1,292	1,100	–192
Tharandt	–1,673	1,049	–624	Hyytiälä	–1,012	674	–338	Sorø	–1,327	1,249	–78
Vielsalm	–1,747	989	–758	Loobos	–1,690	1,616	–74	Vielsalm	–1,411	1,081	–330
Weidenbrunnen	–1,188	1,234	46								

Simulated under Hyytiälä climate conditions

Dense conifers	GPP	Reco	NEE	*Pinus sylvestris*	GPP	Reco	NEE	*Fagus sylvatica*	GPP	Reco	NEE
Aberfeldy	–986	1,225	238	Brasschaat	–1,009	683	–326	Hesse	–1,226	684	–542
Tharandt	–1,510	856	–654	Hyytiälä	–993	694	–299	Sorø	–1,311	1,015	–296
Vielsalm	–,1703	985	–718	Loobos	–1,315	1,118	–197	Vielsalm	–1,301	906	–395
Weidenbrunnen	–1,186	1,133	–53								

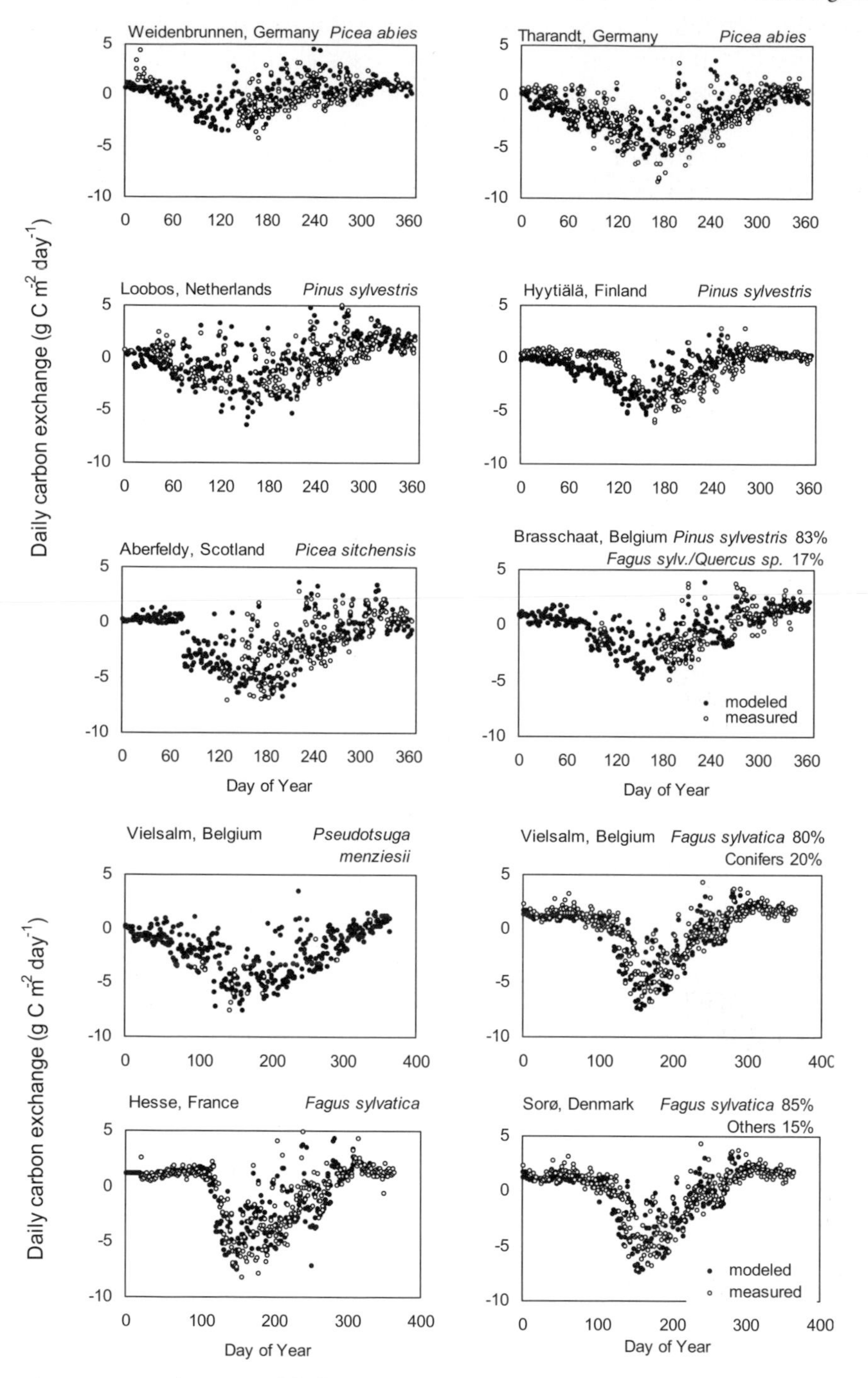

Fig. 8.7. Annual course of daily NEE as measured and modeled with the compensation for aboveground method, both with gap filling (for measurements on days with at least 24 half-hourly observations existing)

ues missing. Vielsalm Douglas fir has few days measured overall because the stand is located at that side of the tower from which the wind least seldom blows. It was, therefore, not included in the comparison. The selected days for the comparison shown in Fig. 8.8 are fairly complete for nine stands and provide good measured estimates of NEE. While the patterns provide information for further study with respect to response at individual sites, the fit of the model is supported as credible.

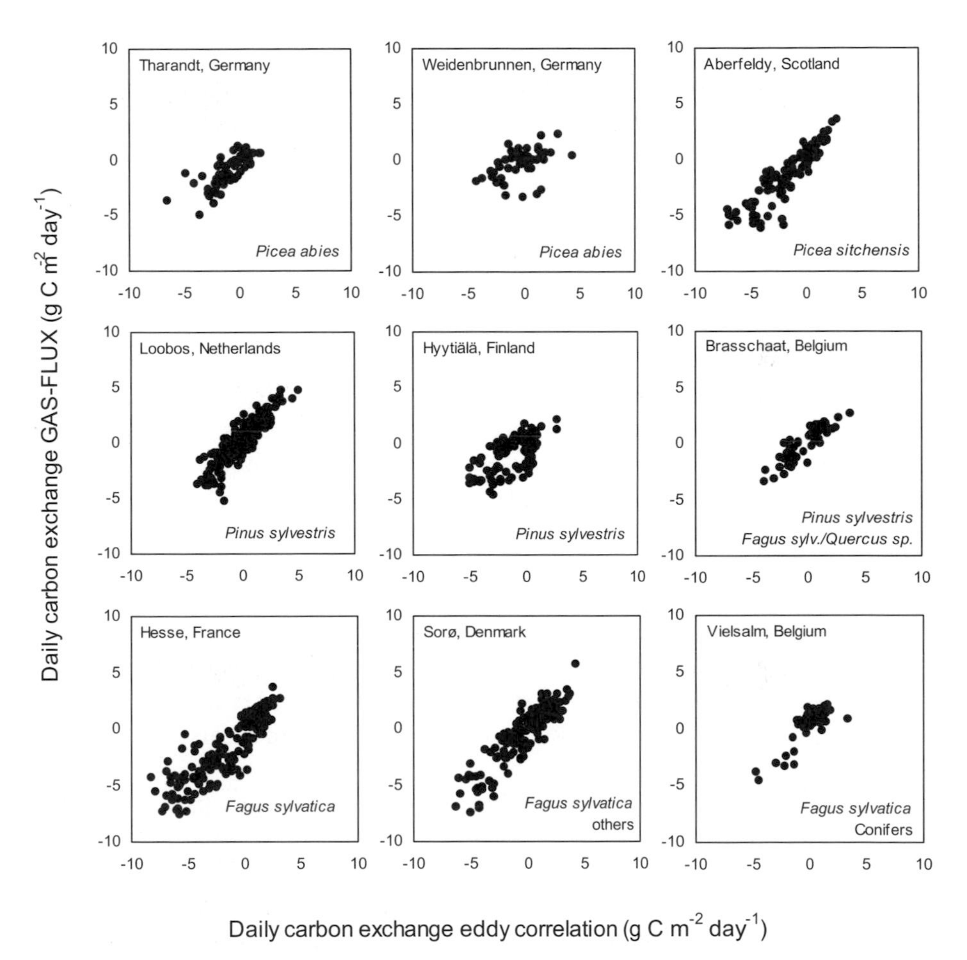

Fig. 8.8. Comparison of the daily integrated carbon dioxide exchange of nine forest canopies as estimated by measurements and the nighttime extrapolation model. Only days are included that are complete or have no more than eight half-hour observations missing. Douglas fir at Vielsalm could not fulfill the criteria and is not included

8.4 Discussion and Future Directions

Differences occur as a result of the different weighting of observations in the process of parameterizing the model, and of gap filling. Long gaps in the NEE data series may be better filled by the model, which has continuous functions over a broad range of conditions; e.g., inadequate function of eddy apparatus during winter makes it difficult to derive robust empirical relationships for gap filling. However, the results demonstrate the following:

1. The proposed methods provide an adequate framework, although (a) LAI estimates that are used require further testing from the standpoint of reproducing seasonal courses of gas exchange, (b) seasonal changes in leaf physiological properties or phenology are not included, (c) soil water availability may in fact decrease during summer at some of these sites and this must be evaluated, (d) the soil model must consider response both to temperature and water content, and (e) a more mechanistic but still simple soil CO_2 efflux model is required in order to better compare function at different sites.
2. The model structure must be reviewed from a variety of perspectives. Given an initial summary of this type, we must now address the degree of complexity that should be used in a variety of applications of network eddy covariance data. This brings us to another point, namely, the testing of this modeling approach for multiple years at each site, or sites with very different vegetation structure.
3. The study shows that we are in fact able to realize the potential offered by a network of eddy covariance measurement sites. However, the sensitivity of carbon balance to accurate parameterization of component subprocesses indicates that much more attention must be paid to measuring those subprocesses independently, in particular, the respiration components. However, LAI and long-term acclimation of physiology, must also be carefully scrutinized in future studies.

   This point will be illustrated further with an example of up-scaling respiration from cuvettes at three sites in comparison with respiration estimates from eddy covariance (Fig. 8.9). The sites where respiration estimates by chamber measurements for leaves (Rleaf), bole (Rbole), and soil (Rsoil) were available were Weidenbrunnen, Brasschaat, and Hesse. For Weidenbrunnen and Hesse, the sum of the three components from up-scaled chamber measurements ($R_{chambers}$) exceeds the estimate for ecosystem respiration by nighttime flux extrapolation method (R_{EC}). At Brasschaat, $R_{chambers}$ is lower (especially during winter periods) than R_{EC}. The annual sum of soil respiration was 680, 660, and 481 g C m^{-2} $year^{-1}$ for Weidenbrunnen, Hesse, and Brasschaat, respectively. At Weidenbrunnen and Hesse, these values compare well with the estimates given in Janssens et al. (this vol., their Table 12.5), with 709 g C m^{-2} $year^{-1}$ at Weidenbrunnen (GE1–47y) and

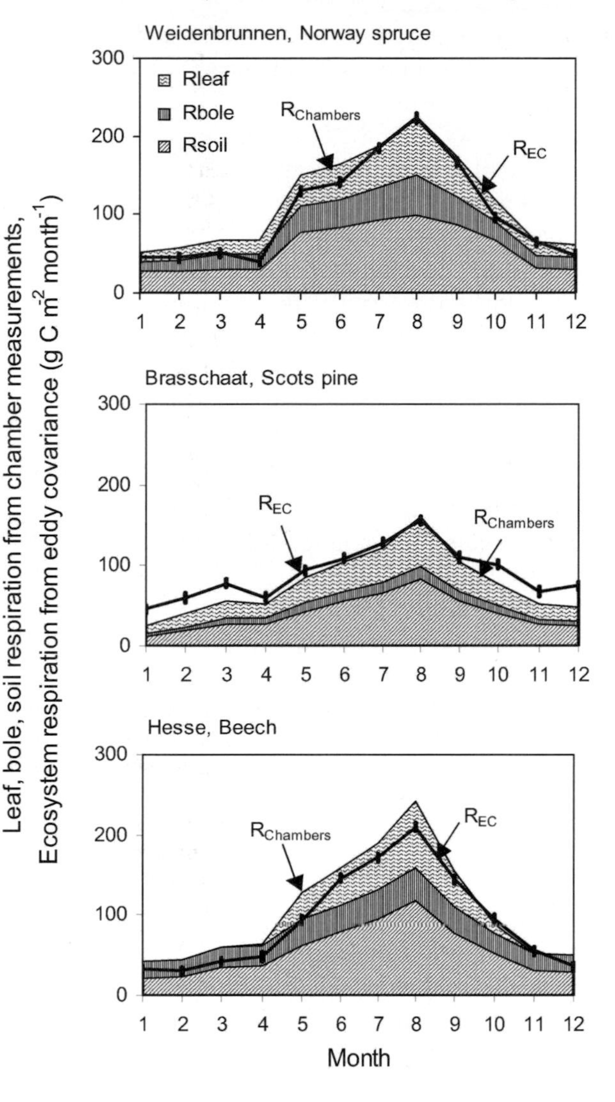

Fig. 8.9. Comparison of the monthly integrated Reco of three sites as estimated by up-scaled chamber measurements of soil, bole, and leaf respiration, and the night-time extrapolation model

685 g C m^{-2} $year^{-1}$ at Hesse (FR1–1997). The value for Brasschaat lays in the range of the four values between 281 and 769 g C m^{-2} $year^{-1}$ reported for the site "BE2" in 1997. Continuous measurements of soil, bole, and leaf respiration over the course of the year would allow parameterization of respiration models for all sites. Such measurements are also necessary to assure a better interpretation of NEE obtained from eddy correlation, e.g., in terms of its major components GPP, and Reco, or in terms of carbon sequestration.

4. Given that a minimum of site-specific information is included in the model formulation, it is satisfying that a first working tool with considerable analytical potential has been easily achieved. While the model can be criticized

because it is validated only for its prediction of GPP (Reco is derived from the eddy covariance measurements), it does provide us with a quantitative hypothesis that is simple enough to be understood by a large community of field researchers. This model allows us in the first approximation to relate processes at smaller scale to total system performance. We see the need to evaluate these processes with greater care, since their influence on system response is apparent.

5. There is potential to derive "functional types" for surface exchange from studies of this type. The response observed within canopy categories is similar enough that it is difficult at this point to judge whether "within category" variability provides us with additional insight except with respect to phenology. Clearly, a better characterization of forest phenology is needed. We must consider as well the time dependent changes in physiological activity of the conifers. Even though the same seasonal course for phenology was initially acceptable for the beech sites under study here, phenology must strongly influence carbon gain of deciduous stands at European continental scale.
6. Comparison of the observations and modeling results from EUROFLUX stands with observations on other continents would indicate whether the forests are "representative" or of special nature due to high levels of N deposition or the specifics of European climate.

8.5 Conclusions

Network data bases such as those assembled in EUROFLUX permit us to analyze ecosystem fluxes along climate gradients and over the seasonal course. Assimilating the data in a model-based study facilitates communication on model and data limitations, as well as optimal exploitation of the data. In this first across-site examination, we obtained a simple model that describes the results of ten EUROFLUX locations.

Experiments like this can advance our understanding on the complexity required, or the degree to which we can simplify models. For use in spatial assessments, we identified the need for improved modules of soil respiration (for drought stressed ecosystems) and seasonal changes in leaf physiological properties or phenology. The analysis helped evaluating bias errors associated with nighttime eddy covariance flux, eventually leading to new experiments and methodological progress to address these problems.

It would be especially useful if the surface exchange "types" we identified and implemented in our model (dense versus open conifers, and deciduous beech forests) could be complemented by agricultural ecosystems. On the base of an effective classification of landscape functional units, we could relate improved model implementations to remotely sensed data of land use, phenology, and meteorology, providing up-scaling tools for use in regional assessments of carbon sequestration.

Acknowledgements. This work was supported by the EUROFLUX project. This study also contributes to the FLUXNET project of the NASA's EOS Validation Project. We gratefully acknowledge the collaboration of numerous researchers in the EUROFLUX project in the acquisition of their field data.

References

Aubinet M, Grelle A, Ibrom A, Rannik Ü, Moncrieff J, Foken T, Kowalski AS, Martin PH, Berbigier P, Bernhofer C, Clement R, Elbers J, Granier A, Grünwald T, Morgenstern K, Pilegaard K, Rebmann C, Snijders W, Valentini R, Vesala T (2000) Estimates of the annual net carbon and water exchange of forests: the EUROFLUX methodology. Adv Ecol Res 30:113–175

Baldocchi DD, Harley PC (1995) Scaling carbon dioxide and water vapor exchange from leaf to canopy in a deciduous forest: model testing and application. Plant Cell Environ 18:1157–1173

Baldocchi D, Valentini R, Running S, Oechel W, Dahlman R (1996) Strategies for measuring and modelling carbon dioxide and water vapour fluxes over terrestrial ecosystems. Global Change Biol 3:159–168

Baldocchi DD, Finnigan J, Wilson K, Paw U KT, Falge E (2000) On measuring net ecosystem carbon exchange over tall vegetation on complex terrain. Bound Layer Meteorol 96:257–291

Ball JT, Woodrow IE, Berry JA (1987) A model predicting stomatal conductance and its contribution to the control of photosynthesis under different environmental conditions. In: Binggins IJ (ed) Progress in photosynthesis research, vol IV.5. Martinus Nijhoff, Dordrecht, pp 221–224

Falge E, Baldocchi D, Olson RJ, Anthoni P, Aubinet M, Bernhofer C, Burba G, Ceulemans R, Clement R, Dolman H, Granier A, Gross P, Grünwald T, Hollinger D, Jensen N-O, Katul G, Keronen P, Kowalski A, Ta Lai C, Law BE, Meyers T, Moncrieff J, Moors E, Munger JW, Pilegaard K, Rannik Ü, Rebmann C, Suyker A, Tenhunen J, Tu K, Verma S, Vesala T, Wilson K, Wofsy S (2001) Gap filling strategies for defensible annual sums of net ecosystem exchange. Agric For Meteorol 107:43–69

Falge EM, Graber W, Siegwolf R, Tenhunen JD (1996) A model of the gas exchange response of *Picea abies* to habitat conditions. Trees 10:277–287

Farquhar GD, Von Caemmerer S (1982) Modelling of photosynthetic response to environment. In: Lange OL, Nobel PS, Osmond CB, Ziegler H (eds) Encyclopedia of plant physiology, vol 12B. Physiological plant ecology II. Water relations and carbon assimilation. Springer, Berlin Heidelberg New York, pp 549–587

Fleck S (2001) Integrated analysis of relationships between 3D-structure, leaf photosynthesis, and branch transpiration of mature *Fagus sylvatica* and *Quercus petraea* trees in a mixed forest stand. PhD Thesis, University of Bayreuth

Goulden ML, Munger JW, Fan S-M, Daube BC, Wofsy SC (1996) Measurements of carbon sequestration by long term eddy covariance: methods and critical evaluation of accuracy. Global Change Biol 2:169–182

Granier A, Ceschia E, Damesin C, Dufrene E, Epron D, Gross P, Lebaube S, Le Dantec V, Le Goff N, Lemoine D, Lucot E, Ottorini JM, Pontailler JY, Saugier B (2000) Carbon balance of a young beech forest over a two-year experiment. Funct Ecol 14:312–325

Harley PC, Tenhunen JD (1991) Modeling the photosynthetic response of C_3 leaves to environmental factors. In: Boote KJ, Loomis RS (eds) Modeling crop photosynthesis – from biochemistry to canopy. Am Soc Agron and Crop Sci Soc Am, Madison, Wisconsin, pp 17–39

Hollinger DY, Kelliher FM, Byers JN, Hunt JE, McSeveny TM, Weir PL (1994) Carbon dioxide exchange between an undisturbed old-growth temperate forest and the atmosphere. Ecology 75:134–150

Lee X (1998) On micrometeorological observations of surface-air exchange over tall vegetation. Agric For Meteorol 91:39–50

Leuning R, Moncrieff J (1990) Eddy-covariance CO_2 measurements using open- and closed-path CO_2 analysers: corrections for analyser water vapor sensitivity and damping of fluctuations in air sampling tubes. Boundary Layer Meteorol 53:63–76

Monteith JL (ed) (1976) Vegetation and the atmosphere, vol 2. Academic Press, New York

Norman JM (1975) Radiation transfer in vegetation. In: de Vries DA, Afgan NH (eds) Heat and mass transfer in the biosphere. Scripta Book, Washington, DC

Norman JM (1980) Interfacing leaf and canopy light interception models. In: Hesketh JD, Jones JW (eds) Predicting photosynthesis for ecosystem models. CRC Press, Boca Raton

Odum EP (1985) Trends expected in stressed ecosystems. BioScience 35:419–422

Pilegaard K, Hummelshøj P, Jensen NO, Chen Z (2001) Two years of continuous CO_2 eddy-flux measurements over a Danish beech forest. Agric For Meteorol 107:29–41

Reiners WA (1983) Disturbance and basic properties of ecosystem energetics. In: Mooney HA, Godron M (eds) Disturbance and ecosystems. Ecological studies 44. Springer, Berlin Heidelberg New York, pp 83–98

Reynolds JR, Wu J (1999) Do landscape structural and functional units exist. In: Tenhunen JD, Kabat P (eds) Integrating hydrology, ecosystem dynamics, and biogeochemistry in complex landscapes. Wiley, Chichester, pp 273–296

Ross J (1975) Radiative transfer in plant communities. In: Monteith JL (ed) Vegetation and the atmosphere, vol 1. Academic Press, New York

Sala A, Tenhunen JD (1996) Simulations of canopy net photosynthesis and transpiration in *Quercus ilex* L. under the influence of seasonal drought. Agric For Meteorol 78:203–222

Sturm N (1998) Steuerung, Skalierung und Umsatz der Wasserflüsse im Hartheimer Kiefernforst (*Pinus sylvestris* L.). Bayreuther Forum Ökologie 63, 190 pp

Sturm N, Köstner B, Hartung W, Tenhunen JD (1998) Environmental and endogenous controls on leaf- and stand-level water conductance in a Scots pine plantation. Ann Sci For 55:237–253

Tenhunen JD, Siegwolf R, Oberbauer SF (1994) Effects of phenology, physiology, and gradients in community composition, structure, and microclimate on tundra ecosystem CO_2 exchange. In: Schulze ED, Caldwell MM (eds) Ecophysiology of photosynthesis. Ecological studies 100. Springer, Berlin Heidelberg New York, pp 431–460

Tenhunen JD, Valentini R, Köstner B, Zimmermann R, Granier A (1998) Variation in forest gas exchange at landscape to continental scales. Ann Sci For 55:1–11

Tenhunen JD, Geyer R, Valentini R, Mauser W, Cernusca A (1999) Ecosystem studies, land-use change, and resource management. In: Tenhunen JD, Kabat P (eds) Integrating hydrology, ecosystem dynamics, and biogeochemistry in complex landscapes. Wiley, Chichester, pp 1–22

Valentini R, Baldocchi DD, Tenhunen JD (1999) Ecological controls on land-surface atmospheric interactions. In: Tenhunen JD, Kabat P (eds) Integrating hydrology, ecosystem dynamics, and biogeochemistry in complex landscapes. Wiley, Chichester, pp 117–145

Williams M, Rastetter EB, Fernandes DN, Goulden ML, Wofsy SC, Shaver GR, Melillo JM, Munger JW, Fan S-M, Nadelhoffer KJ (1996) Modelling the soil-plant-atmosphere continuum in a *Quercus-Acer* stand at Harvard Forest: the regulation of stomatal conductance by light, nitrogen and soil/plant hydraulic properties. Plant Cell Environ 19:911–927

9 A Model-Based Approach for the Estimation of Carbon Sinks in European Forests

D. Mollicone, G. Matteucci, R. Köble, A. Masci,
M. Chiesi, P.C. Smits

9.1 Introduction

Man's knowledge of ecosystems is in constant evolution. This is a consequence of the increased understanding of the underlying biology, but also of the increased collaboration between disciplines that in the past did not interact in the same way and intensity as they do today. In fact, more and more ecosystem scientists are building on what now may be considered solid foundations laid out by disciplines such as biology, ecology, physics, and also computer science (see, for instance, Waring and Running 1998; Patil and Myers 1999). A good understanding of ecosystem processes is vital as it leads to models, which, in turn, provide us ways to predict the possible future behavior of ecosystems under changing conditions. Most models include fluxes of carbon, water, and nitrogen, and are able to provide estimates of gross primary production (GPP) and net primary production (NPP) of carbon per unit area. In the past decade, the numbers of sites where canopy fluxes of carbon and water were measured increased exponentially (Baldocchi et al. 2001) and net ecosystem exchange (NEE) became available as well. Hence, modelers were recently provided with calibration and validation data for the complete process of ecosystem carbon exchange, up to net ecosystem production (NEP=NEE) and even net biome productivity (NBP; see Fig. 9.1).

These developments have led to different views on how ecosystems should be modeled. In the literature, two classes of forest ecosystem models have been proposed (Waring and Running 1998). On the one hand, there are models that focus on physiology and biogeochemistry of the ecosystem and, on the other hand, there are those that use the life-cycle dynamics of trees in ecosystems. Well-known examples of these two classes are FOREST-BGC (Waring and Running 1998) and canopy-gap models (Dale and Rauscher 1994), respectively. It should be noted that all models have their advantages and drawbacks, and so far there is no such thing as a "best model".

Ecological Studies, Vol. 163
R. Valentini (Ed.) Fluxes of Carbon,
Water and Energy of European Forests

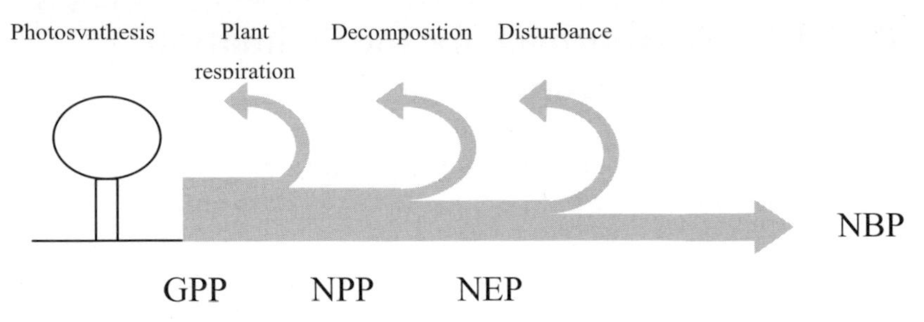

Fig. 9.1. The ecological terms and processes of primary and net productions in the terrestrial carbon cycle. *GPP* Gross primary production; *NPP* net primary production; *NEP* net ecosystem production; *NBP* net biome production. (Modified after Buchmann and Schulze 1999)

A consequence is that different approaches to ecosystem modeling may result in different estimations of carbon sinks. Although there is nothing against using multiple models, a harmonized approach may be desired. At the same time, the scientific community must work on the improvement of its knowledge base, and be prepared to reconsider paradigms as our understanding of ecosystems evolves. An example is the recent integration of remote sensing techniques within ecosystem models. It is our opinion that a flexible model, able to integrate newly acquired knowledge, is preferred. The Biome-BGC model can be considered such a model (Running and Hunt 1993; NTSG 2001).

The studies presented in this chapter are based on the Biome-BGC model for two reasons. First, the model is documented quite well, in terms of both technical documentation and scientific publications; second, the source code of the model is publicly available on the Internet (NTSG 2001). Furthermore, it is also widely used as a benchmark approach in global exercises (e.g., Schimel et al. 1994).

This chapter has two objectives: the first is to validate the Biome-BGC model through a comparative analysis of the model estimations and the flux data produced during the EUROFLUX project experiments (see various chapters, this Vol.), while the second is to report and discuss results that have been obtained by applying the Biome-BGC model on pan-European scale using the carbon budget information system (CBIS). A unique feature of the approach presented in this chapter is the aggregation of ecosystem-level analyses to a pan-European level, based on availability of pan-European databases.

The organization of the remainder of this chapter is as follows; Section 9.2 starts with a brief description of the Biome-BGC model. Section 9.3 reports the findings of the validation tests of Biome-BGC based on EUROFLUX experiments, followed by a description of the methods and data of CBIS (Sect. 9.4). Section 9.5 reports the results obtained with CBIS. Finally, Section 9.6 concludes with a general discussion.

9.2 Biome-BGC Model

Biome-BGC is a computer model that simulates the storage and fluxes of water, carbon, and nitrogen within the vegetation, litter, and soil components of a terrestrial ecosystem. Biome-BGC is primarily a research tool, and many versions have been developed for particular purposes. The numerical terradynamic simulation group (NTSG) of the University of Montana, USA, maintains benchmark code versions for public release, and updates these benchmark versions periodically as new knowledge is gained on the research front (NTSG 2001).

The model is a so-called point model. The term point model means that the point at which the model is applied inherits the values of a region for which these values are a spatial invariant. The model takes as input the mean parameters for a given site (i.e., region of interest) and the output is a mean value valid for that site.

The Biome-BGC model requires three types of information: site parameters, meteorological data, and vegetation constants. The model uses this information in two steps. First, a long-term simulation is carried out to establish an equilibrium in the model. The aim of this long-term simulation (e.g., 3000 years) is to stabilize the carbon and nitrate concentrations in the soil and vegetation, based on the available meteorological data, which is repeated for the duration of the simulation. If this information is already available, for instance by field experiments, the spin-up simulation can be omitted. In a second step, the concentrations found in the long-term simulation are used for the year of interest (in our case 1999), using all the input parameters.

The model can report a wide range of parameters as the output of the simulation, which include, for each day of simulation, leaf area index (LAI), GPP, NPP, and, if specific input parameters are available, NEP and NBP. The model outputs can also be aggregated to obtain coarser time scales (weekly, monthly, yearly intervals).

9.3 Validation of Biome-BGC

Biome-BGC has been tested for the Collelongo beech forest (central Italy, 41°52′N, 13°38′E, altitude 1560 m). In this exercise, the model's input parameters were site-specific, and based mainly on field measurements, but also on literature. At the time of the validation, a complete meteorological dataset for the site was available 1996–1999 and we focused on this 4-year period.

The output parameters considered for validation were LAI, annual GPP, NPP, and NEE of carbon.

9.3.1 Data for Model Validation

NPP was directly measured for all components in 1996 when a destructive biomass sampling campaign was performed (Masci et al. 1998; Scarascia Mugnozza et al. 2000). For the remaining years (1997–1999), NPP was estimated for the different components as follows:

Foliage production was calculated based on LAI measurements (LiCor LAI2000) and litter traps (Cutini et al. 1998).

In 1996, in an experimental plot of 2,000 m^2, dendrometers were placed on 5 % of the trees, selected according to their position in the canopy. Using data from stem analysis carried out on the trees felled in 1996, a regression was tested between percent volume increments and percent increment of cross sectional area at breast height; this regression was then used to estimate stem volume increment in 1997, 1998, and 1999. Finally, stem weight increment was obtained multiplying volume increments by the basal density measured experimentally.

Branches and twigs increments were estimated assuming that, for a short period (few years), the ratio of branch to stem biomass can be considered relatively constant. During the biomass sampling of 1996, we found that branches had increments fairly similar to those of stem, showing thus a higher growing rate (Scarascia Mugnozza et al. 2000). This elevated rate would lead, in few years, to dramatic changes in the ratio of branch to stem biomass; however, data from litter traps showed that the branch necromass that fell every year was also high, thus allowing only negligible changes in the ratio.

Coarse root increment was estimated using a coefficient determined experimentally on six trees felled in 1996. Ring analysis was conducted on stems and coarse roots and the trend of the roots-to-stems increment ratio was reconstructed; this ratio has proved not to change significantly during 90 years before 1996.

Fine root production, measured by sequential coring, was estimated by applying the ratio of fine root production to leaves production as measured in 1996. This could be the weaker point in the estimation of NPP after 1996: nevertheless, peak season fine root biomass measured in other years was comparable to that of 1996.

The beech forest of Collelongo is one of the EUROFLUX sites; hence NEE was measured according to the EUROFLUX protocol (see Chap. 2, this Vol.; Aubinet et al. 2000). Annual GPP was calculated adding total ecosystem respiration to annual NEE. Detailed data, functional relationship, and additional information can be found in Chapter 4 (Granier et al., this Vol., which is dedicated to deciduous forest) and in the comprehensive study on carbon and nitrogen cycling in European forest ecosystems of Schulze et al. (2000).

9.3.2 Validation and Calibration

A preliminary tentative simulation was carried out after a previous "spin-up mode" run, procedure that is normally used to generate the ecosystem state variables in equilibrium with the mean climate of a certain site (Thornton 2000). The results were unrealistic: LAI ranged between 1.2 and 1.3 $m^2 m^{-2}$, NPP averaged 1.4 Mg C $ha^{-1} a^{-1}$, GPP varied from 1.31 to 1.5 Mg C $ha^{-1} a^{-1}$ while NEE was nearly null (from −0.01 to +0.054 Mg C $ha^{-1} a^{-1}$). Furthermore, soil carbon data were extremely underestimated: 27.06 Mg C ha^{-1} compared to 220.3 Mg C ha^{-1} (Persson et al. 2000a).

Successive simulations were performed omitting the "spin-up mode" run and providing the carbon state variables (i.e., leaf carbon, stem carbon, litter carbon, and soil carbon) as site-specific inputs.

The model requires soil and litter carbon to be divided into labile, medium, slow, and recalcitrant fractions. For our site only total litter and total soil carbon were available, so we had to test various combinations of carbon partitioning in the prescribed pools. The best results were obtained after "forcing" most of the carbon into the more slowly decomposable pools. This choice was corroborated by experimental evidence that the soil at Collelongo shows very low rates of carbon mineralization and organic matter decomposition despite the great amount of carbon stored (Cotrufo et al. 2000; Harrison et al. 2000; Matteucci et al. 2000; Persson et al. 2000b). This second series of simulations yielded better results in terms of LAI, NPP, and GPP. Still, NEE remained negative (carbon emission), giving a strong indication of an overestimation of heterotrophic respiration by the model.

In Biome-BGC, the heterotrophic respiration term (R_H) represents the loss of carbon resulting from soil microbial activity. Daily R_H is estimated as a proportion ("respiration fraction") of one of the prescribed soil and litter carbon pools (Kimball et al. 1997). To each one of these pools is attributed a base value of the decomposition rate constant (K-base) coming from laboratory decomposition experiments performed at a given reference temperature. In the model, the K-base values are then modified on the basis of soil water status (Orchard and Cook 1983; Andren and Paustian 1987) and soil temperature (the reference temperature is 25 °C in the model while it was 10 °C in the original Lloyd and Taylor 1994 formulation).

Simulation with additional forcing of soil carbon into the recalcitrant pool proved to be unrealistic and led to a decrease in LAI, NPP, and GPP, probably as a consequence of inadequate nitrogen mineralization which, in the model, is tightly linked to carbon mineralization.

In the model, moisture and temperature constraints are taken into account through climatic data and also litter quality is an input (lignin to cellulose ratio); hence, the only way to include other effects, given the model structure, is to change the partitioning of soil carbon into the pool fractions and the K-base values. As already mentioned, K-base for the different C pools are gener-

Table 9.1. Comparison of simulated and measured values at the EUROFLUX sites of Collelongo

Year	Simulated max. LAI ($m^2\ m^{-2}$)	Measured LAI ($m^2\ m^{-2}$)	Simulated NPP ($Mg\ C\ ha^{-1}\ a^{-1}$)	Measured NPP ($Mg\ C\ ha^{-1}\ a^{-1}$)	Simulated GPP ($Mg\ C\ ha^{-1}\ a^{-1}$)	Measured GPP ($Mg\ C\ ha^{-1}\ a^{-1}$)	Simulated NEE ($Mg\ C\ ha^{-1}\ a^{-1}$)	Measured NEE ($MgC\ ha^{-1}\ a^{-1}$)
1996	5.9	5.4	5.88	6.48	10.08	13.0	1.86	6.6
1997	4.7	5.6	4.66	7.23	8.37	12	0.86	5.7
1998	4.6	5.4	5.85	6.53	9.71	10–13[a]	1.22	4.7–6.6[a]
1999	5.9	5.3	7.36	7.21	12.37	10–13[a]	2.63	4.7–6.6[a]

[a] Range of measured GPP and NEE in 1993–1997.

ally obtained by lab experimentation and, for our site, directly measured values were not available. However, experiments carried out by Harrison et al. (2000) and Persson et al. (2000b) showed that the carbon mineralization rates of bulk soil and litter (^{14}C, and incubation experiment at 15 °C with no water limitations) were very low in the litter and soil of Collelongo.

K-base values for pools of soil organic matter in Collelongo were extrapolated from the above-mentioned experiments carried out at 15 °C (Persson et al. 2000b). A third series of simulations was then performed. The simulations (see Table 9.1) gave a better approximation in terms of NEE, however, NEE was still underestimated with respect to experiment measurements.

9.3.3 Discussion

An ensemble of factors may drive carbon fluxes at site level (Valentini et al. 2000, and currently some of them are not properly considered in models, especially those that aim at a larger-scale application. This oversight could sometimes result in under- or overestimation of relevant variables related to carbon fluxes (NPP, GPP, NEE). It is becoming clearer that to properly address the ecosystem scale, a more detailed representation of process is needed, particularly for factors concerning site-level features.

In the Collelongo beech forest, despite the high organic carbon content of soil (228±18 Mg C ha^{-1}, litter in 0–50 cm soil; Persson et al. 2000a), the mineralization rate as measured by different methods was relatively low (0.8–1.2 Mg C ha^{-1} a^{-1}). The model output for heterotrophic respiration for 1996 and 1997 was 4 and 3.8 Mg C ha^{-1} a^{-1}, respectively, leading to an underestimation of NEE. Nevertheless, NEE values obtained by subtracting the measured heterotrophic respiration in 1996–1997 from the modeled NPP reported in Table 9.1 ranged between 3.7 and 6.4 Mg C ha^{-1} a^{-1}, being comparable to the range of measured NEE (4.7–6.3 Mg C ha^{-1} a^{-1}, 1993–1997).

The average soil carbon mineralization rate of spruce and beech forest ecosystems along a N–S European transect ranged between 1 and 1.5 %, while for Collelongo it was approximately 0.5 % (Persson et al. 2000b; Schulze et al. 2000). This could be explained by snow cover and low soil temperatures during winter, summer drought stress, high percentage of clay in soil, and aluminum content (Persson et al. 2000a), which may protect most of the soil organic matter from decomposition; and litter quality could be another factor that limits decomposition in this ecosystem (Cotrufo et al. 2000). As a consequence, soil organic carbon at Collelongo is probably more resistant to decomposition than we hypothesized; and since a bigger proportion of soil carbon fails to decompose, lower K-bases should have been adopted. Such a reduction could not be done because every additional tentative change of C allocation in the pools or of K-bases led to a reduction of GPP and NPP. Actually, the model shows a certain sensitivity to the heterotrophic respiration:

better estimates of GPP and NPP could be achieved if the model was "allowed" to respire more soil carbon. In other words, the model is not able to simulate a net production as high as in the Collelongo forest with a heterotrophic respiration as low as in the Collelongo soil. This may be because N mineralization is tightly linked to C mineralization. Nevertheless, a test simulation performed after having arbitrarily increased the mineral N pool in the soil produced a sharp increase in heterotrophic respiration together with an increase in GPP and NPP.

Heterotrophic respiration is an ensemble of highly complex processes. There are many variables involved and many of them are site-specific. Though our parameterization is limited by lack of direct data on the repartition of soil carbon between the prescribed pool fractions, some of those data were extrapolated from experiments performed with the site soil. Some parameters that are presently set in the model code as constants may be allowed to change according to certain site features or be treated as input parameters. As an example, the K-base values, excluding the cellulose fraction, can differ according to the site characteristics: lignin, litter-labile carbon pool, the microbial fast, medium, and slow recycling pools, the recalcitrant carbon pool – all may have different compositions and/or be protected within soil aggregates and, consequently, have different decomposition rates. Conversely, our validation exercise for the Collelongo site showed that, if we consider the K-base values proposed by the model as definition values for the different carbon pools, we should force all the soil carbon into the recalcitrant pool, which obviously is not realistic.

It must be underlined that, concerning heterotrophic respiration and the specific decomposition rates, the lack of site-based data for correct parameterization of the model is an important point. In this respect, a better understanding of the respiration processes is necessary, with special consideration of the links between C and N mineralization; further research is needed in that direction.

9.4 Carbon Budget Information System: Methods and Data

The Institute for Environment and Sustainability of the European Commission's Joint Research Center, in collaboration with the DISAFRI of Viterbo University, has set up an information system to provide a harmonized estimate of carbon sinks in the EU 15 member states (Smits 2000). The core of this carbon budget information system (CBIS) is the Biome-BGC model (Running and Hunt 1993; NTSG 2001), which provides estimations of GPP, NPP, NEP, and NBP.

9.4.1 Biome-BGC Model and CBIS

As already mentioned in Section 9.2, the Biome-BGC is a so-called point model. Figure 9.2 visualizes the concept, and shows that the point at which the model is applied inherits the values of a region for which these values are a spatial invariant. The model takes as input the mean parameters for a given site (i.e., region of interest) and the output is a mean value valid for that site. The model can be "spatialized" in that it can be run sequentially or in parallel for any set of regions of interest. Important is that the mean parameters are sufficiently representative of the site.

The Biome-BGC model within CBIS needs for each point (i.e., region) three types of information, represented in a series of 61 parameters. These parameters are related to (1) the physical properties of the region, (2) meteorological data, and (3) ecophysiological constants. A number of them have been obtained through field and literature studies, and some of the parameters can be extracted directly or indirectly from pan-European layers like the CORINE land-cover/land-use database (Annoni et al. 1998), the 1:3 M digital elevation model (DEM), the European soil database, and the meteorological database.

The data flow diagram of CBIS is shown in Fig. 9.3, where it is seen that the data follow the following steps:

Step 1. After the operator gives the command to start the computation of the carbon budget, the information system asks the operator to specify the regions of interest, and if these regions need to be split into smaller regions to obtain homogeneous input parameters for the soil data.

Step 2. All the selected records in the database containing the regions of interest (ROIs) are processed one by one. Step 2 determines which ROI is being analyzed.

Step 3. By overlaying the meteorological grid cells and the ROI, it is determined which meteorological grid cell number is to be used for com-

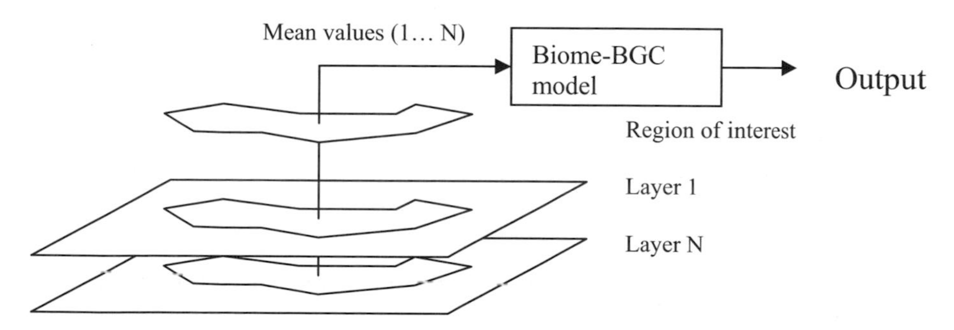

Fig. 9.2. The various layers of CBIS and the Biome-BGC model that can be considered a point model

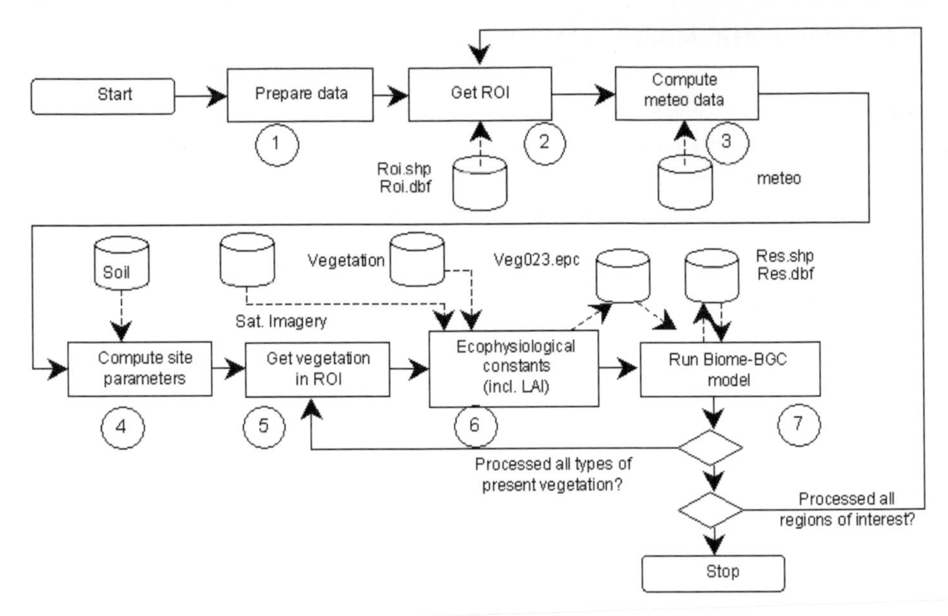

Fig. 9.3. The data flow diagram for the workflow of CBIS

puting the mean meteorological parameters for the ROI. These parameters are computed and stored.

Step 4. Site parameters are computed. Some can be derived directly from the ROI such as longitude, others have to be estimated from pan-European layers like the DEM and the European soil database.

Step 5. The intersection between the ROI and the vegetation coverage determines which of the vegetation types of interest are available, and the physical nature of the surface in the area they cover. A list is created with the available types of vegetation.

Step 6. For each type of vegetation the ecophysiological constants are collected. Ideally, the constants should be as species-specific and local as possible.

Step 7. For the ROI under analysis and the current type of vegetation, the Biome-BGC model is run and the result is assigned to the ROI. Steps 5, 6, and 7 are repeated until all the types of vegetation that are present in the ROI and are of interest to the operator, are processed. After the vegetation is completed, the operator goes back to step 2 until all ROIs are processed.

9.4.1.1 Input Data

It is appropriate to dwell for a moment on two characteristics of geospatial datasets that are important for the understanding the results of any geospatial analysis: (1) the time stamp of the individual sets of data, and (2) the scale of the data. It is especially the aspects related to scale that require some clarification.

Almost all the layers have a different "scale." In the literature the term scale is used in many different ways (Quattrochi and Goodchild 1997). Scale in mapping is mainly a legacy of the use of paper maps, where a distance on the map represents a distance on the surface of the Earth. The term scale does not have an unambiguous meaning in digital geospatial information. Scale may convey a positional accuracy, a dimension of generalization, a minimum mapping unit for polygon coverage, or a pixel size for grid coverage.

In all models, scale is a dimension almost as important as are space and time (De Cola 1997). McNulty et al. (1997) compared the output PnET-IIS, a model similar to Biome-BGC, at three different scales: stand, ecosystem, and regional. They showed that the NPP obtained with PnET-IIS correlates generally well with the measured NPP at ecosystem and regional scales across all sites, but not necessarily with the individual sites due to aggregation of soil and climate data.

In the context of this chapter, scale denotes the minimum mapping unit for vector coverage, and the resolution for grid coverage. For instance, the CORINE land-cover/land-use database has a minimum mapping unit of 25 ha (i.e., scale 1:100,000), and the digital elevation map has a resolution of 1 × 1 km. More information about the influence of the issue of scale to the output of the Biome-BGC model has been presented in Smits and Mollicone (2001).

9.4.1.2 Soil Database

The 1:1,000,000 georeferenced soil database of Europe (Finke et al. 1998) provides precise and harmonized soils information.

The focal point of the database is the soil body. Its genesis is the result of a multitude of factors (climate, parent rock, vegetation, time, etc.) and it is the basic element, the input key for the information system (Hole 1978). The soil body is defined principally by soil attributes. Other objects elaborated from the soil body, i.e., the soilscape and soil region, are used for a better understanding of spatial variability of soils and to provide tools for managing and rationalizing data on the continental scale. The criteria for the geographic delimitation of these objects are not necessarily soil variables, but may also be related to characteristics of soil forming factors: parent material, relief, vegetation, climate, and human influence. The time stamp of the soil database is 1999.

Currently, the parameters that CBIS extracts or estimates from the soil database are (1) the rooting zone soil depth, taking into account the fraction occupied by rocks; and (2) the soil texture (percentage of sand, silt, and clay).

9 4.1.3 Meteorological Database

The meteorological database contains daily meteorological data spatially interpolated on a 50 × 50 km grid cell. The majority of the original observational data originates from nearly 1,500 meteorological stations across Europe, Maghreb countries, and Turkey. The data is received via the Global Telecommunication System (GTS) of the World Meteorological Organization (WMO). Some of the data were obtained from national meteorological services under special copyright and agreements for internal use of the European Commission only.

The whole European area is divided into cells of 50 × 50 km. Each grid cell is georeferenced in latitude/longitude according to the Lambert azimuth projection system. Meteorological parameters are spatially interpolated into the grid system (i.e., each cell contains a set of interpolated meteorological parameters)

The interpolation procedure (Van der Goot 1997) consists of selecting for each grid the best combination of surrounding meteorological stations, the grading being a function of distance to the grid, distance to nearest coastline, altitude difference, and climatic barrier; then in performing for each relevant parameter the average of the observations from the stations, except for rainfall which is taken from the most suitable station.

Daily meteorological parameters are available from 1 January 1975 to the last full month. Within CBIS, the following parameters are used or computed: (1) daily maximum temperature, (2) daily minimum temperature, (3) precipitation, (4) global radiation, (5) potential evapotranspiration, and (6) wind speed at 10 m.

9.4.2 "Clustered" Forest Tree Species Maps

A "pilot tree species criterion" was used for the classification of the European forests. With this approach it was expected that a better accuracy for the model's ecophysiological parameters could be obtained for a given species and region of interest (Köble and Seufert 2001).

The forest tree species maps for the EU15 forest area were compiled by combining spatial information on forest area in different countries and point data on the tree species distribution within the EU15 region.

To define the forest area the CORINE land-cover dataset was used (European Environment Agency 2000). The forest area of the datasets for the EU15 countries was compared with recent forest area statistics (UN-ECE/FAO 2000) for individual countries.

Tree species information was available from International Cooperative Program on Assessment and Monitoring of Air Pollution Effects on Forest (ICP Forest), launched in 1985 under the scope of the Long Range Transboundary Air Pollution (LRTAP) Convention (UN-ECE 1998). Within the ICP Forest program, a transnational survey (Level I) with the objective to document the spatial distribution and development of the forest condition at the European level was initiated. On a 16 × 16 km grid of sample plots (~0.5 ha) crown condition was screened annually. Tree species distribution is also given for the plots. At the moment, site information is available for more than 5,500 plots in Europe and 3,271 sites for EU15.

In Europe, 115 different tree species are distinguished and 111 of them occur in the EU15 area. By interpolating the percentages for each species separately the spatial distribution was assessed.

The maps resulting from overlaying the forest area give the percentage of the species for the 1 × 1 km grids in the forest area. Two examples of maps are presented in Fig. 9.4 for a typical northern European tree (*Pinus sylvestris*) and a Mediterranean species (*Quercus ilex*).

To generate a suitable tool for the CBIS, clustered tree maps were generated in order to obtain a geographical parameterization and a full coverage of the European forest area.

To produce the clustered maps, ecophysiological tree species characteristics and geographical distribution criteria are used to generate 12 tree species

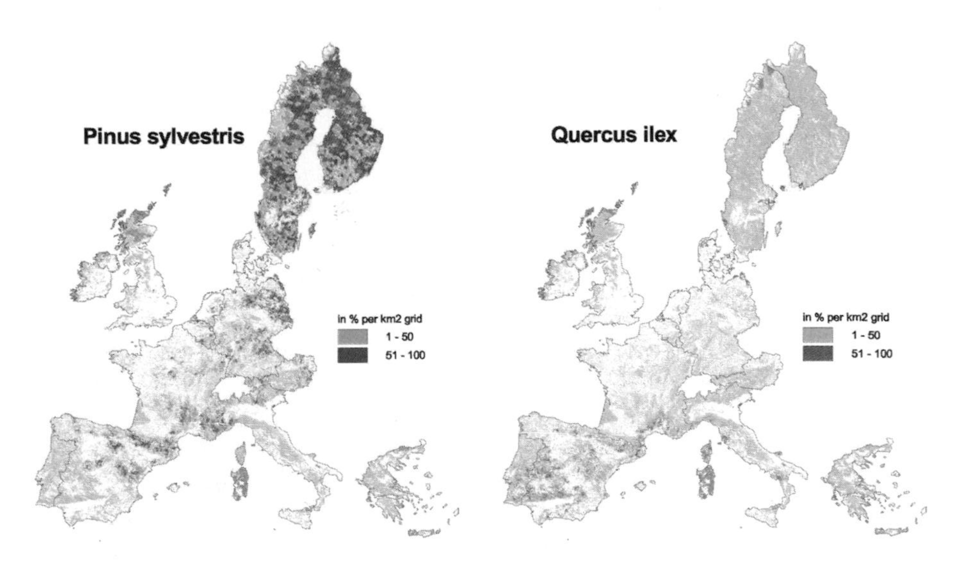

Fig. 9.4. Examples of maps of European forest tree species. The map of *Quercus ilex* also includes *Q. rotundifolia* (Portugal)

Table 9.2. The 12 pilot species and their associated species. The pilot species alone cover 77 % of the total European forest area

	Pilot species *Betula pendula*	*Fagus sylvatica*	*Larix decidua*	*Picea abies*	*Picea sitchensis*	*Pinus halepensis*
Associated species	*Betula pubescens*	*Fagus* spp.	*Larix kaempfery*	*Pinus c embra*	*Pseudo-tsuga m.*	*Pinus pinea*
		Tilia spp.		*Pinus mugo*	*Thuya* sp.	*Pinus can.*
		Acer opalus		*Pinus uncinata*	*Tsuga* sp.	*Pinus cont.*
		Sorbus aucuparia		*Abies spp.*	*Pinus strobus*	*Pinus brutia*
		Ulmus glabra		*Taxus baccata*		

clusters (see Table 9.2). All the groups have a pilot species, which is then used to parameterize the model. The selected pilot species are: *Betula pendula, Fagus sylvatica, Larix decidua, Picea abies, Picea sitchensis, Pinus halepensis, Pinus pinaster, Pinus sylvestris, Populus tremula, Quercus cerris, Quercus ilex, Quercus robur.*

The pilot species alone represents 76 % of the total European forest area.

9.4.3 Ecophysiological Constants of Forest Clusters

To ensure a better model parameterization, the forest cluster pilot species ecophysiology was characterized. No direct measurements of ecophysiological parameters were performed, but rather were retrieved from available databases and literature. The more relevant queried databases were those of the European research project ECOCRAFT, CANIF, and of EUROFLUX itself. Some of the parameters for oaks and pines were obtained from the parameter

Pinus pinaster	*Pinus sylvestris*	*Populus tremula*	*Quercus cerris*	*Quercus ilex*	*Quercus robur*
Pinus nigra		*Populus* spp.	*Quercus pubescens*	*Quercus suber*	*Quercus patraea*
Pinus leucoderermis		*Salix* spp.	*Quercus frainetto*	*Quercus coccifera*	*Quercus rubra*
Juniperus spp.		*Alnus* spp.	*Quercus fructosa*	*Quercus faginea*	*Acer campestre*
Cedrus spp.		*Fraxinus* spp.	*Quercus trojana*	*Quercus rotundifolia*	*Acer pseudo-platanus*
Pinus radiata			*Acer monspes-sulanum*	*Eucalyptus* spp.	*Acer platanoides*
Cupressus sempervirens			*Sorbus spp*	*Arbutus* spp.	*Robinia pseudoaca*
			Castanea sativa	*Erica* spp.	*Ulmus* spp.
			Ostrya carpinifolia	*Laurus nobilis*	*Prunus* spp.
			Juglans spp.	*Ilex aquifolium*	*Platanus orientalis*
			Quercus macrolepis	*Olea europaea*	*Carpinus betulus*
			Cercis siliquastrum	*Phillyrea latifolia*	*Corylus avellana*

list of EC project LTEEF-II. Data for *Betula* were mostly derived from the BOREAS database. For *Larix*, due to the lack of sufficient species-specific data, the default data for "*deciduous needleaf forest*" provided from the Biome-BGC web site (NTSG 2001) were used. When some of the information was not available for a certain pilot species, the solution of using parameters from similar species was employed rather than using default values. The parameters used in the simulations are reported for the 12 pilot species in Table 9.3.

Table 9.3. List of the parameters used in the BCG simulations for the 12 pilot species. The paramet
2000)

Parameter type	Parameter code	*Betula_ spec*	*Fagus_ sylv*	*Larix_ spec*	*Picea_ abies*	*Picea_ sitch*
Photo-synthesis	TfrGpd	0.2	0.2	0.2	0.2	1
	Lfgrs	0.2	0.2	0.2	0.2	1
	AnLfrtf	1	1	1	0.26	1
	AnLwTf	0.7	0.7	0.7	0.7	0.7
	AnWpMtFr	0.01	0.01	0.005	0.005	0.005
	AnFrMtFr	0	0	0	0	0
Carbon allocation	NwFrCnlC	1.2	1.4	1.4	1.4	1.4
	NwSmCnlC	2.2	1	2.2	2.2	2.2
	NwLwCntw	0.16	0.1	0.071	0.071	0.071
	NwctCnsC	0.22	0.36	0.29	0.42	0.29
	CrtGwPrp	0.5	0.5	0.5	0.5	0.5
C/N fractions	CNlvs	22.5	20.3	27	42.73	48
	CNlfltr	57.7	38	120	58.2	58.2
	CNFnRts	48	112	58	36.44	58
	CNLvWd	50	48.2	50	36.44	50
	LfLtrLbPr	0.38	0.39	0.31	0.31	0.31
	LfLtrClPr	0.44	0.44	0.45	0.45	0.45
	LfLtLnPrp	0.18	0.17	0.24	0.24	0.24
	FnRtLbPr	0.34	0.3	0.34	0.34	0.34
	FnRtClPr	0.44	0.45	0.44	0.44	0.44
	FnRtLnPr	0.22	0.25	0.22	0.22	0.22
Decom-position	DdWdClPr	0.77	0.76	0.71	0.71	0.71
	DdWdLnPr	0.23	0.24	0.29	0.29	0.29
	CwItrCff	0.01	0.04	0.01	0.01	0.01
	ClExCff	0.25	0.65	0.51	0.54	0.54
	AsddPrLar	2	2	2.6	2.6	2.6
	CaSfcLa	49	33.3	22	7.4	9.15
	RtShSlSla	2	2.3	2	2	2
Evapo-transpiration	FrLfNRub	0.09	0.08	0.088	0.033	0.033
	MxStmCd	0.01	0.01	0.006	0.005	0.004
	CtrCnd	0	0	0.00006	0.000024	0.00004
	BndLrCnd	0.01	0.02	0.09	0.09	0.09
	LfWrPtSt	−0.34	−0.6	−0.63	−0.6	−0.63
	LfWrPtCt	−2.2	−2.3	−2.3	−2.3	−2.3
	VrPrDfSt	1100	930	610	610	610
	VrPrDfCt	3600		3100	3100	3100
	Text88	705	550	730	556.25	730

been grouped by process type. For parameter codes and units, refer to BGC user guide. (Thornton

Pinus_ halep	*Pinus_ pinas*	*Pinus_ sylv*	*Populus_ trem*	*Quercus_ cerr*	*Quercus_ ilex*	*Quercus robu*
1	1	0.2	0.2	0.2	0.2	0.2
1	1	0.2	0.2	0.2	0.2	0.2
0.61	0.61	0.33	1	1	0.33	1
0.7	0.7	0.7	0.7	0.7	0.7	0.7
0.005	0.005	0.005	0.005	0.005	0.005	0.005
0	0	0	0	0	0	0
0.7	0.7	1	1.2	1	1	1.2
0.63	0.63	2.2	2.2	2.2	1	2.2
0.071	0.071	0.1	0.16	0.1	0.22	0.16
0.16	0.16	0.3	0.22	0.23	0.3	0.22
0.5	0.5	0.5	0.5	0.5	0.5	0.5
59.63	59.63	45	25.06	24	41.8	16.16
192.35	192.35	149	55	49	48	30
45.21	45.21	45	48	42	58	48
45.21	45.21	45	50	50	50	50
0.328	0.328	0.328	0.38	0.39	0.32	0.33
0.447	0.447	0.447	0.44	0.44	0.44	0.47
0.225	0.225	0.225	0.18	0.17	0.24	0.2
0.252	0.252	0.252	0.34	0.3	0.3	0.18
0.495	0.495	0.495	0.44	0.45	0.45	0.48
0.253	0.253	0.253	0.22	0.25	0.25	0.34
0.71	0.71	0.76	0.77	0.76	0.76	0.8
0.29	0.29	0.24	0.23	0.24	0.24	0.2
0.01	0.01	0.041	0.041	0.041	0.05	0.045
0.27	0.27	0.5	0.54	0.7	0.7	0.6
2.6	2.6	2.6	2	2	2	2
12.8	12.8	9.47	32	30	10.47	26
2.52	2.52	2.52	2	2	2	1.49
0.033	0.033	0.04	0.088	0.08	0.06	0.088
0.006	0.006	0.0034	0.006	0.005	0.0038	0.0044
2.5E-06	0.0000025	0.00001	0.00006	0.00001	0.00001	0.00006
0.2	0.2	0.09	0.01	0.01	0.01	0.01
−0.63	−0.63	−0.6	−0.34	−0.6	−0.6	−0.34
−2.3	−2.3	−2.3	−2.2	2.3	−3.2	−1.6
610	610	930	1100	930	1800	1100
4720	4720	4100	3600	4100	4100	5720
498	498	498	550	442	300	550

9.5 Results and Discussion

Following the model results obtained in the site-level validation, in which simulated GPP and NPP usually agreed well with measured data, the CBIS was used to simulate GPP and NPP at European level. In fact, as already discussed, reasonable NEE data could be reached only after the implementation of different model constants for heterotrophic respiration, based on site characteristics (soil carbon). As a version of the model that could use those constants (since geographically variable input is currently not available), we decided not to perform NEE simulations for the European scale.

Model simulations were performed for year 1999 after a spin-up mode run with a training period of 25 meteorological data years as contained in the meteorological database (see Chap. 4.2.4).

In Fig. 9.5, the comparison between modeled and measured GPP is presented for nine EUROFLUX sites. The EUROFLUX data refer to 1997 and 1998 (Valentini et al. 2000). For GPP, model performance showed a general underestimation with significant variation from site to site. For two of the four beech sites (FR1, BE1), modeled GPP was in good agreement with measured values, while for the other two (IT1, DK1), there was an underestimation, particularly relevant for the Italian site. For that site, we have shown above that a proper parameterization resulted in a much better agreement (see Sect. 9.3), while the results of site DK1 can be partially explained by the tendency of the

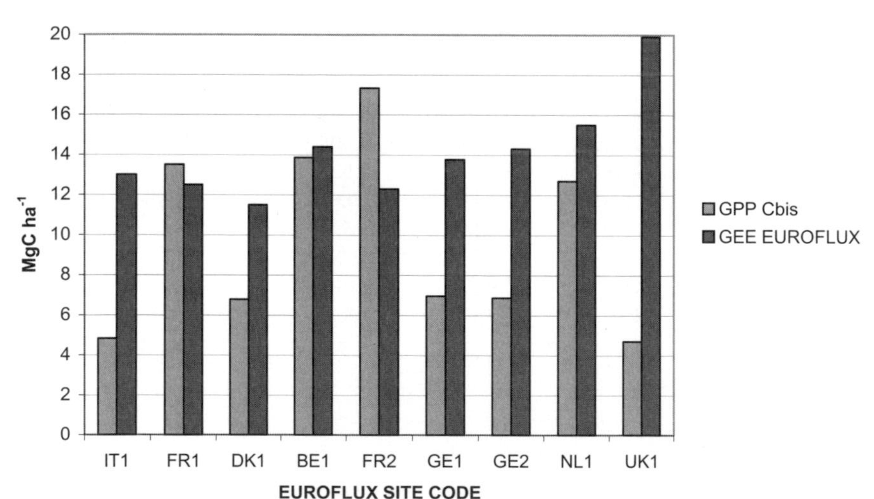

Fig. 9.5. Comparison of gross primary production (*GPP*) estimated with the CBIS and gross ecosystem exchange (*GEE*) measured at nine EUROFLUX sites. (Valentini et al. 2000)

Table 9.4. Net primary production values (Mg C ha^{-1} a^{-1}) for 13 EU member states in 1999. Finland and Sweden not included due to insufficient coverage of the meteorological database. The single values represent the average for a particular *cluster species* in a certain country. In the last column, the total NPP for the countries (Gg C a^{-1}) is reported

Country	*Betula*	*Fagus*	*Larix*	*Picea*	*Pinus h.*	*Pinus p.*	*Pinus s.*	*Populus*	*Quercus c.*	*Quercus i.*	*Quercus r.*	Total NPP
AT	4.68	5.28	7.07	2.00	2.64	2.80	8.08	5.94		0.81		6.6
BE	6.71	6.06	9.28	2.85	5.56	4.74	9.14	7.22		0.69		1.8
DE	4.83	5.52	8.29	3.26	3.31	3.61	8.06	5.18		0.81		25.9
DK		5.44	10.44	1.93		2.44	9.73				3.32	0.54
ES	3.21	3.55	7.26	2.05	1.33	1.48	4.64	1.60	0.88	1.28	0.53	13.1
FR	5.30	5.07	3.94	3.80	6.14	3.19	8.76	4.47	1.68	1.31	2.41	43.8
GR		0.10		0.01	0.74			0.17		0.24		0.28
IR				2.81							8.73	1.1
I	4.23	4.38	6.51	2.52	3.85	4.41	7.27	4.08	2.17	1.35	2.48	28.5
LU	4.66	5.22	8.42	3.24		3.75	7.51			0.63		0.18
NL	4.58	5.56	9.57	3.40	3.93	4.20	7.88			0.88		0.68
PT	2.61				2.28	2.00	7.31	1.77	1.18	1.33	2.09	3.2
UK		6.39		2.03		2.71				0.58	7.64	2.3

model to estimate lower GPP and NPP in colder climates. This may be also partially explain the results for the three spruce sites (GE1, GE2, UK1; but see also discussion of Table 9.4 results).

For site UK1, two factors should be mentioned. Although the site is a Sitka spruce stand, at the relevant location, the forest map does not show that fact. Secondly, the measured data proved to be outliers when compared to other sites (Valentini et al. 2000). It is to be noted that this site is a plantation under intensive management, including fertilization.

Nevertheless, given the different scales from which modeled and measured data have been obtained, the results presented in Fig. 9.5 can be considered promising.

An example of an NPP map for a pilot species as obtained from the CBIS is seen in Fig. 9.6. The map shows *Quercus ilex* NPP at the grid level imposed by the meteorological database, and it is an example of one of the products that can be obtained by the system.

For that particular species cluster, NPP varies between 1 and 7 Mg C ha^{-1} a^{-1}, comparable to the ranges of measured NPP for *Quercus ilex* (Cannell 1982; Matteucci et al. 1995). Furthermore, it is worth noting that, on the average, higher NPP values (4–7 Mg C ha^{-1} a^{-1}) are obtained in the north Tyrrhenian coasts of Italy, in southwest France and along the oceanic coasts of Spain and Portugal, all areas where the climatic conditions, particularly the precipitation

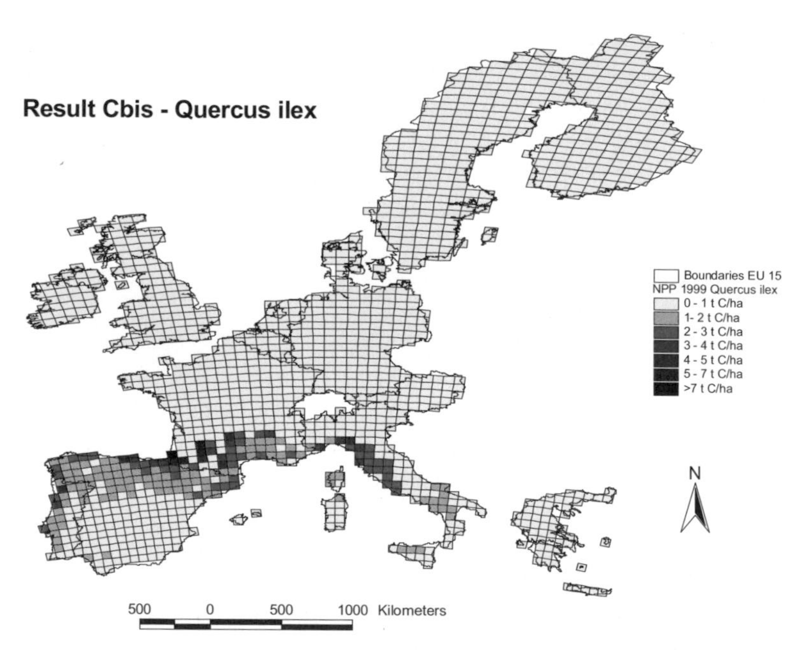

Fig. 9.6. European map of grid scale (50 × 50 km) net primary production rate (Mg C ha^{-1} a^{-1}) of the *Quercus ilex* species cluster for 1999. Differences between this map and that shown in Fig. 9.4 are due to the presence here of the associated species

regimes, are more favorable to the species performance. These results give confidence that model behaviors are consistent with what is known from ecological processes.

Average net primary production results from the Europe-wide model simulation are reported in Table 9.4 for 11 species clusters and 13 EU member states. In this exercise, Finnish and Swedish forests have not been considered because for these countries the meteorological database is too coarse to allow model simulations at ecosystem level. This data gap is currently being filled.

The values reported in Table 9.4 are in the overall range of experimental data reported in the literature (Reichle 1981; Cannell 1982; Scarascia Mugnozza et al. 2000). In some cases, such those of the *Betula*, *Quercus robur,* and *Populus* pilot species, the variation of NPP between countries reasonably represents the climate/NPP relationship for that particular species. This indicates the overall capability of the CBIS to spatialize point data at regional and continental levels.

Average modeled NPPs for *Quercus cerris* (0.9–2 Mg C ha^{-1} a^{-1}) and *Q. Ilex* (0.2–1.3 Mg C ha^{-1} a^{-1}) pilot species are in the low range when compared to other species. Their trend within countries is quite uniform, without particular connection to environmental or climate-specific parameters. Various factors may bear on these results. On the one hand, the two clusters are quite diverse in terms of the species comprising them (see Table 9.2), with the possibility that a quite large ecological gradient is covered: for the case of *Q. cerris* dry to moderately humid climate, while *Q. ilex* is forming different structures, from *macchia* (*maquis*) to forest. On the other hand, it is also possible that the model has less adaptability for representing species of the Mediterranean environment. In any case, among oaks, the more temperate *Quercus robur* pilot species shows a different trend. This is in accord with what has already been described for the model validation experiment, suggesting that for a proper representation of the different forest ecosystems, application of the model at European scale would require consideration of a variable reference temperature in the Lloyd and Taylor equation.

As already discussed regarding Fig. 9.5, for the pilot species *Picea abies* average modeled NPPs are low (1.5–3 Mg C ha^{-1} a^{-1} in the main distribution area) when compared to experimental data (Scarascia Mugnozza et al. 2000). These results, which are not confirmed for other coniferous species in the same geographical area, suggest a check of the ecophysiological constants used for the species.

Grid NPP totals in Gg C are presented in Fig. 9.7, which represent the graphic output of the CBIS at continental scale.

The differences in grid-level NPP (from 0–30 to 370–1500 Gg C a^{-1}) are then the result of the combination of site-level NPP rate (Mg C ha^{-1} a^{-1}), meteorological and soil conditions, and the actual distribution of the species in the 50 × 50 km grid (250,000 ha).

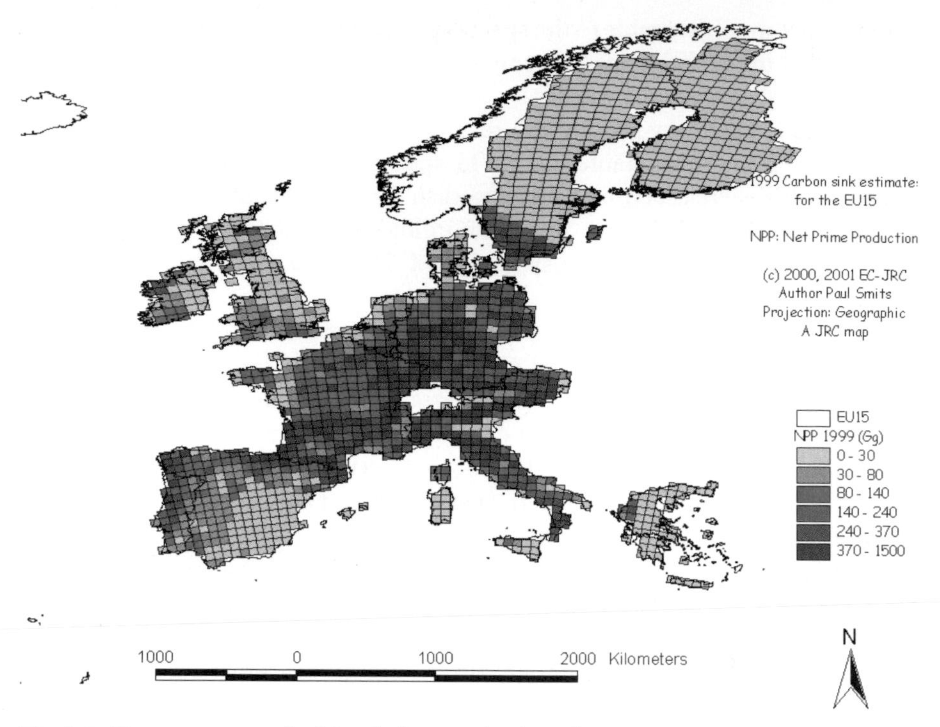

Fig. 9.7. European map of grid scale (50 × 50 km) total net primary roduction (Gg C a^{-1}) estimated with CBIS for the 13 member states of the European Union

In Table 9.5, the NPP simulated for the 11 EU member states (Tg C M S^{-1} a^{-1}) for which it was possible to perform a simulation for the entire land area is compared to the values that the same countries reported to the Intergovernmental Panel on Climate Change (IPCC) in 1999. The carbon sink reports to IPCC may consider different tree components, but they generally include carbon sequestration due to the aboveground growth of trees, which is sometimes expanded to the total growth using empirical factors.

The values presented in Table 9.5 for the two approaches are in the same range for almost all countries. As expected, on average there are larger relative differences between IPCC and CBIS values for the countries in which total NPP is below 5 Tg C a^{-1}.

For Italy, the value reported to IPCC (9.9 Tg C a^{-1}) is substantially different from the CBIS simulation (28.6 Tg C a^{-1}). The national report to IPCC in this instance is likely to have been underestimated, probably referring only to high-stand forests. Still, the forest area of Italy is similar to that of Germany and their CBIS NPP values are comparable.

Table 9.5. Comparison of net primary production as calculated with the CBIS and the values reported by EU member states to IPCC in 1999 (Tg C a^{-1}). EU11: Finland, Greece, Luxembourg, and Sweden not included

	BIOME-BGC results from CBIS	National reporting to IPCC
	(Tg C)	(Tg C)
Austria	6.6	8.1
Belgium	1.9	1.9
Denmark	0.54	1.4
France	43.9	40.0
Germany	25.9	21.7
Ireland	1.1	1.9
Italy	28.6	9.9
Netherlands	0.7	1.0
Portugal	3.2	4.3
Spain	13.1	11.0
United Kingdom	2.4	4.2
EU11	127.9	105.4

9.6 Conclusions

CBIS has proved to be an important research and application tool, with promising prospects. Some critical issues remain to be solved, however, and the presented results should be considered as preliminary.

With respect to the model itself, more detailed understanding of processes is needed, in particular of belowground and soil processes (heterotrophic respiration, soil organic carbon pools, soil and litter carbon recalcitrance, etc.). Furthermore, to apply the model on a large scale for a biome/ecosystem, which has a broad geographical extension, some ecophysiological constants or model parameters should be used as spatially variable inputs. For many of the constants, this would imply intensive research and databasing efforts.

Concerning the CBIS system, the main critical points lay in the database data quality. Despite the fact that the data sources used in our exercise are among the more advanced products available at the European scale, they were not designed to run ecosystem models on a such a scale. As an example, the resolution of the meteorological database is too coarse (50 × 50 km) and not always able to represent climatic variability, especially in mountainous areas. In this respect, the detailed forest map we have employed may represent an example on how it is possible to use and improve existing databases to make them suitable at a more detailed scale.

The CBIS system was also useful as a tool for data quality check of databases. It should be understood that the consideration of single information strata (meteorological or soil data alone) does not allow identification of errors that may occur on a dynamic spatial or temporal scale. During the runs of the CBIS, gaps and failures in the single databases were identified after careful check of anomalous results.

An important characteristic of the CBIS approach is that it provides a homogeneous methodology for the whole of Europe, ensuring at the same time that the bio-geo-chemical parameters are appropriate for the local regions and biomes.

CBIS results were validated by comparison with field measurements obtained from the European project EUROFLUX (Valentini et al. 2000; this Vol.), which project has now been expanded to include more ecosystems (CARBOEUROFLUX). A more detailed validation was obtained for the beech forest of Collelongo, for which detailed NPP and soil carbon mineralization data were available. In the future, results from the projects forming the "CarboEurope" cluster (CarboEurope 2000) will allow validation at various scales.

Future improvements of the system will consider the integration with remote sensing data, which will provide a monitoring and forecasting tool for the carbon cycle in relation to both climate and land-use change. Moreover, this model will be adapted and applied to analyze the EU accession countries and other regions such as Eurasia and the tropical regions.

Finally, we wish to emphasize the unique feature of the approach presented in this chapter, being the aggregation of ecosystem-level analyses to a pan-European level, based on availability of pan-European databases.

Acknowledgements. We thank A. Belward, J.L. Meyer Roux, F. Raes, F. Achard, A.M. Annoni, and G. Seufert for their support. We are grateful to J.M. Terres and L. Montanarella for the collaboration in the use of the European meteorological and soil databases. We also thank the authors of the BIOME-BGC model, P. Thornton and S. Running, for their open data and information policy that allow new applications that can lead to improvements and may identify research needs for a better understanding of the processes involved.

References

Anderson JM (1992) Response of soils to climate change. Adv Ecol Res 22:163–210

Andren O, Paustian K (1987) Barley straw decomposition in the field: a comparison of models. Ecology 68(5):1190–1200

Annoni A, Ehrlich D, Smits P, Montanarella L (1998) Data sets for sustainable management of Europe's regions. In: Proceedings of the 4th EC–GIS Workshop, 24–26 June 1998, Budapest, Hungary, pp 218–234

Aubinet M, Grelle A, Ibrom A, Rannik Ü, Moncrieff J, Foken T, Kowalski AS, Martin PH, Berbigier P, Bernhofer C, Clement R, Elbers J, Granier A, Grünwald T, Morgenstern K, Pilegaard K, Rebmann C, Snijders W, Valentini R, Vesala T (2000) Estimates of the annual net carbon and water exchange of forests: the EUROFLUX methodology. Adv Ecol Res 30:113–175

Baldocchi D, Falge E, Gu L, Olson R, Hollinger D, Running S, Anthoni P, Bernhofer C, Davis K, Evans R, Fuentes J, Goldstein A, Katul G, Law B, Lee X, Malhi Y, Meyers T, Munger W, Oechel W, Paw UKT, Pilegaard K, Schmid HP, Valentini R, Verma S, Vesala T, Wilson K, Wofsy S (2001) FLUXNET: a new tool to study the temporal and spatial variability of ecosystem-scale carbon dioxide, water vapor, and energy flux densities. Bull Am Meteorol Soc 82(11):2415–2434

Bowden RD, Nadelhoffer KJ, Boone RD, Melillo JM, Garrison JB (1993) Contributions of aboveground litter, belowground litter, and root respiration to total soil respiration in a temperate mixed hardwood forest. Can J For Res 23:1402–1407

Buchmann N, Schulze E-D (1999) Net CO_2 and H_2O fluxes of terrestrial ecosystems. Global Biogeochem Cycles 13(3):751–760

Cannell MGR (ed) (1982) World forest biomass and primary production data. Academic Press, London, pp 1–391

Chapman WL, Walsh JE (1993) Recent variations of sea ice and air temperatures in high latitudes. Bull Ann Meteorol Soc 74:33–47

Copley J (2000) Ecology goes underground. Nature 406:452–454

Cotrufo F, Miller M, Zeller B (2000) Litter decomposition. In: Schulze ED (ed) Carbon and nitrogen cycling in European forest ecosystems. Ecological studies 142. Springer, Berlin Heidelberg New York, pp 276–296

Cutini A, Matteucci G, Scarascia Mugnozza G (1998) Estimation of leaf area index with the Li-Cor LAI 2000 in deciduous forests. For Ecol Manage 105:55–65

Dale VH, Rauscher HM (1994) Assessing impacts of climate change on forests: the state of biological modeling. Climate Change 28:65–90

De Cola L (1997) Multiresolution covariation among Landsat and AVHRR vegetation indices. In: Quattrochi DA. Goodchild MF (eds) Scale in remote sensing and GIS. Lewis, Boca Raton, pp 72–91

Epron D, Farque L, Lucot É, Badot P-M (1999) Soil CO_2 efflux in a beech forest: the contribution of root respiration. Ann For Sci 56:221–226

European Environment Agency (2000) CORINE land cover technical guide - addendum 2000. Technical Report no 40, 105 pp

European Forest Institute (2000) Combining geographically referenced Earth observation data and forest statistics for deriving a forest map of Europe. Final report of the contract 15237–1999–08F1EDISPFI

Ewel KC, Cropper WP Jr, Gholz HL (1987) Soil CO_2 evolution in Florida slash pine plantations. II. Importance of root respiration. Can J For Res 17:330–333

Finke P, Hartwich R, Dudal R, Ibanez J, Jamagne M, King D, Montanarella L, Yassoglou N (1998) Georeferenced soil database for Europe. Manual of procedures. Joint Research Centre Report, EUR 18092en

Giardina CP, Ryan MG (2000) Evidence that decomposition rates of organic carbon in mineral soil do not vary with temperature. Nature 400:858–861

Goodenough DG, Chen H, Dyk A, Bhogal AS (2002) The role of multitemporal analysis for forest monitoring and estimation of Kyoto Protocol products. In: Bruzzone L, Smits PC (eds) Analysis of multi-temporal remote sensing images. World Scientific, Singapore

Grace J, Rayment M (2000) Respiration in the balance. Nature 404:819–820

Harrison AF, Harkness DD, Rowland AP, Garnett JS, Bacon PJ (2000) Annual carbon and nitrogen fluxes along the European forest transect, determined using 14C-bomb. In:

Schulze E-D (ed) Carbon and nitrogen cycling in European forest ecosystems. Ecological studies 142. Springer, Berlin Heidelberg New York, pp 237–256

Hole FD (1978) An approach to landscape analysis with emphasis on soils. Geoderma 21:1–23

Janssens IA, Dore S, Epron D, Lankreijer H, Buchmann N, Longdoz B, Montagnani L (2000) Soil respiration: a summary of results from the EUROFLUX sites. In: Valentini R (ed) Biospheric exchanges of carbon, water and energy from European forests. Final Report of the EUROFLUX project, EC, Brussels

Janssens IA, Lankreijer H, Matteucci G, Kowalski AS, Buchmann N, Epron D, Pilegaard K, Kutsch W, Longdoz B, Grünwald T, Montagnani L, Dore S, Rebmann C, Moors EJ, Grelle A, Rannik U, Morgenstern K, Oltchev S, Clement R, Gudmundsson J, Minerbi S, Berbigier P, Ibrom A, Moncrieff J, Aubinet M, Bernhofer C, Jensen N O, Vesala T, Granier A, Schulze E-D, Lindroth A, Dolman AJ, Jarvis PG, Ceulemans R, Valentini R (2001) Productivity overshadows temperature in determining soil and ecosystem respiration across European forests. Global Change Biol 7:269–278

Jenkinson DS, Rayner JH (1977) The turnover of soil organic matter in some of the Rothmsted classical experiments. Soil Sci 123:298–305

Kimball JS, Thornton PE, White MA, Running SW (1997) Simulating forest productivity and surface-atmosphere carbon exchange in the BOREAS study region. Tree Physiol 17:589–599

Kirschbaum MUF (1995) The temperature dependence of soil organic matter decomposition, and the effect of global warming on soil organic C storage. Soil Biol Biochem 6(27):753–760

Kirschbaum MUF (1996) Ecophysiological, ecological, and soil processes in terrestrial ecosystems: a primer on general concepts and relationships. In: Watson RT, Zinyowera MC, Moss RH, Dokken DJ (eds) Climate change 1995. Impacts, adaptations and mitigation of climate change: scientific-technical analyses. Cambridge Univ Press, Cambridge, pp 57–74

Köble R, Seufert G (2001) Novel maps for forest tree species in Europe. In: Proceedings Conference on A changing atmosphere, 8th European Symposium on the Physico-chemical behaviour of atmospheric pollutants, 17–20 September 2001, Torino

Law BE, Ryan MG, Anthoni PM (1999) Seasonal and annual respiration of a ponderosa pine ecosystem. Global Change Biol 5:169–182

Liski J, Ilvesniemi H, Mäkelä A, Westman CJ (1999) CO_2 emissions from soil in response to climatic warming are overestimated – the decomposition of old soil organic matter is tolerant of temperature. Ambio 28(2):171–174

Lloyd J, Taylor JA (1994) On the temperature dependence of soil respiration. Funct Ecol 8:315–323

Masci A, Napoli G, Dore S, Matteucci G Scarascia Mugnozza G (1998) Produzione di biomassa epigea e radicale in una faggeta e in un rimboschimento di abete rosso dell'Appennino abruzzese. In: Borghetti M (ed) La ricerca Italiana in selvicoltura ed ecologia forestale. Atti del I Congresso della SISEF, 4–6 Giugno 1997, Padova, pp 225–232 (Italy)

Matteucci G, Valentini R, Scarascia Mugnozza G (1995) Struttura e funzionalita' di una comunita' vegetale Mediterranea a *Quercus ilex* L. dell'Italia centrale. I. Bilancio del carbonio: variazioni stagionali e fattori limitanti. Studi Trent Sci Nat Acta Biol 69:127–141

Matteucci G, Dore S, Stivanello S, Rebmann C, Buchmann N (2000) Soil respiration in beech and spruce forests in Europe: trends, controlling factors, annual budgets and implications for the ecosystem carbon balance. In: Schulze E-D (ed) Carbon and nitrogen cycling in European forest ecosystems. Ecological studies 142. Springer, Berlin Heidelberg New York, pp 217–236

McNulty SG, Vose JM, Swank WT (1997). Regional hydrologic response of southern pine forests to potential air temperature and precipitation changes. J Am Water Resour Assoc 33:1011–1022

Meir P, Grace J, Miranda A, Lloyd J (1996) Soil respiration in a rainforest in Amazonia and in cerrado in central Brazil. In: Gash JHC, Nobre CA, Roberts JM, Victoria RL (eds) Amazonian deforestation and climate. Wiley, New York, pp 319–329

Motavalli PP, Palm CA, Parton WJ, Elliot ET, Frey SD (1994) Comparison of laboratory and modeling simulation methods for estimating soil carbon pools in tropical forest soils. Soil Biol Biochem 26:935–944

NTSG (2001) Numerical terradynamics simulation group, University of Montana, USA. On-line: http://www.forestry.umt.edu/ntsg/

Palmer Winkler J, Cherry RS, Schlesinger WH (1996) The Q_{10} relationship of microbial respiration in a temperate forest soil. Soil Biol Biochem 8(28):1067–1072

Orchard VA, Cook F, (1983) Relationship between soil respiration and soil moisture. Soil Biol Biochem 15:447–453

Patil GP, Myers WL (1999) Guest editorial: Environmental and ecological health assessment of landscapes and watersheds with remote sensing data. Ecosyst Health 5(4):221–224

Persson T, Van Oene H, Harrison A.F, Karlsson P, Bauer G, Cenry J, Coûteaux M-M, Dambrine E, Högberg P, Kjøller A. Matteucci G, Rubedeck A, Schulze E-D, Paces T (2000a) Experimental sites in the NYPHYS/CANIF project. In: Schulze E-D (ed) Carbon and nitrogen cycling in European forest ecosystems. Ecological studies 142. Springer, Berlin Heidelberg New York, pp 14–48

Persson T, Karlsson P S, Seyferth U, Sjöberg RM, Rudebeck A. (2000b) Carbon mineralization in European forest soils. In: Schulze ED (ed) Carbon and nitrogen cycling in European forest ecosystems. Ecological studies 142. Springer, Berlin Heidelberg New York, pp 257–275

Phillipson J, Putman R J, Steel J, Woodell SRJ (1975) Litter input, litter decomposition, and the evolution of carbon dioxide in a beech woodland-Wytham woods, Oxford. Oecologia 20:203–217

Post WM, Emanuel WR, Zinke PJ, Stangenberger AG (1982) Soil carbon pools and world life zones. Nature 298:156–159

Quattrochi DA, Goodchild MF (eds) (1997) Scale in remote sensing and GIS. CRC Press, Boston

Reichle DE (1981) Dynamic properties of forest ecosystems. IBP 23, Cambridge Univ Press, Cambridge

Running SW, Hunt ER Jr (1993) Generalization of a forest ecosystem process model for other biomes, BIOME-BGC, and an application for global-scale models. In: Ehleringer JR, Field C (eds) Scaling physiological processes: leaf to globe. Academic Press, San Diego, pp 141–158

Scarascia Mugnozza G, Bauer G, Persson H, Matteucci G, Masci A. (2000) Tree biomass, growth and nutrient pools. In: Schulze E-D (ed) Carbon and nitrogen cycling in European forest ecosystems. Ecological studies 142. Springer, Berlin Heidelberg New York, pp 49–62

Schulze E-D (ed) Carbon and nitrogen cycling in European forest ecosystems. Ecological studies 142. Springer, Berlin Heidelberg New York, pp 1–491

Schulze E-D, Högberg P, Van Oene H, Persson T, Harrison AF, Read D, Kjøller A, Matteucci G (2000) Interactions between the carbon- and nitrogen cycle and the role of biodiversity: a synopsis of a study along a north–south transect through Europe. In: Schulze E-D (ed) Carbon and nitrogen cycling in European forest ecosystems. Ecological studies 142. Springer, Berlin Heidelberg New York, pp 468–491

Schimel DS, Braswell BH, Holland EA, McKeown R, Ojima DS, Painter TH, Parton WJ, Townsend AR (1994) Climatic, edaphic, and biotic controls over storage and turnover of carbon in soils. Global Biogeochem Cycles 8:279–293

Smits PC (2000) The harmonization of European spatial data sets to initiate and run models to generate a European carbon budget for 1999. EC-JRC final report of contract 16436-2000-07 F1EI ISP IT. Ispra, Italy, Oct 2000

Smits P, Mollicone D (2001). On the use of pan-European data sets for the estimation of carbon sinks. In: Belward A. Binaghi E, Brivio PA, Lanzarone GA, Tosi G (eds) International Workshop on Geo-spatial knowledge processing for natural resource management, Proceedings. European Commission SPI 01.81

Sørenson LH (1981) Carbon–nitrogen relationships during the humification of cellulose in soils containing different amounts of clay. Soil Biol Biochem 13:313–321

Tate KR, Ross DJ, O'Brien BJ, Kelliher FM (1993) Carbon storage and turnover, and respiratory activity, in the litter and soil of an old-growth southern beech (*Nothofagus*) forest. Soil Biol Biochem 11(25):1601–1612

Thornton PE (2000) User's guide for Biome-BGC, Version 4.1.1. July 2000. Report of the numerical terradynamic simulation group, School of Forestry, University of Montana, Missoula 59812 MT, USA. On-line: http://www.forestry.umt.edu/ntsg

Trumbore S (2000) Age of soil organic matter and soil respiration: radiocarbon constraints on belowground C dynamics. Ecol Appl 10(2):399–411

UN-ECE (1998) International co-operative programme on assessment and monitoring of air pollution effects on forests. Manual on methods and criteria for harmonized sampling, assessment, monitoring and analysis of the effects of air pollution on forests. On-line. http://www.icpforests.org

UN-ECE/FAO (2000) Forest resources of Europe, CIS, North America, Australia, Japan and New Zealand. UN-ECE/FAO contribution to the global forest resources assessment 2000. Geneva Timber and Forest Study Papers, no 17, 445 pp

UNFCCC (2001) United Nations framework convention on climate change. On-line. http://www.unfccc.int

Valentini R, Matteucci G, Dolman AJ, Schulze E-D, Rebmann C, Moors EJ, Granier A, Gross P, Jensen NO, Pilegaard K, Lindroth A, Grelle A, Bernhofer C, Grünwald T, Aubinet M, Ceulemans R, Kowalski AS, Vesala T, Rannik Ü, Berbigier P, Loustau D, Gudmundsson J, Thorgeirsson H, Ibrom A, Morgenstern K, Clement R, Moncrieff J, Montagnani L, Minerbi S, Jarvis PG (2000) Respiration as the main determinant of carbon balance in European forests. Nature 404:861–865

Van der Goot E (1997) Technical description of interpolation and processing of meteorological data in CGMS. JRC Internal Report

Waring RH, Running SW (1998) Forest ecosystems – analysis at multiple scales. Academic Press, San Diego

Xu M, DeBiase T, Qi Y, Goldstein A, Liu Z (2001) Ecosystem respiration in a young ponderosa pine plantation in the Sierra Nevada mountains, California. Tree Physiol 21:309–318

10 Factors Controlling Forest Atmosphere Exchange of Water, Energy, and Carbon

A.J. Dolman, E.J. Moors, T. Grunwald, P. Berbigier, C. Bernhofer

10.1 Introduction

Forests play an important role in the water and energy balance of the land surface. It has been known since the early studies of Horton (1919) and Rutter (1975) that the water use of forests can be considerably higher than that of vegetation of different structure and height. Subsequent work has elucidated the main factors influencing this behavior (e.g., Shuttleworth and Calder 1979; Shuttleworth 1989). They showed that the combination of a high aerodynamic roughness, with a relative low and strongly controlled surface resistance, was the main cause for high evaporation rates from wet canopies and somewhat low transpiration rates from dry canopies. They also suggested that care must be given to the separate modeling of dry and wet canopy evaporation, and that total evaporation could not simply be derived from equations relating total evaporation to a net radiation estimate. Building on that work, numerous modeling studies at regional and global scale have provided evidence that the interaction of the forests with the atmosphere is a major component in shaping regional to global climate and weather (Nobre et al. 1991; Blyth et al. 1994). Forests not only use more water by evaporating more, they also influence the rainfall patterns and magnitude at regional and global scales by increasing the low level moisture convergence (supply of moisture through horizontal advection in the lower layers of the atmosphere).

The controls of forest water use are essentially the same as those that influence mesoscale weather: the relatively low albedo, the aerodynamically rough surface, and the tight physiological control on stomata. The first two of these are related to the structural characteristics of the tall and dense canopies that absorb considerable amounts of solar radiation relatively easily. The tall irregular canopies of forests furthermore generate increased turbulence, which improves the efficiency of the transport processes over forests substantially. The stomatal control of forests relates to the balance between the strength of the stomatal and aerodynamic resistance in the exchange pathway, with the

Ecological Studies, Vol. 163
R. Valentini (Ed.) Fluxes of Carbon,
Water and Energy of European Forests

stomatal, or canopy conductance presenting the larger resistance. This phenomenon has also been put in the framework of the omega coupling factor (Jarvis and McNaugthon 1986), which effectively describes the coupling of the canopy to the atmosphere by taking the ratio of the dry canopy evaporation to saturated canopy evaporation.

The purpose of this chapter is to provide an overview of what is currently known about the control mechanisms for surface atmosphere exchange of forests, the general trend of the water and energy fluxes and then to analyze the data obtained by the EUROFLUX project. Analyzing these data shows a number of remarkable generalities and discrepancies, that are useful for parameterizing studies (Dolman 1993) and studies of regional climate (Blyth et al. 1994) and CO_2 sequestration (e.g., Martin et al. 1998).

10.2 Radiation Balance

The radiation balance of a forest is primarily determined by the shortwave albedo and the emissivity of the surface. Table 10.1 shows the values for total shortwave hemispheric reflectance, the albedo for a number of EUROFLUX sites. In general, the values are between 0.08 and 0.11. This range is smaller than that quoted by Jarvis et al. (1976) who list values for coniferous forests between 0.08 and 0.14. There is some variation between species and between sites. The pine and spruce forests have generally values of 0.08, deciduous forest tend towards higher values, e.g., 0.11 for the Danish beech forest. The dense spruce stands are clearly more efficient in trapping the incident light and have the lowest values of albedo. On the basis of these results, a suitable value for spruce and pine would be 0.08 and for foliated deciduous 0.11. Unfoliated larch or deciduous trees have a somewhat higher albedo, on the order of 0.13. These values are lower than those quoted by Monteith and Unsworth (1990). This may reflect a difference in measurement techniques, but a more likely explanation is that the longer observation period from the EUROFLUX forests shows a more representative average than previous "campaign"-based measurements. The albedo shows a marked diurnal trend, related to the elevation cycle of the sun. This is shown in Fig. 10.1, where for four Dutch forests the relation of albedo to solar elevation is plotted. The largest difference in elevation sensitivity is to be expected by the smoothest canopy. A smooth canopy will trap the light less efficiently at low solar elevations. For the tall larch and mixed forests this results in a relatively small difference between reflectance at low and high solar elevations.

The incoming shortwave radiation is mostly absorbed by the canopy and re-emitted as longwave radiation. The balance of incoming and reflected shortwave and longwave radiation is the net radiation. There is a clear empirical relation between incoming shortwave radiation and net radiation caused

Table 10.1. Estimates of shortwave albedo and regression constants for the relation of net to solar radiation

Site	Species	Albedo	A (W m^{-2})	b
Sarrebourg	*Fagus sylvatica*	0.11		
1997		0.14	–24.2	0.76
1998			–22.5	0.79
1999			–17.8	0.77
Bordeaux	*Pinus maritima*			
1996			–25.2	0.78
1997			–36.5	0.80
1998			–33.8	0.80
Norunda	*Pinus sylvestris*			
1996			–33.8	0.79
1997			–42.0	0.82
1998			–29.8	0.8
Bayreuth	*Picea abies*	0.085		
1996			–4.6	0.91
1997			–21.4	0.85
1998			–25.9	0.79
Tharandt	*Picea abies*	0.075		
1996			–30.4	0.78
1997			–34.9	0.82
1998			–35.7	0.82
1999			–37.0	0.82
Loobos	*Pinus sylvestris*	0.081		
1996			–22.6	0.80
1997			–25.6	0.82
1998			–22.0	0.82
Aberfeldy	*Picea abies*			
1997			–15.4	0.67
1998			–15.7	0.70
Viesalm				
1996			–12.8	0.72
1997			–20.4	0.80
1998			–17.4	0.84
Brasschaat	*Pinus sylvestris*	0.05		
1997			–28.0	0.80
1998			–25.0	0.80
Lille Boegeskov	Beech	0.105		
Fleditenbos	*Populus* spp.	0.089		
		0.091		
Kampina	Mixed	0.10		
		0.13		
Bankenbos	*Larix*	0.125		
		0.013		

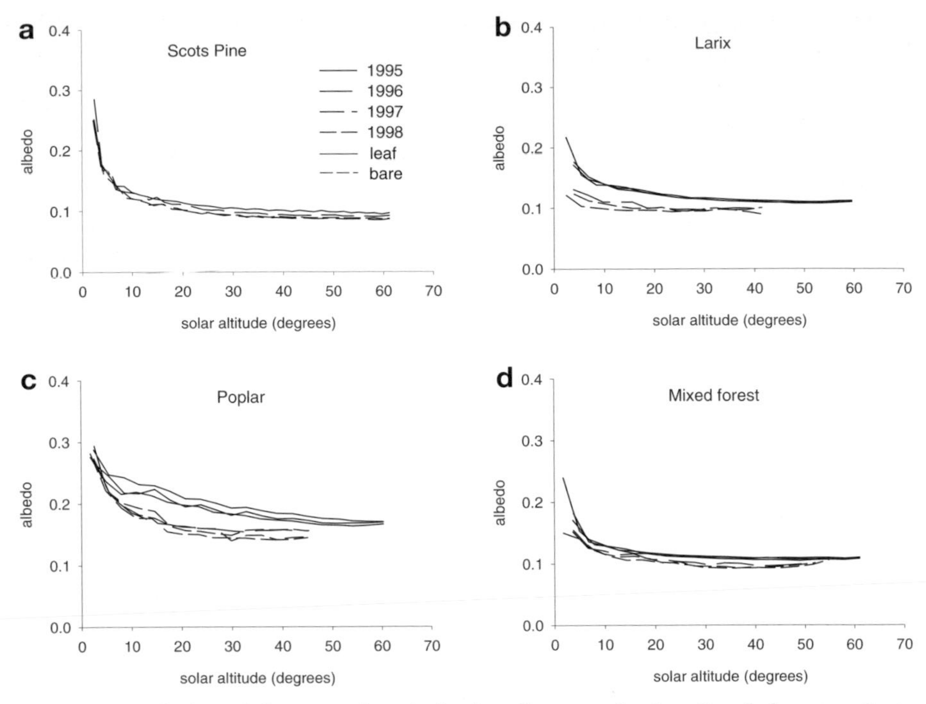

Fig. 10.1. Variation of shortwave hemispheric reflectance for four Dutch forests: **a** Scots pine; **b** Larch **c** Poplar; **d** mixed deciduous, pine

by the albedo, surface energy balance, and emissivity. This relation can be expressed as

$$R_n=a+bR_g$$

where R_n is net radiation, R_g incoming solar shortwave downward radiation, and a and b dimensionless parameters that depend primarily on latitude. Table 10.1 gives values for the two parameters for several of the EUROFLUX forests. The values were based on a relation of half-hourly radiation values. It can easily be shown that the regression coefficient of Eq. (1) relates to the albedo, α, as $b=1-\alpha$, however, the regression coefficients tend to be of order 0.7 to 0.8, indicating a much higher albedo. The data shown here combine various effects that may confound this simple theoretical relationship. For instance, no effort is made to distinguish between cloudy or sunny days, where a different balance between diffuse and direct radiation will alter the reflectance. The average value for b is 0.80 ($\pm$0.01), which is the same as that quoted by Jarvis et al. (1976) for coniferous forest. Apparently, this relation is remarkably constant and may be used in practical applications where a need exists to calculate net radiation when only estimates of sunshine hours or shortwave radiation are available. The physical interpretation of the intercept is that it corresponds to the average longwave radiation balance of the site. The average longwave balance of these sites is -25 W m^{-2} (± 8). There is, however, considerably more variation in the intercept, both between years and between sites, reflecting differences primarily in surface temperature of the forests. Unexpectedly, the data do not show a clear geographical relation to the value of the intercept. Monteith and Unsworth (1990) quote an average value of 60 W m^{-2} as longwave loss to the atmosphere for grass in Germany (Hamburg). The EUROFLUX forests lose on average less than half that value because of the lower surface temperature of forests compared to grassland. For photosynthesis the photosynthetic photon flux density is important. At the three sites where this quantity was measured reliably, PPFD was found to be on average 1.92 times the incoming shortwave radiation.

10.3 Turbulent Exchange

The rough canopy of forests is one of the most important characteristics determining the exchange of water heat and carbon between the forest and the atmosphere. This can be shown by writing the aerodynamic part of the resistance pathway, r_a between an arbitrary reference level z above the canopy and the source/sink height within the canopy as a function of wind speed, u, and friction velocity, u_*:

$$r_a = u(z)/u^2_*$$

The values u and u_* are related by the aerodynamic roughness length z_0 and zero plane displacement height d in the logarithmic wind profile under neutral stability conditions:

$$u = u_*/k\ (\ln\ (z-d)/z_{0)}$$

where k is von Karman's constant. For most crops the roughness length is a function of the height of the canopy and to a lesser extent of the amount of leaf and spacing of the vegetative elements. To be able to determine the roughness length from measurements, either another level of measurement of the momentum flux ($\tau = u_*^2$), or several levels of wind speed measurements are required. In the latter case, care has to taken when relating cup anemometer measurements with those of a sonic anemometer.

For the EUROFLUX sites the values of z_0 and d are given in Table 10.2. Roughness length and displacement height are mostly related through simple linear approximations with height. Shuttleworth (1989) reviewed some of the earlier work and concluded that for z_0 a fraction of 0.086 and for d 0.76 are appropriate for forests. The EUROFLUX results give a slightly higher values

Table 10.2. Height, leaf area index, aerodynamic roughness, and zero plane displacement height for several EUROFLUX forests

Site	Species	Height	LAI	z_0	d
Brasschaat	Pine	21	3	1.5	16
Sarrebourg	Beech	12.7	5.7		
Viesalm	Mixed	35		3	
Loobos	Pine	15.1	1.90	1.5	8.1
Bankenbos	Larch	22.8	1.9	2.4	12.5
				2.2	13
Fleditenbos	Poplar	16.2	3.9	1.7	9.1
				1.2	8.9
Kampina	Mixed	17.5	3.5		
Tharandt	Spruce	25	7.2	1.5	18.5
Hythalla	Pine	13	3	1.2	9
Lille Boegeskov	Beech	25	4.75	1.6	19
Weidenbrunnen	Spruce	19	5.5	2.5	10.5
Aberfeldy	Spruce	9.8	6.4	0.6	3.4

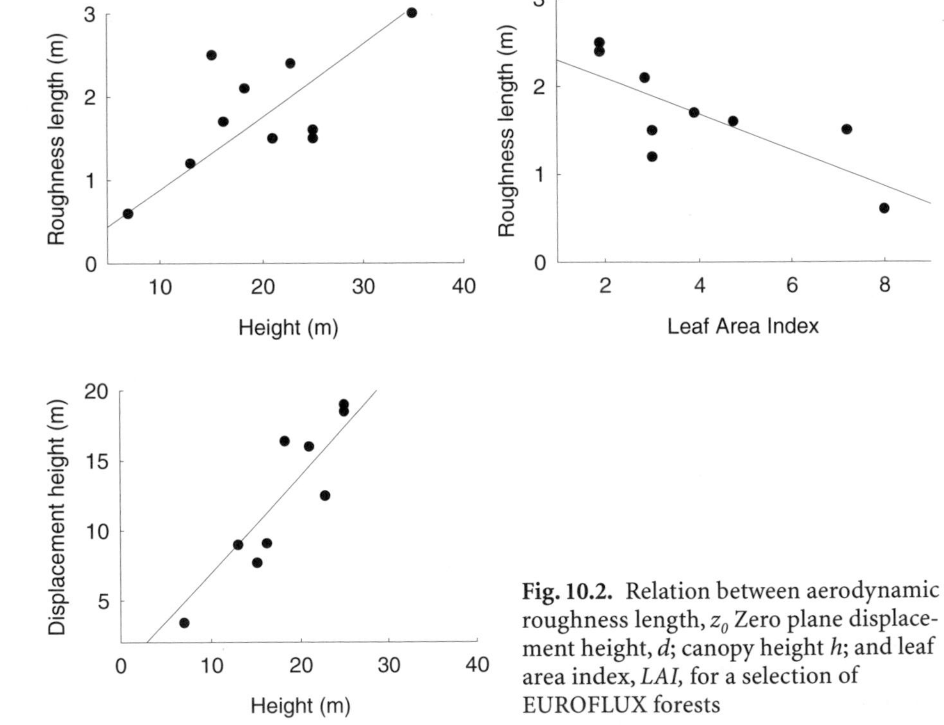

Fig. 10.2. Relation between aerodynamic roughness length, z_0 Zero plane displacement height, d; canopy height h; and leaf area index, *LAI*, for a selection of EUROFLUX forests

for z_0 of 0.088 and a similar value for displacement height (0.76) (Fig. 10.2). However, the scatter is considerable and measurement differences between sites and individual site errors cannot be completely ruled out.

Figure 10.2 also shows the relation of the two turbulent parameters with leaf area index (LAI). The relation is less clear, but it would appear that the canopies with high LAI are somewhat more smooth than those with lower LAI. This inverse relation with LAI, and the relation with height may give scope for developing new estimation methods for roughness length.

10.4 Water Use of Forests

The water use of forests is generally regarded as higher than that of grasslands or agricultural crops under similar climatic conditions. Analysis has shown that it is essential to distinguish between wet and dry canopy evaporation. Because of the large leaf area of forests a considerable amount of precipitation is intercepted. The aerodynamic roughness of the forests makes turbulent transfer a very efficient transfer mechanism for water vapor, even under conditions of marginal net radiation: at the extreme the canopy behaves as a well ventilated wet bulb. When the canopy is dry, the physiological control by stomata presents the main resistance in the system, and transpiration is constrained by the physiology of the trees.

10.4.1 Wet Canopy Evaporation

Table 10.3 lists the average interception loss, defined as the relative difference between gross precipitation and throughfall, for several EUROFLUX forests. This quantity was not measured routinely at all sites, hence it can only be given for a few forests.

Table 10.3. Interception loss as a fraction of gross precipitation for Euroflux forests

Site	Species	Interception loss (%)
Brasschaat	Pine	22–30
Sarrenbourg	Beech	25
Loobos	Pine	26
Bankenbos	Larch	23
Fleditenbos	Poplar	20
Kampina	Mixed	30
Tharandt	Spruce	45

Pine forests have interception losses of around 25 %; the spruce forest at Tharandt shows almost double that, 45 %. This is the range commonly observed for forest interception loss. A simple relationship is not apparent between interception loss and the leaf area of these forests (see Table 10.2). A linear relation of LAI with the saturation storage capacity is often used to calculate saturation storage capacity. The absence of a relation between annual average interception loss as a percentage of LAI brings into question the use of those simple relations in climate models.

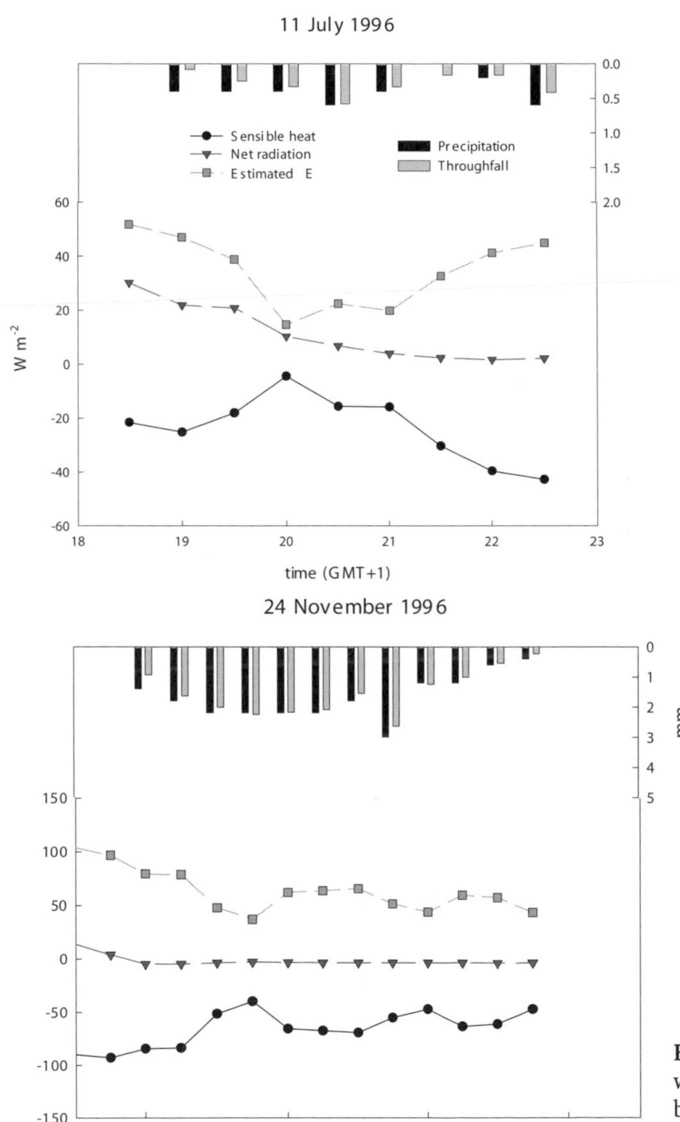

Fig. 10.3. Examples of wet canopy energy balance for a Dutch poplar stand in summer and winter

A further noteworthy feature of these results is the high interception loss during winter in deciduous forests. This has been found before (Dolman 1987) and suggest that it is the efficiency of aerodynamic transfer rather than the saturation storage capacity that determines interception losses of leafless canopies in the winter. During these periods, the net available radiative energy is low. Energy to sustain the wet canopy evaporation is then obtained from the overlying warmer air through a downward negative sensible heat flux. This is shown in Fig. 10.3 for the case of a Dutch poplar forest in the winter. Due to unreliability of the latent heat flux measurements in wet periods, the latent heat flux is estimated from the residual of the energy balance in these plots. The plots show clearly that the sensible heat turns negative during rainfall events, and that, with low or zero net radiation, a positive latent heat flux is maintained. Also shown is precipitation and throughfall. Note that to obtain a closed water and energy balance, the release of stored energy in the canopy below crown space and the amount of water stored on the canopy need to be taken into account.

10.4.2 Dry Canopy Evaporation

The water use of forests has received considerable attention in land-use planning since Horton's (1919) early studies (e.g., Newson and Calder 1989). Shuttleworth and Calder (1979) presented the physical analysis that explained most of the differences in water use by forest and agricultural crops and grassland. One of their conclusions was that forest with dry canopy conditions (transpiration) hardly ever evaporated at potential rates as well watered grassland would do. Stomatal closure in forests occurs as soon as humidity deficits in the air increase. This causes Bowen ratios of forests to be above one on average, implying that a substantial part of the energy is transferred as sensible heat. The EUROFLUX forests also behave according to this general pattern.

In Fig. 10.4, the evaporative fraction, here defined as $\alpha=\lambda E/(\lambda E+H)$ is plotted against normalized evaporation for two pine forests and the Danish beech forest. The sum of the fluxes is used as a substitute for available energy. Thus, the ratio expresses more adequately the partitioning of the turbulent fluxes independent of measurement errors in net radiation. The normalization is achieved through dividing by the maximum rate. These evaporation-space diagrams show the general behavior of evaporation in these forests. Extrapolation of the upper envelope to zero evaporation on the graphs yields the maximum evaporative fraction. For these three forests this value is 0.85, which indicates that under these conditions 85 % of the available energy is transferred into evaporation, the remaining 15 % lost as sensible heat. These conditions would prevail under low atmospheric humidity deficits and suggest that the approximate value of the Priestley-Taylor coefficient for these forests may

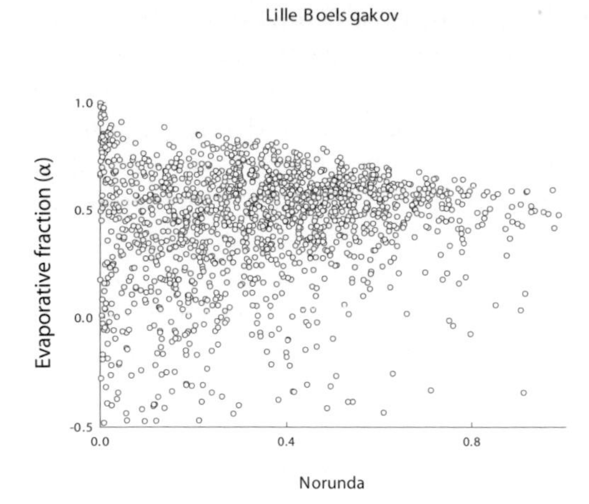

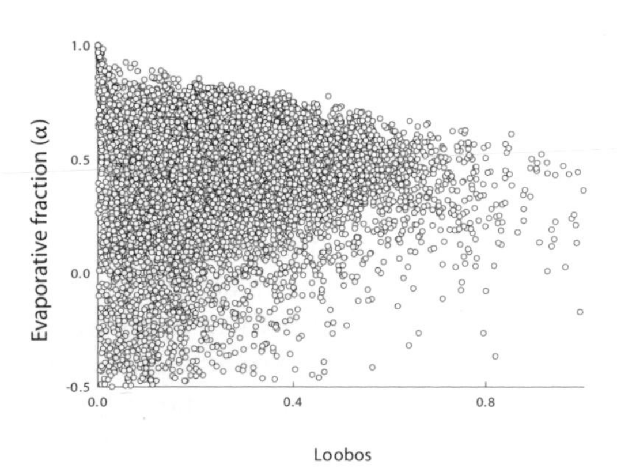

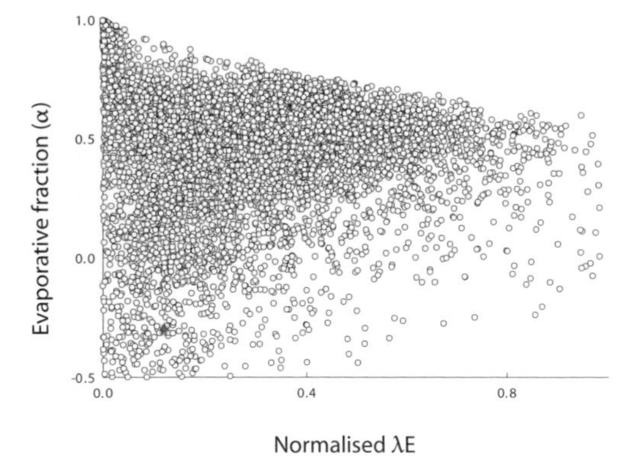

Fig. 10.4. Relation between the evaporative fraction and latent heat flux for a pine forest in the Netherlands, a spruce forest in central Sweden, and a beech forest in Denmark

be close to 0.85. At high evaporation values, the evaporative fraction has been found to decrease to 0.5 for the beech forest and to a similar value for the pine forests. For vegetation with higher aerodynamic resistance, such as grasslands, the slope would be less steep and high evaporative fractions maintained under high evaporation (radiation). The nature of the slope confirms the tight coupling and response of forest surface conductance to atmospheric humidity deficit. Indeed, the linearity of the upper envelope for these forests is striking.

The diagrams also suggest a marked variability in surface energy balance partitioning during the growing season. On average the evaporative fraction would appear to be around 0.5, but clearly stomatal control can produce variation between 0.25 and 0.75.

The control by stomata is the main cause for variation in the surface energy balance of forests. At the canopy level this behavior translates with the leaf boundary layer resistances into the surface conductance. The surface conductance can be obtained by inverting the Penman-Monteith equation (see Stewart 1988):

$$g_s^{-1}=(\lambda E\Delta/c_p\ \beta-1)/G_a+\varrho\delta q/E$$

Figure 10.5 shows the behavior of the average daily conductance throughout 1997 for the pine site in Loobos, Holland. Error bars give an indication of the amount of data used in calculating the mean. Also shown is precipitation. There are some high values of conductance in the winter that are related to wet canopy and soil conditions. Even during the winter in this forest evaporation can reach 1–1.5 mm/day. Once the growing season has started the conductance becomes truly physiological. During drier periods, the average conductance appears to decline somewhat, while just after rainfall increases may be noted. This phenomenon turns out to be a response to increasing humidity deficits in the air rather than in the soil. The average mean summer daily value is around 7 mm s^{-1}.

Previous analysis of the behavior of surface conductance has shown that the humidity deficit in the air tends to constrain the conductance and that shortwave radiation increases the conductance until a maximum value is reached (Jarvis 1976; Dolman et al. 1988). These features are used to parameterize the surface conductance in various climate and hydrological models. Figure 10.6 shows the behavior of the surface conductance in the Dutch pine forest. The conductance decreases sharply with increasing specific humidity deficit. For comparison a similar graph for spruce (Tharandt) is also shown. The pine forest shows a stronger response to atmospheric humidity deficit than the spruce as is shown by the steeper decline at humidity deficits around 7 g kg^{-1}. In general, this response leads to a more conservative water use of the Scots pine forest compared to spruce.

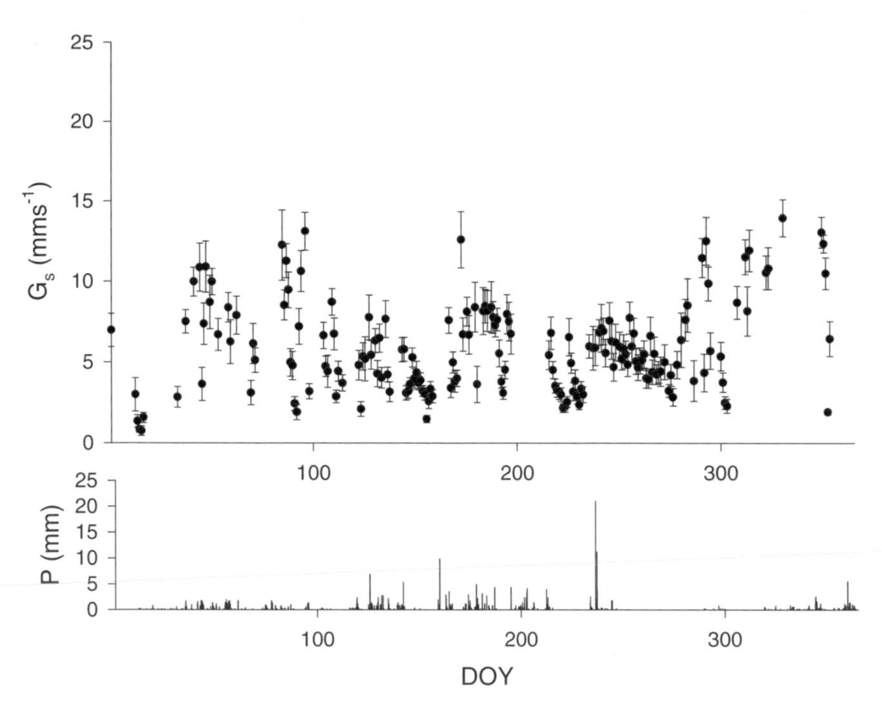

Fig. 10.5. Daily average surface conductance of a Dutch Scots pine forest in 1997. Precipitation is also shown

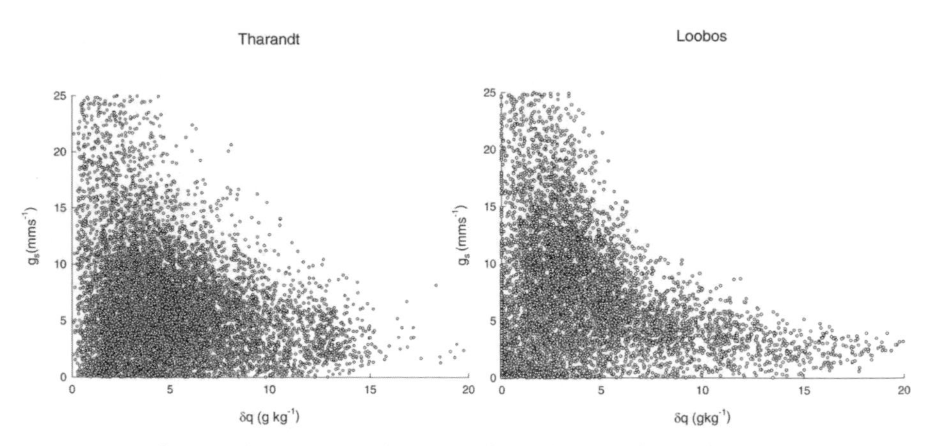

Fig. 10.6. Surface conductance as a function of specific humidity deficit for a Dutch pine forest and a German spruce forest

10.5 Surface Conductance Photosynthesis Relations

It has been established (that at the leaf level a strong relationship exist between stomatal conductance and net assimilation (e.g., Leuning 1995).

$$g_{s=}a+b\ F_{nee}h_s/c_i$$

where g_s is the surface conductance, i.e., the integrated overall conductance of the surface as shown before, including canopy and soil; F_{nee} is the half-hourly net ecosystem flux (or leaf level assimilation minus respiration in the original formulation); h_s is the surface relative humidity; and c_i is the internal CO_2 concentration of the fictional big leaf. The values a and b are regression parameters of a linear fit of g_s against $F_{nee}h_s/c_i$.

This relation is now forming the basis for a new generation of conductance models which link the carbon cycle directly with the water cycle (Collatz et al. 1991; Cox et al. 1998). The relation between conductance and photosynthesis is not fully understood in causal and physiological terms, and therefore remains empirical. It is of interest to note that in some ways this relation can be interpreted as a climate-normalized water-use efficiency, since for evaporation the parameter g_s is used, effectively normalizing the climatological influence on evaporation.

It is of some interest to explore the possibility of the existence of this relationship at ecosystem level instead of leaf level. At ecosystem level all sources of respiration, autotrophic and heterotrophic are lumped together with the canopy assimilation to produce the NEE. At this level of integration, however, detailed radiation extinction models and canopy scaling methods may not be necessary to calculate g_s if NEE is related to the overall surface conductance g_s. Furthermore, the predictive power of such a relationship may be very useful for scaling the tower measurements across larger areas with the aid of remote sensing. Also, the performance of scaled models based on leaf level and soil respiration can be checked against such an ecosystem-level relation that may represent an adequate signature of the relation between water and carbon fluxes at the ecosystem level. Deriving this "big leaf NEE-g_s model" can be achieved by using the observed friction velocities, wind speed and stability corrections to calculate the aerodynamic transport resistance. The surface values of temperature, humidity and CO_2 concentration are obtained by inverting the flux equations.

Figures 10.7 and 10.8 show the final result of these calculations for the big leaf NEE-g_s relation, where the data have been grouped by classes of $F_{nee}h_s/C_i$ for the pine forest near Bordeaux and the EUROFLUX forests, respectively. Only data during the growing season are used (May–September). Using data outside the mean growing season reduces the correlation between NEE and g_s as the latter becomes less dependent on the physiological behavior of the system and more on the surface and soil wetness.

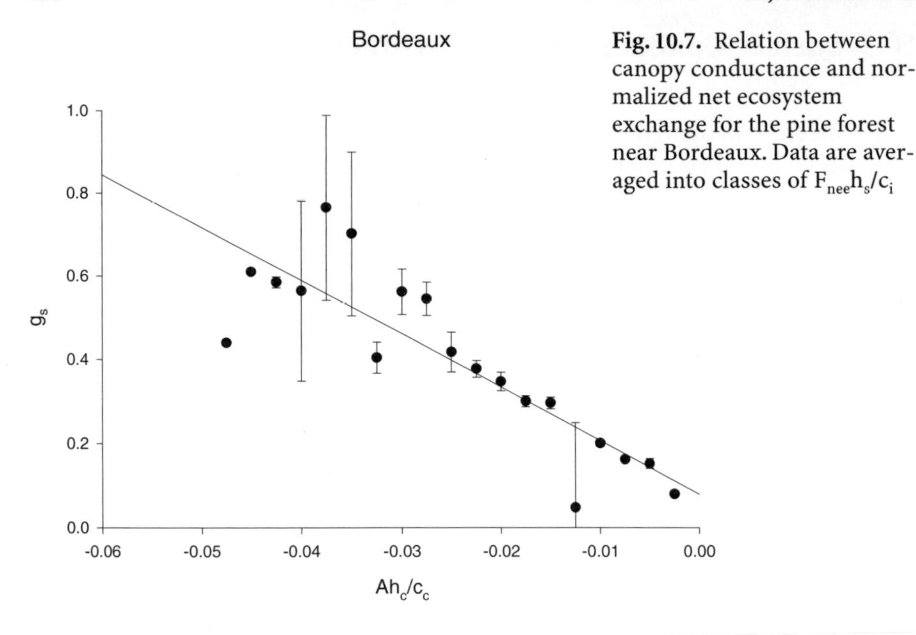

Fig. 10.7. Relation between canopy conductance and normalized net ecosystem exchange for the pine forest near Bordeaux. Data are averaged into classes of $F_{nee}h_s/c_i$

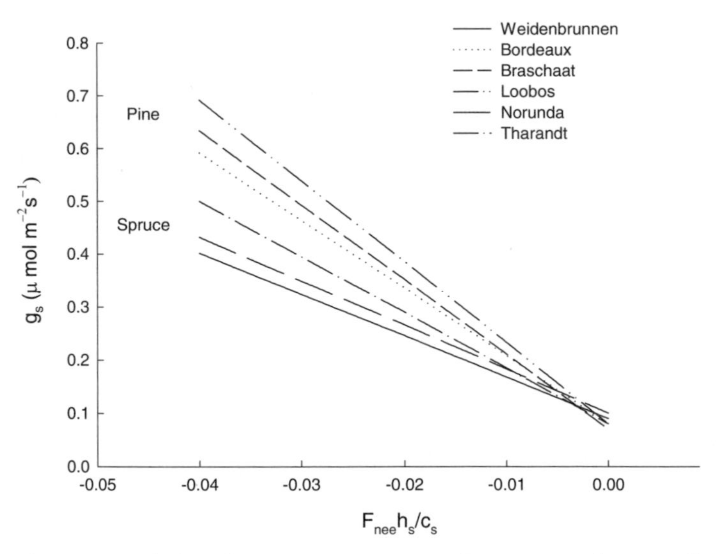

Fig. 10.8. Relation between canopy conductance and normalized net ecosystem exchange for the EUROFLUX forests. Data are averaged into classes of $F_{nee}h_s/c_i$

Table 10.4. Regression parameters of the g_s–$F_{nee}h_s/c_i$ relation

Site	Species	A	B
Weidenbrunnen	Picea	-7.8	0.09
Bordeaux	Pinus maritima	–12.8	0.08
Brasschaat	Pinus sylvestris	–14.1	0.07
Loobos	Pinus sylvestris	–15.3	0.08
Castel Porziano	*Quercus ilex*	-4.0	0.4
Norunda	Picea	-8.3	0.1
Tharandt	*Picea*	–10.5	0.08

The first thing to note (Table 10.4) is that for all the seven forests analyzed the data show a convincingly linear relationship up to $F_{nee}h_c/c_c$~0.03, with a slope varying from -7.8 to –15.3 (leaving out the anomalous low value of Castel Porziano). There is little variation between the sites and species. The pine locations at Loobos and Brasschaat are very similar, even though the Brasschaat canopy is higher and would appear more closed than the Loobos site. The slope of the pine site near Bordeaux is also similar but shows more scatter at the lower values of $F_{nee}h_s/c_i$. The two spruce sites in Germany are again similar, although data problems with Weidenbrunnen do not permit extrapolation beyond $F_{nee}h_c/c_c$~–0.04. In general, the spruce sites have a somewhat lesser slope.

The second thing to note is a general leveling off beyond $F_{nee}h_c/c_c$~–0.03 of the two Mediterranean sites, the *Quercus ilex* at Castel Porziano and the *Pinus maritima* at Bordeaux (see Fig. 10.7). This is likely to have been caused by soil moisture deficits occurring during the summer period. These have the effect of reducing the conductance, thereby changing the slope of the F_{nee}-g_s relation. There is also more scatter around the mean in these values, indicating that different conductances may result from similar values of $F_{nee}h_s/c_i$.

10.6 Conclusions

This chapter has analyzed some of the main controls on forest evaporation and radiation exchange. The data obtained in EUROFLUX have proved to be useful for determining some general trends. In the growing season on average the net radiation is 80 % of incoming shortwave radiation for the forests studied in this chapter. The average longwave balance is negative and amounts to 25 W m^{-2}. There is no clear geographical trend in this number with a pine forest at Bordeaux having almost the same balance as a spruce forest in Norunda, Sweden.

The shortwave albedo varies between 0.08 and 0.13. On the basis of the current study, a suitable value for spruce would be 0.08, for pine 0.08, and for leafed deciduous 0.12. Unfoliated larch or deciduous trees have a somewhat higher albedo, on the order of 0.13

The aerodynamic roughness length of forests varies with canopy height according to the relation z_0=0.088 × mean tree height. The zero plane displacement is 0.76 times the mean tree height. As a second-order effect, the leaf area index appears to show an inverse relation with roughness length.

The interception losses of the EUROFLUX forests vary between 20 and 45 %. Scots pine has an average of 22 %, the dense spruce forest lose 45 % of the annual gross precipitation by interception evaporation. Also in winter relatively high percentages of interception loss were found for deciduous forests.

The evaporative fraction of the EUROFLUX forests shows a maximum of 0.85, with an average value closer to 0.5. There appears to be little variation in the relation of evaporative fraction with evaporation. Analysis of surface conductance shows a strong response to humidity deficit and the well-known asymptotic relationship with shortwave radiation.

An analysis of net ecosystem exchange of CO_2 versus surface conductance shows a strong linear behavior. The slope of the relation appears to be similar between forest species, although pine forests tend to have larger slopes than spruce forests. The intercept of this relation is related to the average ecosystem respiration rate and has values between 0.07 and 0.1. The remarkable similarity in these relationships between species show that it may be important to exploit these in predictive global models. It furthermore suggests that there is a rather conservative optimized relation between water use and regulation of CO_2 exchange. This is further shown in Fig. 10.8 where the linear regressions of the coniferous EUROFLUX forests are plotted. The lines clearly fall into two groups, with the spruce forests showing a smaller slope than the three pine forests. It should be noted that these regressions are only valid for well watered forests up to $F_{nee}h_s/c_i=-0.04$, as is shown in Fig. 10.7; the slope decreases in the Mediterranean forest, probably due to increasing soil moisture deficits.

The variability in driving forces of the water and energy fluxes appears to be primarily species-related. Site-dependent characteristics and climatic differences will however generate rather different total annual fluxes, as is for instance shown by the analysis of the annual NEE by Valentini et al. (2000).

References

Blyth EM, Dolman AJ, Noilhan J (1994) The effect of forests on mesoscale rainfall: an example from HAPEX-MOBILHY. J Appl Meteorol 33:445–454

Collatz GJ, Grivet C, Ball JT, Berry JA (1991) Physiological and environmental regulation of stomatal conductance, photosynthesis, and transpiration: a model that includes a laminar boundary layer. Agric Forest Meteorol 54:107–136

Cox PM, Huntingford C, Harding RJ (1998) A canopy conductance model for use in a GCM land surface scheme. J Hydrol 212, 213:79–94

Dolman AJ (1987) Summer and winter rainfall interception in an oak forest. Predictions with an analytical and a numerical simulation model. J Hydrol 90:1–9

Dolman AJ (1993) A multiple source land surface energy balance model for use in GCMs. Agric Forest Meteorol 65:2–45

Dolman AJ, Stewart JB, Cooper JD (1988) Predicting forest transpiration from climatological data. Agric Forest Meteorol 42:339–353

Horton RE (1919) Rainfall interception. Mon Weather Rev 47:603–623

Jarvis PG (1976) The interpretation of the variations in leaf water potential and stomatal conductance found in the field. Philos Trans R Soc Lond B273:593–610

Jarvis PG, McNaughton KG (1986) Stomatal control of transpiration: scaling up from leaf to region. Adv Ecol Res 15:1–49

Jarvis PG, James GB, Landsberg JJ (1976) Coniferous forests. In: Monteith JL (ed) Vegetation and the atmosphere, vol II. Academic Press, London, pp 171–240

Leuning R (1995) A critical appraisal of a combined stomatal-photosynthesis model. Plant Cell Environ 18:339–357

Martin P, Valentini R, Jaqcues M, Fabbri K, Galati D, Quarantino R, Moncrief JB, Jarvis P, Jensen NO, Lindroth A, Grelle A, Aubinet M, Ceulemans R, Kowalski AS, Vesala T, Keronen P, Matteucci G, Granier A, Berbingier P, Lousteau D, Shulze ED, Tenhunen J, Rebmann C, Dolman AJ, Elbers JE, Bernhofer C, Grunwald T, Thorgeirson H (1998) New estimate of the carbon sink strength of EU forests integrating flux measurements, field surveys and space observations. Ambio 27:582–584

Monteith JL, Unsworth MH (1990) Principles of environmental physics. Arnold, London

Newson MD, Calder IR (1989) Forests and water resources: problems of prediction on a regional scale. Philos Trans R Soc B324:283–298

Nobre CA, Sellers PJ, Shukla Y (1991) Amazonian deforestation and climate change. J Climate 4:957–988

Rutter AJ (1975) The hydrological cycle in vegetation. In: Monteith JL (ed) Vegetation and the atmosphere, vol I. Academic Press, London, pp 111–154

Shuttleworth WJ (1989) Micrometeorology of temperate and tropical forest. Philos Trans R Soc B324:299–334

Shuttleworth WJ, Calder IR (1979) Has the Priestley Taylor equation any relevance to forest evaporation? J Appl Meteorol 18:634–638

Stewart JB (1988) Modelling surface conductance of pine forests. Agric For Meteorol 43:19–35

Valentini R, Matteucci G, Dolman AJ, Schulze E-D, Rebmann C, Moors EJ, Granier A, Gross P, Jemsen NO, Pilegaard K, Lindroth A, Grelle A, Bernhofer C, Grunwald T, Aubinet M, Ceulemans R, Kowalski AS, Vesala T, Rannik U, Berbigier P, Lousteau D, Gudmundson J, Thorgeirson H, Ibrom A, Morgenstern K, Clement R, Moncrieff J, Montagni L, Minerbi S, Jarvis PG (2000) Respiration as the main determinant of carbon balance in European forests. Nature 404:861–865

11 The Carbon Sink Strength of Forests in Europe: a Synthesis of Results

R. Valentini, G. Matteucci, A.J. Dolman, E.-D. Schulze, P.G. Jarvis

11.1 Introduction

The terrestrial sink for carbon is estimated to be of the order of 2±1 Gt C - y^{-1} (IPCC 2000) However the accumulation of CO_2 in the atmosphere as documented by atmospheric stations around the globe for the past 40 years can vary by a factor of two from one year to the next, which is equivalent to several Gt C per year. Such changes reflect interannual shifts in the carbon uptake of land and oceans of the same magnitude as the average uptake itself. Fossil fuel emissions changes tend to be smooth in time: year-to-year variations are less than 4% of the total. Both ocean data and global ocean carbon models suggest that the air-sea carbon fluxes are rather stable. The land biosphere may thus explain most of the observed CO_2 interannual growth rate variation. Several studies, utilizing different techniques, have shown that, in the northern hemisphere, the terrestrial biosphere is currently absorbing carbon (Dixon et al. 1994; Myneni et al. 1997). Most, if not all, of these methods depend on indirect estimates of the carbon fluxes, like isotopic analysis and inversion methods from CO_2 concentrations measurements (Ciais et al. 1995), remote sensing (Myneni et al. 1997), growth trend analysis (Dixon et al. 1994; Kauppi et al. 1992), and modeling. All these methods provide the necessary global and continental scale perspective for carbon balance calculations. However, these studies suffer from uncertainties in the assumptions used, for instance in the inverse modeling studies the anthropogenic sources and sinks are frequently prescribed "a priori" and they lack adequate parameterization of the carbon balance at local scales. Their use in addressing small temporal and spatial changes in the carbon balance is therefore rather limited.

The net carbon exchange of terrestrial biota is the result of a delicate balance between uptake (photosynthesis) of vegetation and losses by anthropogenic fossil fuel emissions, autotrophic and heterotrophic respiration, and natural and human-induced disturbances involving several temporal and spatial scales.

Ecological Studies, Vol. 163
R. Valentini (Ed.) Fluxes of Carbon,
Water and Energy of European Forests

At ecosystem scale the balance is between uptake (photosynthesis) and loss (respiration), and shows a strong diurnal, seasonal, and annual variability. Under favorable conditions, during daytime the net ecosystem flux is dominated by photosynthesis, while during night, and for deciduous ecosystems in leafless periods, the systems loses carbon by respiration. The influence of climate and phenology can in some cases shift a terrestrial ecosystem from a sink to a source of carbon (Goulden et al. 1998; Lindroth et al. 1998).

Global and continental scale techniques are of limited use in addressing such processes.

Indeed, one of the major effects of land-use changes, including the afforestation, reforestation, and deforestation of land, is a change in soil organic matter, both as buildup and decomposition (Schlesinger 1997). The changes in stocks of soil carbon in a 4–5 year period are unfortunately within the errors of the survey techniques used for most ecosystems. Remote sensing techniques also appear inadequate for such purposes, since they have limited capability for estimating below-canopy processes such as soil respiration.

In this context, the direct long-term measurement of carbon fluxes by the eddy covariance technique (Baldocchi et al. 1988, Aubinet et al. 2000) offers a distinct possibility of assessing the carbon sequestration rates of forests and of land-use changes activities at the local scale. The technique can also provide a better understanding of the vulnerability of the carbon balance of ecosystems to climate variability, and at the same time can be used to validate ecosystem models and provide parameterization data for land surface exchange schemes in global models (Baldocchi et al. 1996). This chapter presents a synthesis of results in the EUROFLUX network concerning the carbon source/sink strength of European forests.

11.2 Spatial Distribution of Carbon Fluxes Across Europe

The EUROFLUX results for 1996–1998 show a sink strength of up to 6.6–6.7 t C ha^{-1} y^{-1} for two forests in Southern Europe and for a Sitka spruce plantation in Scotland and indicate that European old boreal forests are close to equilibrium and may switch from being a source one year to a sink of carbon in the next. Within the same biome, younger stands may still gain carbon, although at a lower rate than temperate and Mediterranean forests and fast-growing plantations (Table 11.1). Despite the wide range of species composition, stand structure, soils, tree age, site disturbance history, and year to year variability, a consistent latitudinal trend in NEE is found (Fig. 11.1) (Valentini et al. 2000).

Indeed a multivariate statistical analysis, based on different procedures (forward selection, stepwise selection, and maximum R-square improvement; SAS/STAT User's Guide 1989) on the effect of the single factors (latitude, precipitation, ecosystem type, elevation, mean annual temperature, age, management type, LAI) in explaining variation in NEE, shows that latitude is the most significant single variable (r^2=0.55, P <0.001). Latitude is not (per se) a phenomenological driving variable, however, it is a good proxy for the actions of a multiplicity of factors (e.g., radiation balance, length of growing season, frost events, regime of disturbances etc.). The trend indicates that high latitude forests generally show lower and more variable carbon sequestration rates than low latitude forests. The several forests growing within 50° and 52°N show a pronounced variability, with NEE ranging from an uptake of less than1 t C ha^{-1} y^{-1} (points 21-23, Table 11.1) to 6.6 t C ha^{-1} y^{-1} (points 1-2). In this latitudinal band, the variability can be related to stand, soil, and climate characteristics, ranging from continental to maritime. An intensively managed fast-growing fertilized spruce plantation (points 19 and 20, United Kingdom 1) falls off the latitudinal trend, with a higher uptake of carbon than more continental stands located at similar latitude.

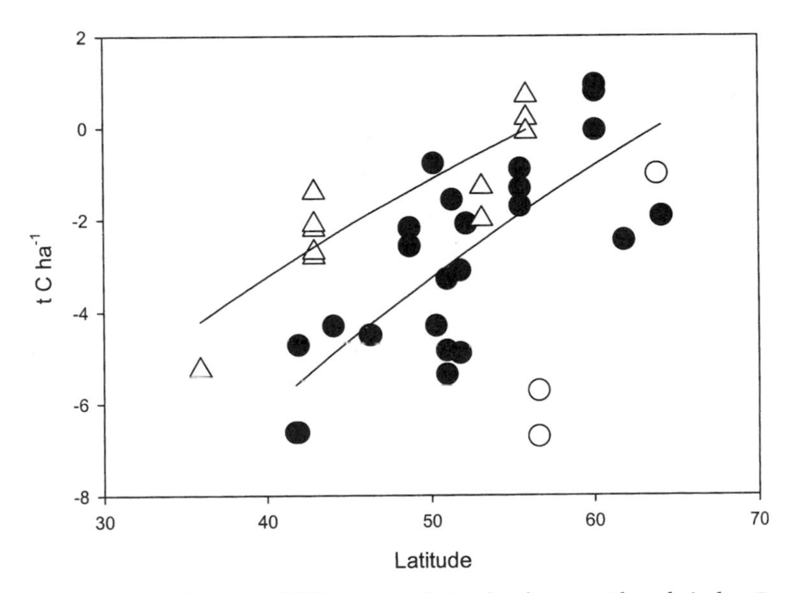

Fig. 11.1. Distribution of NEE across a latitudinal range. Closed circles: European Forest data, Open circles: artificial plantations (see text), Open triangles: North American forest data.

Directly measured data cover between 60 and 95 % of the yearly half-hour periods. Fluxes in stable conditions (usually u* < 0.2 m s^{-1}) and data gaps have been filled through site-based functional relationships using meteorological variables, such as global or photosynthetic active radiation (PAR) during the day and soil or air temperature during the night. Small gaps (a few half-hours) during single days were filled by simple interpolation. Site Denmark 1 (16-18) and Italy ext. (4) filled all the gaps by interpolation.

Table 11.1. Characteristics of the EUROFLUX sites

Site	Period of observation	NEE ($t C ha^{-1} y^{-1}$)	RE ($t C ha^{-1} y^{-1}$)
1 Italy 2	1997	-6.6	
2 Italy 1	28/06/96-27/06/97	-6.6	6.4
3 France 1	13/07/96-12/07/97	-4.3	8.0
4 Italy ext.	1998	-4.5	4.45
5 France 2	01/05/96-30/04/97	-2.2	7.9
6	1997	-2.6	9.9
7 Germany 1	01/05/97-30/04/98	-0.77	13
8 Belgium 1	Aug 96 - Jul 97	-4.3	10.1
9 Germany 2	1996	-3.3	8.3
10	1997	-4.8	9.5
11	1998	-5.4	9.7
12 Belgium 2	1997	-1.57	
13 Germany ext	1996	-3.1	8.3
14	1997	-4.9	9.6
15 Netherland 1	1997	-2.1	13.4
16 Denmark 1	1997	-0.9	10.6
17	01/06/96-31/05/97	-1.7	9.7
18	01/06/97-31/05/98	-1.3	11.1
19 Unit.King. 1	1997	-6.7	13.2
20	1998	-5.7	13.5
21 Sweden 1	1995	0.9	13.4
22	1996	-0.05	12.4
23	1997	0.8	14
24 Finland 1	1997	-2.45	7.6
25 Iceland 1	1997	-1	6.1
26 Sweden 2	1997	-1.9	10.65
2a Italy 1	1993-94	-4.7	5.4

11.3 Interpreting the Spatial Distribution of Carbon Sinks/Sources

Interestingly, a similar trend has been found by plotting the carbon flux data coming from North American sites, representing a range of different ecosystems distributed across a latitudinal belt similar to the European ones. It can be seen that at the same latitude, North American sites uptake less carbon than the corresponding ones in Europe, indicating the effects of continentality on climate. Indeed, at the same latitude in North America climate is colder and shows more pronounced extremes. On a similar ground it can be shown that the length of growing season is playing an important role in determining the sink strength of forest ecosystems Baldocchi et al. (2000 presented an example for deciduous forests of both the European and North American continent where the length of growing season is related to changes in the annual carbon uptake. More difficult is derivation of a similar relationships for conifers for which the length of growing season is not well defined.

Despite the large variation of NEE, gross primary production (GPP) is rather conservative across sites and latitude, indicating that other components of the carbon balance are responsible for the observed variation in NEE(Table 11.1). It is noteworthy that the young spruce plantation has the largest values of GPP, indicating strong stimulation of photosynthesis, while the young poplar plantation (point 25) that has not yet attained canopy closure in a colder climate at 64°N shows the smallest GPP.

The observed variation in NEE across sites can be explained by the relative importance of ecosystem respiration (RE) in relation to NEE. Indeed as it is shown in Fig.11.2 , NEE is linearly related to ecosystem respiration RE. Generally, our data show that while GPP tends to be constant across sites, annual ecosystem respiration increases with latitude despite the general decrease of mean annual air temperature (Table 11.1). This trend indicates that factors other than mean annual temperature control ecosystem respiration. In forests, total ecosystem respiration tends to be dominated by root and microbial soil respiration. Boreal soils contain a larger amount of soil organic matter (SOM) as a labile fraction that is readily decomposed (Kirschbaum 1995; Vogt et al. 1995) than do temperate soils (Anderson 1992), The effective temperature sensitivity (Q_{10}) of SOM decomposition is much higher in colder than in warmer climates, and temperature increases in cold regions are likely to affect decomposition rates more than net primary productivity. There is also evidence that northern latitudes have warmed by more than 4 °C degrees, while southern latitudes have warmed less (Chapman and Walsh 1993). This may have resulted in non-steady-state conditions for SOM that can explain relative enhancement of respiration in the north compared to the south. In this respect, land-use change and site history can also play a role. For example, site Sweden 1 (points 21–23) is losing carbon as a result of past soil

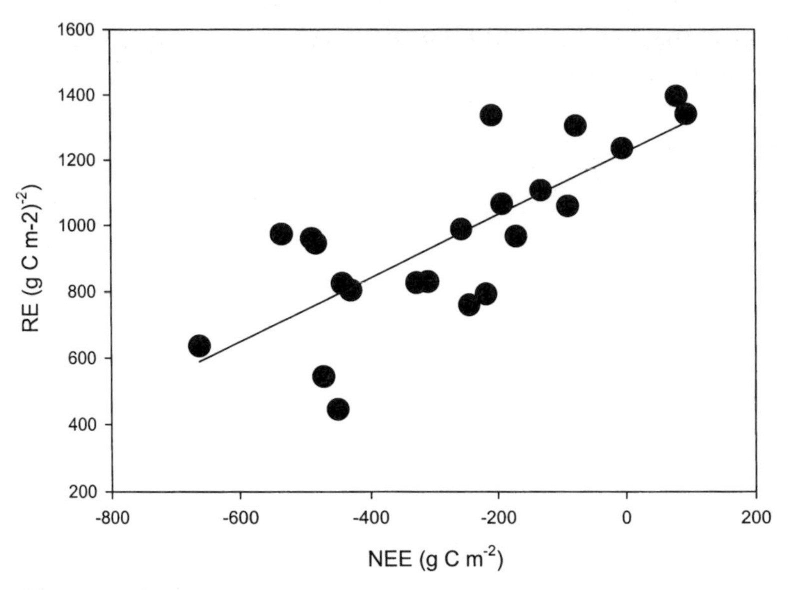

Fig. 11.2. Relation between Ecosystem Respiration (RE) and Net Ecosystem Exchange (NEE) for the Euroflux forests

drainage, while the high respiration rates of the maritime spruce plantation may be linked to preparation of the site by plowing, to the favorable Atlantic climate, and to fertilization. Furthermore, the relatively low rates of respiration of the southern sites may be the result of drought limitations to soil respiration (Schlenter and Van Cleve 1985; Burton et al. 1998) (see also Chap. 3). The carbon balance is ultimately a delicate equilibrium between the two large fluxes in photosynthesis and respiration, and this appears to be particularly true for boreal European ecosystems, making them very vulnerable to disturbances in climate. Indeed, annual variability for these high latitude sites is very pronounced, as shown by the remarkable variation in NEE from year to year: warm winters tend to switch old boreal stands from a sink to a source of carbon by increasing the annual amount of respiration (see site Sweden 1 in Table 11.1). In other boreal ecosystems, year-to-year changes in timing the soil's spring thaw play a major role in the carbon balance.

The direct flux estimates of carbon exchange provide a useful tool for understanding the overall carbon balance processes of terrestrial ecosystems. Indeed, partial accounting of carbon dynamics can easily lead to erroneous conclusions. For example, plant biomass is currently increasing in all the EUROFLUX sites, even though some of these sites have a carbon budget close to neutral and one is losing carbon on a yearly basis. Similarly, the increases in plant growth at northern latitudes estimated by remote sensing of normalized difference vegetation index (NDVI) (Myneni et al. 1997) must be examined critically in the light of these results, as the authors also pointed out in their conclusions when they referred to the need for considering ecosystem respiration. Forest inventory-based carbon balance estimates should also be carefully examined in relation to comprehensive carbon budget accounting (Nabuurs et al. 1997).

In general, an integrated multi-scale approach involving tower flux measurements, remote sensing analysis, inversion modeling, and ecosystem inventories is required to survey the carbon metabolism of the terrestrial biosphere (Running 1999).

References

Anderson JM (1992) Responses of soils to climate change. Adv Ecol Res 22:163–210

Aubinet M, Grelle A, Ibrom A, Rannik Ü et al (2000) Estimates of the Annual Net Carbon and Water Exchange of Forests: the EUROFLUX Methodology Adv. Ecol. Research, 30:113–175

Baldocchi DD, Hicks BB, Meyers TP (1988) Measuring biosphere-atmosphere exchanges of biologically related gases with micrometeorological methods. Ecology 69:1331–1340

Baldocchi DD, Valentini R, Running S, Oechel WC, Dahlman R (1996) Strategies for measuring and modelling carbon dioxide and water vapour fluxes over terrestrial ecosystems. Global Change Biol 2:159–167

Burton AJ, Pregitzer KS, Zogg GP, Zak DR (1998) Drought reduces root respiration in sugar maple forests. Ecol Appl 8:771–778

Chapman WL, Walsh JE (1993) Recent variations of sea ice and air temperatures in high latitudes. Bull Am Meteorol Soc 74:33–47

Chen WJ, Black TA, Yang PC, Barr AG et al (1999). Effects of climatic variability on the annual carbon sequestration by a boreal aspen forest. Global Change Biol 1:41–53

Ciais P, Tans PP, Trolier M, White JWC, Francey RJ (1995) A large Northern hemisphere terrestrial CO_2 sink indicated by $^{13}C/^{12}C$ ratio of atmospheric CO_2. Science 269:1098–1102

Dixon RK, Brown S, Houghton RA, Solomon AM, Trexler MC, Wisniewski J (1994a) Carbon pools and flux of global forest ecosystems. Science 263:185–190

Fan S, Gloor M, Mahlman J, Pacala S, Srmiento J, Takahashi T, Tans P (1998) A large terrestrial carbon sink in North America implied by atmospheric and oceanic carbon dioxide data and models. Science 282:442–446

Goulden ML, Munger JW, Fan S-M, Daube BC, Wosfy WC (1996) Measurements of carbon sequestration by long-term eddy covariance: methods and critical evaluation of accuracy. Global Change Biol 2:169–181

Goulden ML, Wofly SC, Harden JW, Trumbore SE et al (1998) Sensitivity of boreal forest carbon balance to soil thaw. Science 279:214–217

Hanson PJ, Wullschleger SD, Bohlmann SA, Todd DE (1993) Seasonal and topographic patterns of forest flora CO_2 efflux from an upland oak forest. Tree Physiol 13:1–15

IGBP Terrestrial Carbon Working Group (1998) The terrestrial carbon cycle: implications for the Kyoto protocol. Science 280:1393–1394

IPCC (2000) Land use, land-use change, and forestry; Special report. Cambridge University press, 377 pp

Kauppi PE, Mielikäinen K, Kuusela K (1992) Biomass and carbon budget of European forests, 1971 to 1990. Science 256:70–74

Kirschbaum MU (1995) The temperature dependence of soil organic matter decomposition, and the effect of global warming on soil organic C storage. Soil Biol Biochem 6:753–760

Law BE, Ryan MG, Anthoni PM (1999) Seasonal and annual respiration of a ponderosa pine ecosystem. Global Change Biol 5:169–182

Lindroth A, Grelle A, Morén A-S (1998) Long-term measurements of boreal forest. Global Change Biol 4:443–450

Moncrieff JB, Malhi Y, Leuning R (1996) The propagation of errors in long-term measurements of land atmosphere fluxes of carbon and water. Global Change Biol 2:231–240

Myneni RB, Keeling CD, Tucker CJ, Asrar G, Nemani RR (1997) Increased plant growth in the northern high latitudes from 1981 to 1991. Nature 386:698–702

Nabuurs GJ, Pavinen R, Sikkema R, Mohren GMJ (1997) The role of European forests in the global carbon cycle–a review. Biomass Bioenergy 13:345–358

Oechel WC et al (1993) Recent change of arctic tundra ecosystems from a net carbon dioxide sink to a source. Nature 361:520–523

Running SW (1998) A blueprint for improved global change monitoring of the terrestrial biosphere. Earth Observer 10:8–12

Running SW, Baldocchi DD, Turner D, Gower ST, Bakwin P, Hibbard K (1999) A global terrestrial monitoring network, scaling tower fluxes with ecosystem modeling and EOS satellite data. Remote Sensing of the Environment. 70:108–127

Schlenter RE, Van Cleve K (1985) Relationship between CO_2 evolution from soil, substrate temperature, and substrate moisture in four mature forest types in interior Alaska. Can J For Res 15:97–106

Schlesinger WH (1997) Biogeochemistry: an analysis of global change. Academic Press, San Diego, CA, pp 161–165

Schulze ED, Heimann M (1998) Carbon and water exchange of terrestrial systems. In: Galloway J, Melillo J (eds) Asian change in the context of global change. IGBP Book series. Cambridge Univ Press, Cambridge, pp 145–161

Scott Denning A, Fung IY, Randall D (1995) Latitudinal gradient of atmospheric CO_2 due to seasonal exchange with the land biota. Nature 376:240–243

Valentini R; Matteucci G; Dolman AJ; Schulze ED et al (2000). Respiration as the main determinant of carbon balance in European forests. Nature 404, 861–865

Villar R, Held AA, Merino J (1994) Dark leaf respiration in light and darkness of an evergreen and a deciduous plant species. Plant Physiol 107:421–427

Vogt KA, Vogt DJ, Brown S, Tilley JP et al (1995) Dynamics of forest floor and soil organic matter accumulation in boreal, temperate, and tropical forests. In: Lal R, Kimble J, Levine E, Stewart BA (eds) Soil management and greenhouse effect. CRC Press, Boca Raton, pp 159–178

12 Climatic Influences on Seasonal and Spatial Differences in Soil CO_2 Efflux

I.A. Janssens, S. Dore, D. Epron, H. Lankreijer, N. Buchmann, B. Longdoz, J. Brossaud, L. Montagnani

12.1 Modeling the Temporal Variability of Soil CO_2 Efflux

The efflux of CO_2 from the soil is characterized by large seasonal fluctuations due to seasonal changes in root and microbial respiration. Although several biotic and abiotic factors influence root and microbial activity (see Chap. 3), the control exerted by temperature, and in some cases moisture, is usually dominant. In the absence of water stress, variation in soil temperature accounts for most of the seasonal and diurnal variation in soil CO_2 efflux. Where water stress frequently occurs, soil CO_2 efflux may not be correlated with soil temperature, but with its moisture content (Rout and Gupta 1989). Thus, CO_2 release from the soil appears to respond to temperature or moisture, whichever is most limiting at the time of measurement (Schlentner and van Cleve 1985).

Interest in the rate-controlling factors of soil CO_2 efflux is growing because of the potential for climate change to increase the flux of CO_2 from the ecosystems to the atmosphere (Raich and Potter 1995). Information on the relationship between soil CO_2 efflux and its driving variables is also needed for the development of models of value for the assessment of climate-change effects and for the interpretation of the processes involved.

Soil CO_2 efflux has been successfully modeled with process-based models that simulate root and microbial respiration separately (Simunek and Suarez 1993; Freijer and Leffelaar 1996; Fang and Moncrieff 1999). These one-dimensional models use Fick's diffusion law to describe the transport of CO_2 in soil, dependent on soil characteristics, soil water content, and temperature. However, because soil CO_2 efflux can be successfully modeled using only the temperature and moisture relationships (Keith et al. 1997; Epron et al. 1999a; Janssens et al. 1999; Buchmann 2000; Longdoz et al. 2000), empirical models are most frequently used to simulate soil CO_2 efflux.

Ecological Studies, Vol. 163
R. Valentini (Ed.) Fluxes of Carbon, Water and Energy of European Forests

12.1.1 Temperature Responses

Microbial communities and plant root systems are particularly sensitive to changes in soil temperature (Killham 1994). Both specific respiration rates and microbial and root biomass are positively affected by elevated temperature, and in most ecosystems a positive correlation between soil CO_2 efflux and temperature is observed (Singh and Gupta 1977), leading to large seasonal fluctuations in the flux rates. This positive relationship between soil CO_2 efflux and temperature was also observed in most of the EUROFLUX forests.

There is, however, no consensus on the exact form of the relationship (Lloyd and Taylor 1994). Soil CO_2 efflux has been modeled using linear (Witkamp 1966; Anderson 1973), power (Kucera and Kirkham 1971), and sigmoid (Schlentner and van Cleve 1985; Janssens et al. 1999; Matteucci et al. 2000) relationships with temperature (Table 12.1). However, exponential rela-

Table 12.1. List of relationships between soil CO_2 efflux and soil temperature that have been applied in empirical models

Relationship	Equation[a]	Comments
Linear	$SR = a + b \times Temp$	
Power function	$SR = a \times Temp^b$	
Sigmoid	$SR = a + \left(\frac{1}{b + c\left(\frac{10 - Temp}{10}\right)} \right)$	
Exponential	$SR = a \times e^{b \times Temp}$	
Q_{10} function	$SR = SR_{ref} \times Q_{10}\left(\frac{Temp - Temp_{ref}}{10}\right)$	
Arrhenius type	$SR = SR_{10} \times e\left(Ea \times \left(\frac{Temp - 283.15}{283.15 \times Temp \times R_g}\right)\right)$	Temp in K R_g = gas constant = 8.314 J mol^{-1} K^{-1} Ea = Activation Energy $\left(\mathrm{J\ mol^{-1}}\right) = a \times \left(\frac{Temp}{Temp - 227.13}\right)$

[a] *SR*, soil CO_2 efflux; *Temp*, soil temperature; SR_{ref}, soil CO_2 efflux at reference temperature; $Temp_{ref}$, reference temperature; SR_{10}, soil CO_2 efflux at 10 °C; *a*, *b*, and *c* are constants.

tionships, especially the Q_{10} relationship, are more frequently used to predict respiration rates from temperature (Peterjohn et al. 1994; Raich and Potter 1995; Boone et al. 1998; Davidson et al. 1998; Epron et al. 1999a; Buchmann 2000; Morén and Lindroth 2000).

The median value of the reported Q_{10} values for forest soil CO_2 efflux is 2.4, but the range is very broad (Schleser 1982; Raich and Schlesinger 1992; Kicklighter et al. 1994; Kirschbaum 1995). The use of the Q_{10} relationship often has been criticized because the Q_{10} factor itself decreases with increasing temperature and depends on soil moisture conditions (Howard and Howard 1993). This problem may be mitigated by using a temperature-dependent (McGuire et al. 1992) or a moisture-dependent Q_{10} factor (Carlyle and Than 1988). However, Lloyd and Taylor (1994) reported that the assumption of an exponential Q_{10} relation between soil respiration and soil temperature is invalid and systematically leads to underestimated fluxes at low temperatures, and overestimated fluxes at high temperatures. They found that soil respiration was better described by an Arrhenius-type relationship in which the activation energy decreases with increasing soil temperature.

12.1.2 Sensitivity of Empirical Models to the Type of Temperature Regression

To test the use of different temperature response functions, we fitted three of the equations listed in Table 12.1 (i.e., Q_{10}, Arrhenius, and sigmoid) to data from the BE-2 site (Table 12.2). With the parameterized temperature response functions we then simulated soil CO_2 efflux from the measured soil temperatures (Fig. 12.1).

In general, all regressions fitted the data well (Table 12.2, Fig. 12.1). Although the different regressions did not result in significant differences in the total annual soil CO_2 efflux (Table 12.2), we observed large deviations during the year. In mid-summer, when soil temperature exceeded the temperature range observed during the fitting exercise, the Arrhenius and Q_{10} relationships produced significantly higher estimates of soil CO_2 efflux than the sigmoid relationship (Fig. 12.1). At temperatures below 5 °C the opposite was observed. These differences highlight the importance of measuring soil CO_2 efflux in the widest possible temperature range in order to obtain an optimal parameterization of the models. However, high soil temperature often coincides with low moisture availability, confounding the temperature response of soil CO_2 efflux. Therefore, it may be difficult to get the necessary data, especially at drier sites, and an irrigation experiment during warm, drought-stressed periods could be very informative.

The sigmoid response function provided the best fit at the BE-2 site. However, this was not the case at other sites. At the FR-1 site, the sigmoid function showed the weakest fit with the data, and at the GE-1 site an exponential func-

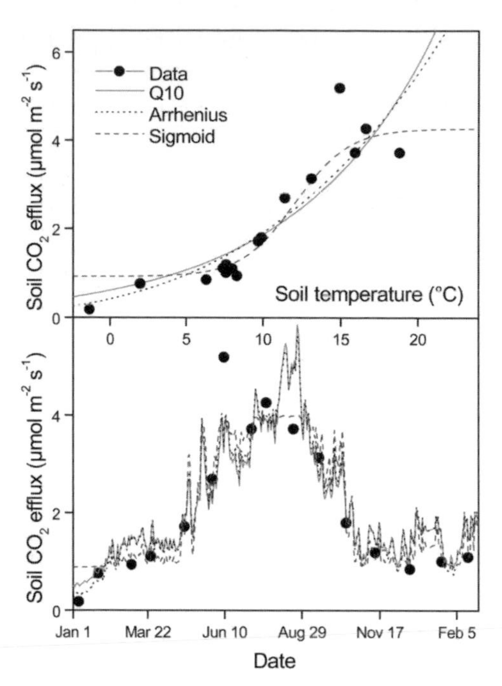

Fig. 12.1. *Top panel* Plot of the temperature response functions (fitted to the data from the BE-2 oak site) that were applied to simulate the annual soil CO_2 efflux rates in Table 12.2. Soil temperature was measured at 5 cm. *Bottom panel* Time series of the measured and simulated soil CO_2 efflux rates in the BE-2 oak site. Regressions and parameters are given in Tables 12.1 and 12.2. Data shown are for 1997 and the beginning of 1998

Table 12.2. List of the regressions that were fitted to the data from the BE-2 site and the obtained parameter values. Regressions are listed in Table 12.1. Total annual flux is in g C m^{-2} $year^{-1}$, range is in µmol CO_2 m^{-2} s^{-1}, R2adj is adjusted R^2

Regression	Parameter values	Total annual flux	Range	R2adj	*n*
Q_{10}	SR_{ref}=1.87, Q_{10}=2.87	825	0.47–5.85	0.685	23
Arrhenius	SR_{10}=1.91, a=1.65 × 10^4	820	0.27–5.37	0.732	23
Sigmoid	a=0.888, b=0.320, c=891	819	0.89–4.00	0.832	23

tion was found to fit the measured fluxes best. Whichever model is selected to simulate soil CO_2 efflux from soil temperature, the annual totals are not likely to differ significantly. However, inappropriate models may introduce significant errors in the estimated fluxes during the year.

12.1.3 Moisture Responses

Soil moisture may negatively affect soil CO_2 efflux rates when it becomes either very high (poor aeration and reduced CO_2 diffusivity) or too low (desiccation stress). In drought-stressed ecosystems, soil CO_2 efflux usually peaks in spring and after rain events (Matteucci et al. 2000). Soil temperature is often poorly correlated with soil CO_2 efflux (Fig. 12.2), and soil moisture may become the best predictor of soil CO_2 efflux (Rout and Gupta 1989; Holt et al. 1990; Keith et al. 1997).

The shape of the moisture response curve and the moisture content at which maximum respiration occurs depend on an array of site-specific factors such as soil texture and structure, amount and type of organic matter, and soil temperature (Howard and Howard 1993). Nonetheless, the response of soil CO_2 efflux to soil moisture has been successfully described using linear (Kowalenko et al. 1978; Rout and Gupta 1989; Holt et al. 1990; Epron et al. 1999a), exponential (Keith et al. 1997; Davidson et al. 1998), power (Skopp et al. 1990), Gompertz (Janssens et al. 1999), and first- (Hanson et al. 1993), and second-degree inverse polynomial functions (Bunnell et al. 1977; Schlentner and van Cleve 1985; Carlyle and Than 1988; Table 12.3).

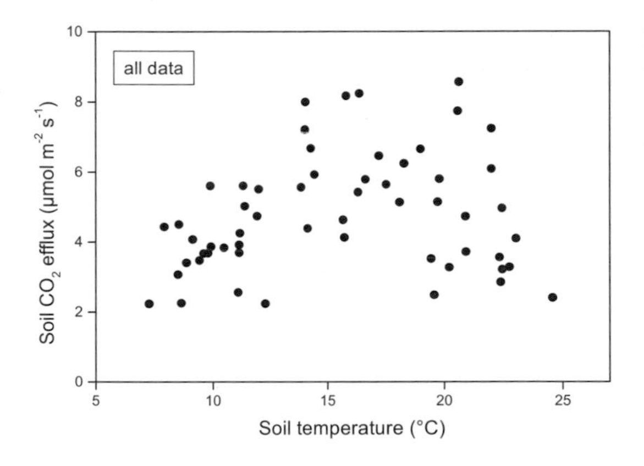

Fig. 12.2. The temperature relationship of soil CO_2 efflux in the IT-2 site. Soil temperature was measured at 5 cm depth

Table 12.3. List of moisture response functions of soil CO_2 efflux that have been applied in empirical models. $f(M)$ = moisture response function, M = moisture content, and a, b, and c are constants

Relationship	Equation	Comments
Linear	$f(M)=a+b\times M$	
Exponential	$f(M)=a\times e^{b\times M}$	
Power	$f(M)=a\times M^b$	
Gompertz function	$f(M)=e^{-e^{(a-b\times M)}}$	
First-degree inverse polynomial	$f(M)=\frac{a\times b\times M}{(a\times M)+b}$	
Second-degree inverse polynomial	$f(M)=\frac{M}{a+M}\times\frac{b}{b+M}\times c$	a = M at half field capacity b = M at half water holding capacity c = scaling factor

12.1.4 Sensitivity of Empirical Models to the Type of Moisture Regression

We selected the IT-2 site to test a number of models for simulating the moisture dependence, because this was the driest of the EUROFLUX sites and soil CO_2 efflux was not correlated with soil temperature (Fig. 12.2). Thus, variability in the flux simulations from different moisture regressions were expected to be large at this site.

First, a Q_{10} temperature relation was derived, using data that were not limited by soil moisture, i.e., when soil moisture was above 10 vol% (Fig. 12.3, top panel). With the obtained Q_{10} function we then normalized all data for the influence of temperature and examined the dependence on soil moisture (Fig. 12.3, bottom panel). The normalized data can easily be separated into two groups: above a volumetric moisture content of 10% (solid circles in Fig. 12.3) they appear to be uncorrelated with moisture, while below that threshold (open circles) a strong correlation is observed.

To test different moisture response curves, we fitted four of the equations listed in Table 12.3 (i.e., linear, Gompertz, and first- and second-degree inverse polynomials) to the data normalized for temperature (Fig. 12.3, Table 12.4). Except for the linear function, all regressions had similar shapes and they all fitted the normalized data rather well. Thus, as long as the mois-

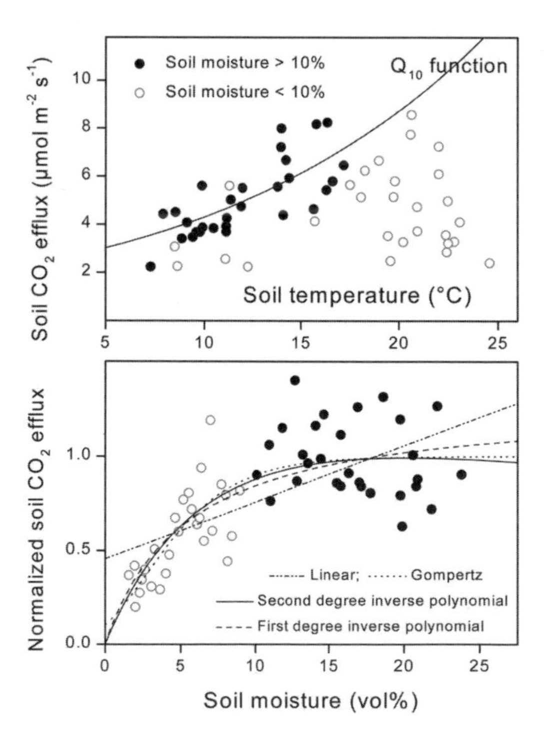

Fig. 12.3. *Top panel* Temperature relationship of soil CO_2 efflux in the IT-2 site. The Q10 function (*solid line*) was fitted to the data without water stress (*solid circles*, soil moisture above 10 vol%). Soil temperature was measured at a 5 cm depth. *Bottom panel* Plot of the moisture response functions and the normalized soil fluxes versus soil moisture at the IT-2 site. The moisture regressions and parameterization are given in Tables 12.3 and 12.4. Soil moisture was measured in the upper 15 cm with a TRIME system (IMKO GmbH, Germany)

Table 12.4. List of the regressions that were fitted to the data from the IT-2 site and the obtained parameter values. Regressions are listed in Table 12.3. R^2adj is adjusted R^2

Regression	Parameter values	R2adj	*n*
Linear	a=0.457, b=0.03	0.425	56
Gompertz	a=0.824, b=0.308	0.612	56
First-degree inverse polynomial	a=0.244, b=1.29	0.597	56
Second-degree inverse polynomial	a=20.2, b=20.2, c=3.97	0.601	56

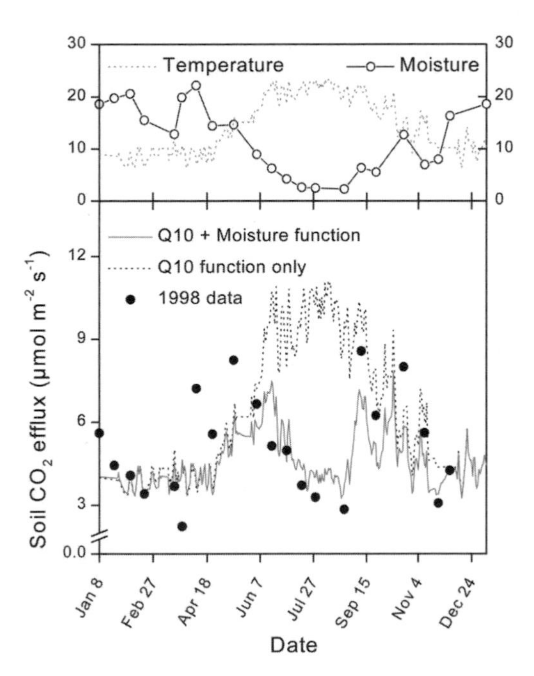

Fig. 12.4. *Top panel* Time series (1997) of mean daily soil temperature and soil moisture at the IT-2 site. Soil temperature was measured at 5 cm, soil moisture in upper 15 cm. *Bottom panel* Time series (1997) of the measured and simulated soil CO_2 efflux rates at the IT-2 site. The Q_{10} function is given in Fig. 12.3. The moisture regression applied in this simulation was a first-degree inverse polynomial (see Tables 12.3 and 12.4 for equation and parameter values)

ture response functions accurately fit the data below the threshold where moisture stress occurs, and approximate the data above that threshold, no significant deviations in the simulated fluxes are to be expected.

As in most drought-stressed ecosystems, soil CO_2 efflux at the IT-2 site peaked in spring and autumn, was low in winter because of moderate temperatures, and diminished (in comparison with spring and autumn) in summer when drought occurred (Fig. 12.4). The Q_{10} function alone [$SR=4.3 \times 2.03^{(Temp-10/10)}$] significantly overestimated soil CO_2 efflux in dry periods. When including a moisture response function (first degree inverse polynomial, see Table 12.4), the empirical model fitted the data better and explained about 60 % of the temporal variability (R^2=0.579, P=0.006, n=21).

Because predictions of soil CO_2 efflux based solely on temperature do not account for reductions due to moisture limitation, they are likely to overestimate soil CO_2 efflux when drought occurs. At the IT-2 site, the reduction in total annual soil CO_2 efflux induced by drought was 26 % (Fig. 12.4).

Even in forests with less pronounced drought stress, where drought occurs infrequently or moderately, the inclusion of moisture regressions in empirical models will increase their fit with the data. Moderate drought may also partly explain the inter-annual variability in soil CO_2 efflux (Epron et al. 1999a).

12.1.5 Additional Comments on Empirical Models

Empirical models based on soil temperature and moisture usually explain about 60–90 % of the temporal variability in soil CO_2 efflux. The unexplained variability may be due to several factors. For instance, the annual pattern of above- and belowground litterfall may not coincide with that of temperature. Rapid decomposition of the labile components of this fresh litter may thus confound the climate dependency of soil CO_2 efflux (Trumbore et al. 1996).

Some of the variability might also be related to the seasonal changes in the basal rates and temperature sensitivity (Q_{10} coefficient) of respiration that have been observed in forests (Hagihara and Hozumi 1991; Goulden and Crill 1997; Lavigne et al. 1997). Temporal changes in the basal ecosystem respiration rates may be due to the fluctuating root and microbial biomass. Although root production in most tree species in temperate climates is related to soil temperature, root growth often peaks in early summer and diminishes in midsummer due to unfavorable moisture conditions (Lyr and Hoffmann 1967). Also, microbial biomass may vary considerably during the year, usually peaking in spring or summer (Wardle 1998). The annual pattern of root and microbial biomass might therefore be unrelated to that of soil temperature, and increase the scatter in the relationship between soil CO_2 efflux and temperature, thus reducing the predictive power of the empirical models. In addition, roots and microbes may have different temperature sensitivities; temporal decoupling of root and microbial processes might therefore enhance the variability in the temperature response of soil CO_2 efflux.

Another source of unexplained variability in the temperature response may be related to the temperature profile within the soil. Whereas the soil temperature measurements used in the models (usually in the upper soil layers) are coupled to air temperature, temperatures lower in the soil (0.5 m) are not. Although most respiratory activity occurs in the upper soil layers, forests may have significant CO_2 production in the deeper layers (Trumbore et al. 1995). Respiration occurring farther down in the soil will thus be poorly correlated with the near-surface temperature, and confound the temperature response of soil CO_2 efflux.

Besides the natural phenomena mentioned above, an inappropriate measurement depth of soil temperature may also add to the unexplained temporal variability in soil CO_2 efflux. If the measurement depth is too deep (if most respiratory activity occurs above it) then soil temperature will lag behind soil CO_2 efflux, on both a daily and an annual time scale. If the measurement depth is too close to the surface, the opposite will be observed. As a result of this time lag, hysteresis will occur when plotting the complete diurnal or annual cycle of soil CO_2 efflux versus soil temperature: morning (or spring) fluxes will differ from the evening (or fall) fluxes at the same temperature (Janssens et al. 1998). Inappropriate measurement depths will therefore artificially increase the variability in the relationship between the measurements of soil CO_2 efflux and temperature.

To cover the widest possible temperature range, most empirical models are based on in situ measurements of soil CO_2 efflux from all seasons. At this large time scale, respiration is affected not only by temperature and moisture, but also by the changing size of the root and microbial biomass. Thus, when applied to simulate diurnal fluctuations, empirical models using data obtained in both winter and summer may overestimate diurnal fluctuations, because daily changes in root and microbial biomass are much smaller than seasonal changes. This problem may be overcome by smoothing either the model inputs or the outputs over longer time periods. The proper smoothing period will vary from site to site, but can be retrieved by fitting the model simulations to the diurnal patterns of soil CO_2 efflux (Janssens et al. 2001a).

12.2 Spatial Variability Among the EUROFLUX Forests

Large differences in total annual soil CO_2 efflux were observed among the different EUROFLUX forests (Table 12.5). As was shown in the previous section, the use of different temperature and moisture regressions to simulate these fluxes does not contribute to these differences. However, no standard methodology for measuring soil CO_2 efflux was applied in the EUROFLUX network (see Chap. 3). Different methodologies typically result in different fluxes (Norman et al. 1997; Le Dantec et al. 1999; Janssens et al. 2000), and the reader should bear in mind that this lack of standardization is likely to have contributed to the variability found in most of the relationships with the influencing biotic and abiotic factors. None the less, most of the spatial variability in soil CO_2 efflux is likely to be related to differences in climate, vegetation, and site characteristics.

A positive trend was observed in the correlation between soil CO_2 efflux and soil pH, while higher litter layer C/N ratios tended to have a negative

Table 12.5. Overview of soil CO_2 efflux rates and a number of related fluxes and site characteristics in the EUROFLUX sites. TSR = Total annual soil C efflux (g C m^{-2} $year^{-1}$); SR_{10} = soil CO_2 efflux at 10 °C in µmol m^{-2} s^{-1} (for IT1 and IT2 in the absence of drought stress); Tsoil = mean annual soil temperature, at or near a depth of 5 cm (°C); Precip = total annual precipitation (mm); pHsoil = pH_{KCl} of upper mineral soil; TSC = total soil C content (kg C m^{-2}); C/Nlit = C/N ratio of litter layer; TER = total annual ecosystem respiration (g C m^{-2} $year^{-1}$); NPP = aboveground net primary productivity (g C m^{-2} $year^{-1}$)

Site+year	Code	TSR	SR_{10}	Q_{10}	Tsoil	Precip	pHsoil	TSC	C/Nlit	TER	NPP
IT1-(1996–1997)	X	879	4.1	2.2	2.5	1180	5.6	22.7	20.7	636	312
IT2-(1997)	Y	1456	4.3	2.0	14.5	–	5.8	–	–	–	793
ITex-(1998)	A	1379	4.9	3.4	4.0	–	4.4	–	20.7	–	995
BE1Douglas-(1997)	B1	–	1.4	3.0	8.1	792	–	–	–	–	–
BE1beech-(1997)	B2	844	2.6	2.4	8.1	792	–	–	–	1095	–
BE2pine1-(1997)	C1	281	0.7	2.6	10.6	662	2.6	14.4	27.1	–	–
BE2pine2-(1997)	C2	338	0.8	2.4	10.6	662	2.7	14.9	25.3	–	–
BE2oak1-(1997)	C3	578	1.2	3.6	10.6	662	2.6	–	21.3	–	–
BE2oak2-(1997)	C4	769	1.9	3.1	10.6	662	3.5	–	19.6	–	–
FR1-(1996)	F1	509	1.6	3.4	9.1	672	4.8	–	13.4	793	402
FR1-(1997)	F2	685	1.8	3.8	9.6	871	4.8	–	13.4	988	556
FR1-(1998)	F3	713	2.2	4.0	9.4	–	4.8	–	13.4	1235	364
DK-(1996)	D1	370	1.3	6.3	5.8	–	4.4	19.6	17.5	967	–
DK-(1997)	D2	460	1.3	2.8	9.1	510	4.4	19.6	17.5	1107	–
DK-(1998)	D3	–	1.7	5.0	7.4	–	4.4	19.6	17.5	–	–
GE1–47y-(1998)	G1	709	2.4	2.4	6.1	–	3.6	–	22.0	1373	1060
GE1–87y-(1998)	G2	740	2.6	3.2	6.1	–	–	–	–	–	–
GE1–111y-(1998)	G3	859	3.0	2.9	6.1	–	–	–	–	–	–
GE1–146y-(1998)	G4	624	2.1	2.4	6.1	–	–	–	–	–	534
GE-Kiel-(1997)	H	–	1.6	3.9	7.6	–	3.5	–	23.0	742	594
SE1-(1995)	S1	1250	–	–	–	437	3.4	–	–	1341	1247
SE1-(1996)	S2	1220	4.3	4.8	4.8	393	3.4	–	–	1236	1241
SE1-(1998)	S3	1080	3.4	2.8	5.7	–	3.4	–	–	–	–

effect on soil fluxes. In this chapter, however, we focus on the effects of precipitation, temperature, and NPP on the variability in annual soil CO_2 efflux among the EUROFLUX forests.

12.2.1 Effect of Precipitation

Globally, soil CO_2 efflux correlates significantly with annual precipitation (Raich and Schlesinger 1992). Among the EUROFLUX forests, however, this relation was not observed (Fig. 12.5). It should be understood that annual precipitation is a poor estimate of moisture availability. In the drought-stressed Mediterranean sites, e.g., precipitation is relatively large, but occurs in winter. Thus, moisture availability is quite high in winter, but extremely low in summer, which annual precipitation does not indicate. We believe that the actual evapotranspiration (ET) rate, or the ratio of actual/potential ET, provides better estimates of moisture availability, but these were not available.

12.2.2 Effect of Soil Temperature

Although temperature was positively correlated with soil CO_2 efflux in most of the sites, we found no positive trend among the different forests (Fig. 12.6). With the exception of the IT-2 site (Y in Fig. 12.6), total annual soil CO_2 efflux was even higher in the colder sites (Table 12.5). The enhanced fluxes in the colder sites may be explained by a combination a different factors. Firstly, in

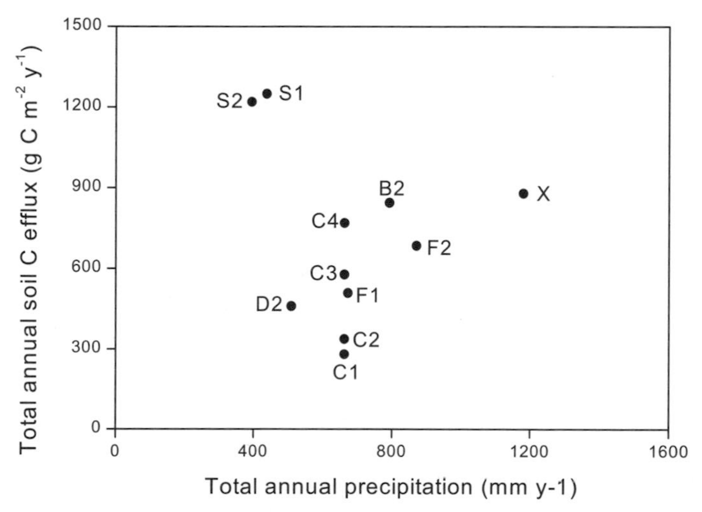

Fig. 12.5. Total annual soil CO_2 efflux versus total annual precipitation in the different EUROFLUX sites. Codes for different sites are explained in Table 12.5

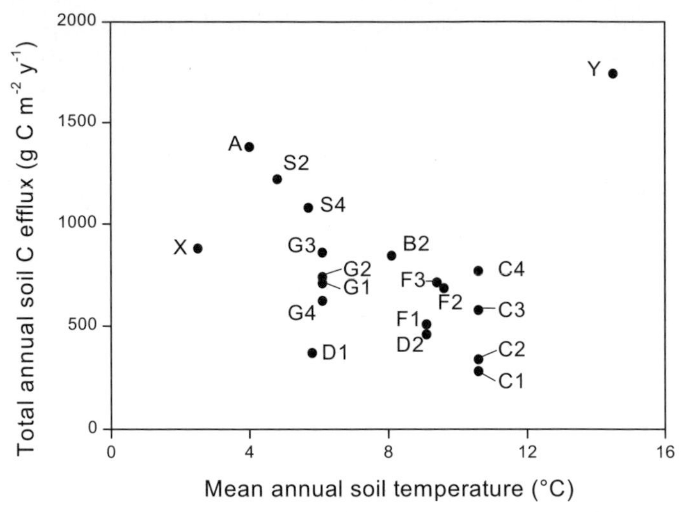

Fig. 12.6. Total annual soil CO_2 efflux versus mean annual soil temperature in the different EUROFLUX forests. Codes for different sites are explained in Table 12.5

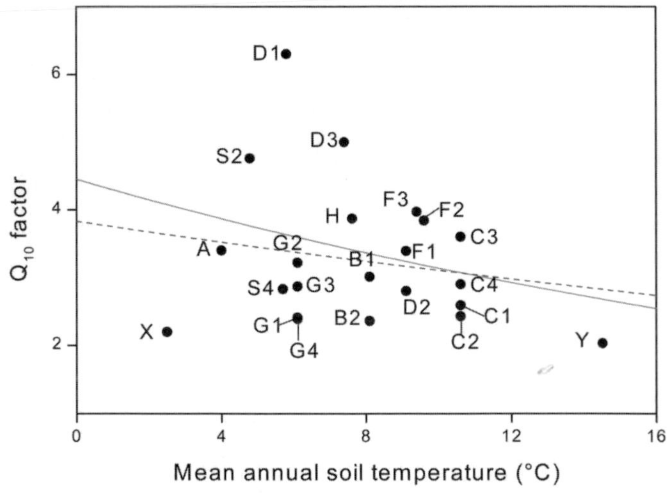

Fig. 12.7. Temperature sensitivity (Q_{10} factor) of soil CO_2 efflux versus mean annual soil temperature in the different EUROFLUX forests. Codes for different sites are explained in Table 12.5. The *plotted lines* represent negative exponential functions fitted to the entire data set (*dashed line*), and to the data set without the drought-stressed sites (*solid line*). See text for discussion of parameter values

agreement with previous studies (Schleser 1982; Kirschbaum 1995), we observed a (non-significant) negative trend in the temperature sensitivity of soil CO_2 efflux with increasing temperature (Fig. 12.7). Recent increases in global temperature will therefore have stimulated soil CO_2 efflux more at low than at high temperatures.

Secondly, soil CO_2 efflux is positively correlated with productivity (Raich and Schlesinger 1992), as is extensively discussed in Janssens et al. (2001b). Recent increases in temperature and atmospheric CO_2 concentrations have stimulated NPP directly, through enhanced photosynthetic rates, and indirectly, through the lengthening of the growing season (Myneni et al. 1997). Because temperature increases are larger in colder regions, productivity is stimulated more, and thus also soil CO_2 efflux is expected to be enhanced more.

Thirdly, soil drainage at the Swedish site (S2 and S4 in Fig. 12.6) has resulted in increased aeration that may have stimulated decomposition of native SOM.

Fourthly, enhanced decomposition rates in response to increasing temperature results in the release of more nutrients. This fertilizing effect will stimulate productivity (Schimel et al. 1996), but will probably have a stronger effect in the cold and nutrient-limited northern ecosystems than in the nitrogen-saturated ecosystems of western Europe.

Fifthly, the use of different methodologies is also likely to have contributed to this observation. Soil respiration at the two coolest sites was measured with the PP-Systems SRC-1 soil chamber, which may overestimate soil fluxes (Chap. 3; Le Dantec et al. 1999; Janssens et al. 2000).

Nevertheless, even among forests from temperate climates, we did not observe the expected positive relationship between annual soil CO_2 efflux and soil temperature (Fig. 12.6). The use of different techniques to measure soil CO_2 efflux (Chap. 3) is probably an important source of variability, but even among sites with identical measurement systems, no positive relationship with temperature was detected. Other factors that may have confounded the relation with temperature are differences in vegetation cover, site productivity, soil acidity and texture, quality and quantity of soil organic matter, and, of course, drought stress.

Both temperature sensitivity (Q_{10}) and basal rate (SR_{10}, efflux at 10 °C) of soil CO_2 efflux tended to decrease with mean annual soil temperature (Figs. 12.7, 12.8). A negative exponential function [$y=a \times \exp^{(b \times x)}$] was fitted to the data. The negative trend in Q_{10} with increasing soil temperature (dashed line in Fig. 12.7) was very weak and not statistically different from zero (a=3.8, b=-0.021, R^2=0.041, P=0.364, n=22). Exclusion of the two drought-stressed Mediterranean sites, where the Q_{10} factor was derived from a selected data set (without water stress), made the fitted relationship slightly more negative (solid line in Fig 12.7), but still not significantly different from zero (a=4.45, b=-0.035, R^2=0.062, P=0.292, n=20). A negative relationship between the Q_{10} values of soil respiration and soil temperature was also reported by Schleser (1982) and Kirschbaum (1995), and probably originates from the larger seasonal fluctuations of root and microbial biomass in colder climates. Wardle (1998) reviewed the literature on the temporal changes in microbial biomass and found that the most northern sites showed the highest

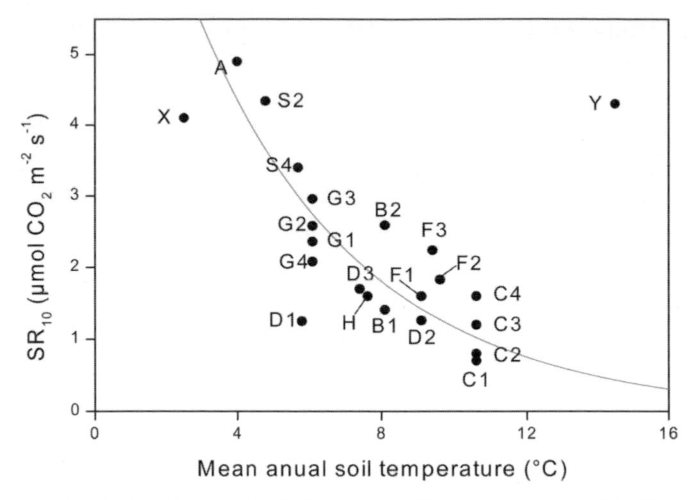

Fig. 12.8. Basal rate of soil CO_2 efflux (SR_{10}, flux rate at 10 °C) versus mean annual soil temperature in the different EUROFLUX forests. Codes for different sites are explained in Table 12.5. The *plotted line* represents the negative exponential function fitted to the data set without the drought-stressed sites. See text for discussion of parameter values

temporal variability. Ecosystems where the soil does not freeze over winter (with lower mortality due to frost) showed only small differences in microbial biomass during the year. The same probably holds for root biomass. Because the temporal changes in soil CO_2 efflux are highly dependent on root and microbial biomass, these observations may explain why soil CO_2 efflux is more sensitive to temperature in cooler climates.

The calculated Q_{10} values depend on the amplitude of the reference temperature (Kicklighter et al. 1994); thus, they are partly determined by the depth of the soil temperature measurements. Lower measurement depths have a smaller temperature range, and therefore a higher temperature sensitivity (Q_{10}). At shallower depths, soil temperature covers a broader range, which will result in a lower temperature sensitivity. Thus, small differences in the measurement depth may have introduced some variability in Fig. 12.7.

Except for the drought-stressed IT-2 site (Y), a decrease in base respiration rate with annual temperature was observed (Fig. 12.8). When excluding the two drought-stressed sites (X and Y), the exponential function fitted the data rather well and the decrease was significantly different from zero ($a=10.5$, $b=-0.22$, $R^2=0.691$, $P<0.001$, $n=20$). The decrease in base respiration rates in warmer sites is likely due to acclimation of roots and microorganisms to local climate. For the cooler climates, 10 °C is at the high end of the soil temperature range, whereas in the temperate region it is only slightly above the annual mean. This, combined with the higher temperature sensitivity of soil CO_2 efflux in the cooler climates, probably explains the sharp increase in base respiration rate at lower temperature.

12.2.3 Effect of Site Productivity

We assumed aboveground NPP (net primary productivity) to be a good representative of site productivity. In their review, Raich and Potter (1995) reported a significant positive correlation between soil CO_2 efflux and NPP on the global scale. This positive correlation between the total annual soil CO_2 efflux rates and NPP was also found among the EUROFLUX forests (P=0.09), as is shown in Fig. 12.9. This is not unexpected, since forests with high NPP are likely to have enhanced root activity and higher litter production, both resulting in high soil CO_2 efflux rates. For more information on the relation between soil respiration and productivity, the reader is referred to Janssens et al. (2001b).

However, increases in soil CO_2 efflux (e.g., due to temperature increases) may also stimulate NPP. Due to the enhanced decomposition, more nutrients will be released from the soil organic matter that become available for the trees. Because boreal forests are generally nutrient limited (Tamm 1985; Linder 1987), enhanced decomposition may thus fertilize these sites and stimulate NPP. The high NPP at the Swedish site might therefore be partly related to the high soil CO_2 efflux rates observed at that site (Table 12.5).

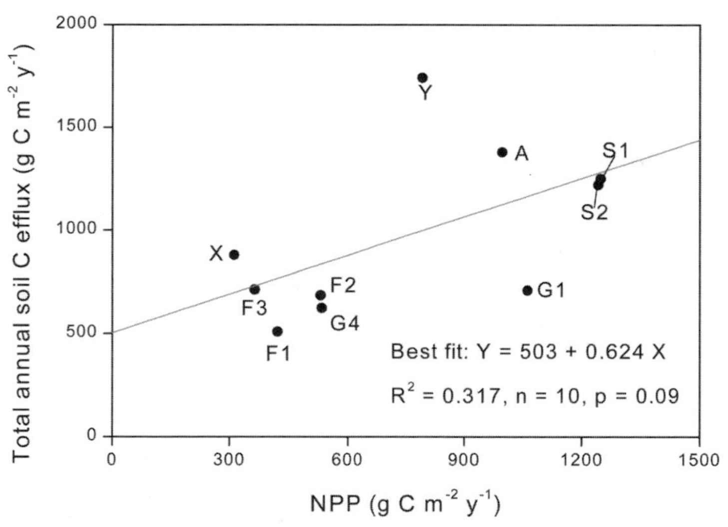

Fig. 12.9. Total annual soil CO_2 efflux rates versus aboveground NPP in the different EUROFLUX forests. Codes for different sites are explained in Table 12.5

12.3 Importance of Roots in Soil CO_2 Efflux

There is large uncertainty concerning the relative contribution of heterotrophic and root respiration to total soil CO_2 efflux. Reported estimates of the contribution of root respiration in forests average around 45 % (Landsberg and Gower 1997), but the estimates range from 22 % (Tate et al. 1993) to 90 % (Thierron and Laudelout 1996). Some of this variability is natural and may be related to differences in vegetation and/or soil type. Forests growing on soils with low organic carbon content will have a higher relative contribution of root respiration than primeval forests with a significantly larger soil carbon pool. Recently Boone et al. (1998) reported that the temperature sensitivity of root respiration is much higher than that of heterotrophic respiration. This was also found in the FR-1 site, where the Q_{10} value of root respiration was 3.86 and that of heterotrophic soil respiration was 2.34. These differences in temperature sensitivity are probably related to the larger temporal variability in root biomass compared to microbial biomass, and to the lower soil moisture content in summer, which would affect microbial respiration more than root respiration, because trees can extract water from deeper soil layers. Because of the different temperature sensitivities, the contribution of root respiration to total soil CO_2 efflux is likely to be higher in summer than in winter. This seasonal pattern was indeed observed at the FR-1 site (Fig. 12.10). Estimates of the relative contribution of root respiration obtained in different seasons are thus likely to be different, which could also add to the variability of the estimates found in literature.

Another, and probably larger source of variability in the estimates of the contribution of root respiration to the total soil CO_2 efflux, is the application

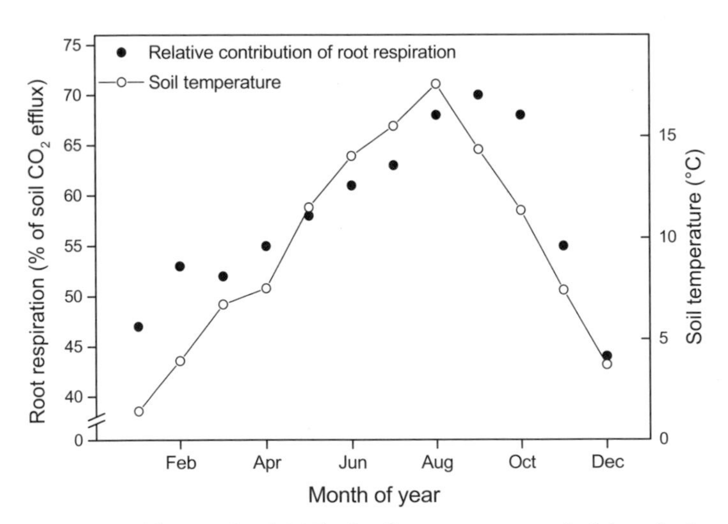

Fig. 12.10. Time series (1997) of soil temperature and of the relative contribution of root respiration to soil CO_2 efflux at the F-2 site

of different methodologies. Estimates of the contribution of root respiration have been obtained by scaling-up measurements of root respiration and comparing these with measured soil CO_2 efflux rates (Gansert 1994), by comparing soil CO_2 efflux in trenched and control plots (Ewel et al. 1987; Bowden et al. 1993; Boone et al. 1998; Epron et al. 1999b), by girdling trees and comparing soil CO_2 efflux below them with control plots (Högberg et al. 2001), by subtracting heterotrophic respiration estimated via laboratory incubations from the in situ measurements of soil CO_2 efflux (Edwards and Sollins 1973; Ewel et al. 1987; Thierron and Laudelout 1996), by extrapolating the regression of soil CO_2 efflux versus root biomass to zero root biomass (Kucera and Kirkham 1971; Behera et al. 1990), and from the ^{13}C and/or ^{14}C fingerprint of the respired CO_2 (Swinnen et al. 1994; Högberg and Ekblad 1996; Lin et al. 1999). The last method estimates rhizosphere respiration without disturbing the soil or the roots, and might therefore produce the best reckoning of the source of CO_2.

At the FR-1 beech site the contribution of root respiration to soil carbon efflux was estimated by comparing soil CO_2 efflux from small (2 × 1.5 m) trenched plots to efflux from undisturbed control areas. The treeless trenched plots were established by digging a trench (1 m depth) around each, lining the trench with a polyethylene film and filling it back. Soil CO_2 efflux was measured every 2 to 4 weeks in 1997 (Epron et al. 1999b). An empirical model ($y=a\,\theta_v\,e^{bT}$), fitted to the soil CO_2 efflux data was used to calculate annual soil CO_2 efflux from soil temperature (T) and soil water content (θ_v). The annual soil carbon efflux were 680 g C m^{-2} $year^{-1}$ in the main plot and 510 g C m^{-2} $year^{-1}$ in the trenched plots. Since trenching strongly influenced soil water content by eliminating tree transpiration, the annual soil carbon efflux on the trenched plots was corrected for differences in soil water content between trenched and control plots. In addition, respiration in the trenched plots was corrected for the decomposition of roots that were killed following trenching (Epron et al. 1999b). Thus root respiration was estimated to account for 60 % of soil C release (410 g C m^{-2} $year^{-1}$), representing 40 % of total ecosystem respiration and 30 % of gross primary productivity at this site.

At the BE-2 Scots pine site, the contribution of root respiration in summer was estimated indirectly by extrapolating the relation between soil CO_2 efflux and root biomass underneath the soil collar (n=21) to the Y-intercept (no roots). Thus, root respiration was estimated to account for 53 % of total soil CO_2 efflux when undergrowth was present and for 75 % when undergrowth was absent (Janssens and Ceulemans, unpubl.). The difference between the two cases was not related to differences in root respiration but to the enhanced heterotrophic respiration when undergrowth was present.

12.4 Conclusions

1. Soil temperature explained most of the temporal variability in soil CO_2 efflux in the majority of the EUROFLUX forests, but not in the drought-stressed sites. Empirical models explained 70–90 % of the seasonal variability in forests with limited drought stress, while they only explained 60–70 % in the drought-stressed Mediterranean sites.
2. Estimates of total annual soil CO_2 efflux by empirical modeling are not sensitive to the type of temperature or moisture regression. However, within a single year, large differences are found between the different regression functions.
3. We found no positive relationship between total annual soil CO_2 efflux and total precipitation or mean annual soil temperature among the different EUROFLUX forests.
4. Annual soil CO_2 efflux was positively correlated with NPP, suggesting that site productivity was more important than local climate in determining the differences in soil CO_2 efflux observed among the EUROFLUX forests. A positive trend was also observed with soil pH, while higher litter layer C/N ratios tended to have a negative effect on soil CO_2 efflux.
5. The relative contribution of root respiration to total soil CO_2 efflux was found to be higher in summer than in winter in the FR-1 site, which was probably related to the higher temperature sensitivity of roots.

Acknowledgements. I.A.J. is indebted to the Fund for Scientific Research – Flanders for a post-doctoral fellowship.

References

Anderson JM (1973) Carbon dioxide evolution from two temperate deciduous woodland soils. J Appl Ecol 10:361–378

Behera N, Joshi SK, Pati DP (1990) Root contribution to total soil metabolism in a tropical forest soil from Orissa, India. For Ecol Manage 36:125–134

Boone RD, Nadelhoffer KJ, Canary JD, Kaye JP (1998) Roots exert a strong influence on the temperature sensitivity of soil respiration. Nature 396:570–572

Bowden RD, Nadelhoffer KJ, Boone RD, Melillo JM, Garrison JB (1993) Contributions of aboveground litter, belowground litter, and root respiration to total soil respiration in a temperate mixed hardwood forest. Can J For Res 23:1402–1407

Buchmann N (2000) Biotic and abiotic factors regulating soil respiration rates in *Picea abies* stands. Soil Biol Biochem 32:1625–1635

Bunnell FL, Tait DEN, Flannagan PW, van Cleve K (1977) Microbial respiration and substrate weight loss. I. A general model of the influence of abiotic variables. Soil Biol Biochem 9:33–40

Carlyle JC, Than UB (1988) Abiotic controls of soil respiration beneath an eighteen-year-old *Pinus radiata* stand in south-eastern Australia. J Ecol 76:654–662

Davidson EA, Belk E, Boone RD (1998) Soil water content and temperature as independent or confounded factors controlling soil respiration in a temperate mixed hardwood forest. Global Change Biol 4:217–227

Edwards NT, Sollins P (1973) Continuous measurement of carbon dioxide evolution from partitioned forest floor components. Ecology 54:406–412

Epron D, Farque L, Lucot E, Badot P-M (1999a) Soil CO_2 efflux in a beech forest: dependence on soil temperature and soil water content. Ann For Sci 56:221–226

Epron D, Farque L, Lucot E, Badot P-M (1999b) Soil CO_2 efflux in a beech forest: the contribution of root respiration. Ann For Sci 56:289–295

Ewel KC, Cropper WP, Gholz HL (1987) Soil CO_2 evolution in Florida slash pine plantations. II. Importance of root respiration. Can J For Res 17:330–333

Fang C, Moncrieff JB (1999) A model for soil CO_2 production and transport. 1. Model development. Agric For Meteorol 95:225–236

Freijer JI, Leffelaar PA (1996) Adapted Fick's law applied to soil respiration. Water Resour Res 32:791–800

Gansert D (1994) Root respiration and its importance for the carbon balance of beech saplings (*Fagus sylvatica* L.) in a montane beech forest. Plant Soil 167:109–119

Goulden ML, Crill PM (1997) Automated measurements of CO_2 exchange at the moss surface of a black spruce forest. Tree Physiol 17:537–542

Hagihara A, Hozumi K (1991) Respiration. In: Raghavendra AS (ed) Physiology of trees. Wiley, New York, pp 87–110

Hanson PJ, Wullschleger SD, Bohlman SA, Todd DE (1993) Seasonal and topographic patterns of forest floor CO_2 efflux from upland oak forest. Tree Physiol 13:1–15

Högberg P, Ekblad A (1996) Substrate-induced respiration measured *in situ* in a C_3-plant ecosystem using additions of C_4-sucrose. Soil Biol Biochem 28:1131–1138

Högberg P, Nordgren A, Buchmann N, Taylor AFS, Ekblad A, Högberg MN, Nyberg G, Ottosson-Löfvenius M, Read DJ (2001) Large-scale forest girdling shows that current photosynthesis drives soil respiration. Nature 411:789–792

Holt JA, Hodgen MJ, Lamb D (1990) Soil respiration in the seasonally dry tropics near Townsville, North Queensland. Aust J Soil Res 28:737–747

Howard DM, Howard PJA (1993) Relationships between CO_2 evolution, moisture content and temperature for a range of soil types. Soil Biol Biochem 25:1537–1546

Janssens IA, Barigah ST, Ceulemans R (1998) Soil CO_2 efflux rates in different tropical vegetation types in French Guiana. Ann For Sci 55:671–680

Janssens IA, Meiresonne L, Ceulemans R (1999) Mean soil CO_2 efflux from a mixed forest: temporal and spatial integration. In: Ceulemans R, Veroustraete F, Gond V, Van Rensbergen J (eds) Forest ecosystem modelling, upscaling and remote sensing. SPB Academic Publishing, The Hague, pp 19–33

Janssens IA, Kowalski AS, Longdoz B, Ceulemans R (2000) Assessing forest soil CO_2 efflux: an in situ comparison of four techniques. Tree Physiol 20:23–32

Janssens IA, Kowalski AS, Ceulemans R (2001a) Forest floor CO_2 fluxes estimated by eddy covariance and chamber-based model. Agric For Meteorol 106:61–69

Janssens IA, Lankreijer H, Matteucci G, Kowalski AS, Buchmann N, Epron D, Pilegaard K, Kutsch W, Longdoz B, Grünwald T, Montagnani L, Dore S, Rebmann C, Moors EJ, Grelle A, Rannik Ü, Morgenstern K, Clement R, Oltchev S, Gudmundsson J, Minerbi S, Berbigier P, Ibrom A, Moncrieff J, Aubinet M, Bernhofer C, Jensen NO, Vesala T, Granier A, Schulze E-D, Lindroth A, Dolman AJ, Jarvis PG, Ceulemans R, Valentini R (2001b) Productivity and disturbance overshadow temperature in determining soil and ecosystem respiration across European forests. Global Change Biol 7:269–278

Keith H, Jacobsen KL, Raison RJ (1997) Effects of soil phosphorus availability, temperature and moisture on soil respiration in *Eucalyptus pauciflora* forest. Plant Soil 190:127–141

Kicklighter DW, Melillo JM, Peterjohn WT, Rastetter EB, McGuire AD, Steudler PA (1994) Aspects of spatial and temporal aggregation in estimating regional carbon dioxide fluxes from temperate forest soils. J Geophys Res 99:1303–1315

Killham K (1994) Soil ecology. Cambridge Univ Press, Cambridge

Kirschbaum MU (1995) The temperature dependence of soil organic matter decomposition, and the effect of global warming on soil organic C storage. Soil Biol Biochem 27:753–760

Kowalenko CG, Ivarson KC, Cameron DR (1978) Effect of moisture content, temperature and nitrogen fertilization on carbon dioxide evolution from field soils. Soil Biol Biochem 10:417–423

Kucera CL, Kirkham DL (1971) Soil respiration studies in tall grass prairies in Missouri. Ecology 52:912–915

Landsberg JJ, Gower ST (1997) Applications of physiological ecology to forest management. Academic Press, San Diego

Lavigne MB, Ryan MG, Anderson DE, Baldocchi DD, Crill PM, Fitzjarrald DR, Goulden ML, Gower ST, Massheder JM, McCaughey JH, Rayment M, Striegl RG (1997) Comparing nocturnal eddy covariance measurements to estimates of ecosystem respiration made by scaling chamber measurements at six coniferous boreal sites. J Geophys Res 102:28977–28985

Le Dantec V, Epron D, Dufrêne E (1999) Soil CO_2 efflux in a beech forest: comparison of two closed dynamic systems. Plant Soil 214:125–132

Lin G, Ehleringer JR, Rygiewicz PT, Johnson MG, Tingey DT (1999) Elevated CO_2 and temperature impacts on different components of soil CO_2 efflux in Douglas-fir terracosms. Global Change Biol 5:157–168

Linder S (1987) Responses to water and nutrition in coniferous ecosystems. In: Schulze ED, Zwölfer HE (eds) Potentials and limitations of ecosystem analysis. Springer, Berlin Heidelberg New York, pp 180–202

Lloyd J, Taylor JA (1994) On the temperature dependence of soil respiration. Funct Ecol 8:315–323

Longdoz B, Yernaux M, Aubinet M (2000) Soil CO_2 efflux measurements in a mixed forest: impact of chamber disturbances, spatial variability and seasonal evolution. Global Change Biol 6:907–917

Lyr H, Hoffmann G (1967) Growth rates and growth periodicity of tree roots. Int Rev For Res 2:181–236

Matteucci G, Dore S, Rebmann C, Stivanello S, Buchmann N (2000) Soil respiration in beech and spruce forests in Europe: trends, controlling factors, annual budgets and implications for the ecosystem carbon balance. In: Schulze E-D (ed) Carbon and nitrogen cycling in European forest ecosystems. Springer, Berlin Heidelberg New York, pp 217–236

McGuire AD, Melillo JM, Joyce LA, Kicklighter DW, Grace AL, Moore B III, Vorosmarty CJ (1992) Interactions between carbon and nitrogen dynamics in estimating net primary productivity for potential vegetation in North America. Global Biogeochem Cycles 6:101–124

Morén A-S, Lindroth A (2000) CO_2 exchange at the floor of a mixed boreal pine and spruce forest. Agric For Meteorol 101:1–14

Myneni RB, Keeling CD, Tucker CJ, Asrar G, Nemani RR (1997) Increased plant growth in the northern latitudes from 1981 to 1991. Nature 386:698–702

Norman JM, Kucharik CJ, Gower ST, Baldocchi DD, Crill PM, Rayment M, Savage K, Striegl RG (1997) A comparison of six methods for measuring soil-surface carbon dioxide fluxes. J Geophys Res 102:28771–28777

Peterjohn WT, Melillo JM, Steudler PA, Newkirk KM (1994) Responses of trace gas fluxes and N availability to experimentally elevated soil temperatures. Ecol Appl 4:617–625

Raich JW, Potter CS (1995) Global patterns of carbon dioxide emissions from soils. Global Biogeochem Cycles 9:23–36

Raich JW, Schlesinger WH (1992) The global carbon dioxide flux in soil respiration and its relationship to vegetation and climate. Tellus 44B:81–99

Rout SK, Gupta SR (1989) Soil respiration in relation to abiotic factors, forest floor litter, root biomass and litter quality in forest ecosystems of Siwaliks in northern India. Acta OEcologica/OEcologica Plantarum 10:229–244

Schimel DS, Braswell BH, McKeown R, Ojima DS, Parton WJ, Pulliam W (1996) Climate and nitrogen controls on the geography and timescales of terrestrial biogeochemical cycling. Global Biogeochem Cycles 10:677–692

Schlentner RE, van Cleve K (1985) Relationships between soil CO_2 evolution from soil, substrate temperature, and substrate moisture in four mature forest types in interior Alaska. Can J For Res 15:97–106

Schleser GH (1982) The response of CO_2 evolution from soils to global temperature changes. Z Naturforsch 37a:287–291

Simunek J, Suarez DL (1993) Modeling of carbon dioxide transport and production in soil. 1. Model development. Water Resources Res 29:487–497

Singh JS, Gupta SR (1977) Plant decomposition and soil respiration in terrestrial ecosystems. Bot Rev 43:449–528

Skopp J, Jawson MD, Doran JW (1990) Steady-state aerobic microbial activity as a function of soil water content. Soil Sci Soc Am J 54:1619–1625

Swinnen J, Van Veen JA, Merckx R (1994) Rhizosphere carbon fluxes in field-grown spring wheat: model calculations based on ^{14}C partitioning after pulse-labelling. Soil Biol Biochem 26:171–182

Tamm CO (1985) The Swedish optimum nutrition experiments in forest stands – aim, methods, yield, results. J R Swed Acad Agric For 17:9–29 (in Swedish, English summary)

Tate KR, Ross DJ, O'Brien BJ, Kelliher FM (1993) Carbon storage and turnover, and respiratory activity, in the litter and soil of an old-growth southern beech (*Nothofagus*) forest. Soil Biol Biochem 25:1601–1612

Thierron V, Laudelout H (1996) Contribution of root respiration to total CO_2 efflux from the soil of a deciduous forest. Can J For Res 26:1142–1148

Trumbore SE, Davidson EA, Barbosa de Camargo P, Nepstad DC, Martinelli LA (1995) Belowground cycling of carbon in forests and pastures of Eastern Amazonia. Global Biogeochem Cycles 9:515–528

Trumbore SE, Chadwick OA, Amundson R (1996) Rapid exchange between soil carbon and atmospheric carbon dioxide driven by temperature change. Science 272:393–396

Wardle DA (1998) Controls of temporal variability of the soil microbial biomass: a global-scale synthesis. Soil Biol Biochem 30:1627–1637

Witkamp M (1966) Decomposition of leaf litter in relation to environment, microflora and microbial respiration. Ecology 47:194–201

13 Conclusions: The Role of Canopy Flux Measurements in Global C-Cycle Research

R. Valentini, G. Matteucci, A.J. Dolman, E.-D. Schulze

13.1 Introduction

The carbon cycle is central to the Earth System, being inextricably coupled with climate, the water cycle, the nutrient cycles, and the production of biomass by photosynthesis on land and in the oceans. Over the past century, the Earth's carbon cycle experienced large perturbations. Since the beginning of the industrial revolution, the mean global carbon dioxide (CO_2) concentration has risen from about 280 ppm to over 368 ppm (Conway et al. 1994; Keeling and Whorf 2002). The worldwide rise in atmospheric CO_2 concentration is occurring due to an imbalance between the rate at which anthropogenic and natural sources emit CO_2 (by burning fossil fuel and respiring) and the rate at which biospheric and oceanic sinks remove CO_2 from the atmosphere by photosynthesis and physiochemical processes. Superimposed on the overall trend regarding CO_2 is a record of great interannual variability of sources and sinks in the rate of growth of atmospheric CO_2. Typical values are on the order of 0.5 to 3 ppm year^{-1}. On a mass basis, these values correspond to 1 and 5 Gt C year^{-1}, respectively. Potential sources of year-to-year changes in CO_2 remain a hot topic of debate. Such variation has been attributed to El Niño/La Niña events, which cause regions of droughts or excessive rainfall (Conway et al. 1994; Keeling et al. 1995), and alterations in the timing and length of the growing season (Myneni et al. 1997; Randerson et al. 1997).

The global C cycle is additionally altered by dramatic changes in the cover and composition of the land surface due to the needs of the growing human population, as well as by excessive exploitation of natural vegetation, the result being large-scale deforestation and desertification. Globally, the land surface is a not a negligible factor in the carbon and nitrogen cycles. About 20 % of the current atmospheric CO_2 concentration can be attributed to human influence, and the amount of nitrogen fixated has changed by almost 60 % due to human action. Land cover and its use thus plays an important part in biogeochemical cycles. Many agricultural lands have been trans-

Ecological Studies, Vol. 163
R. Valentini (Ed.) Fluxes of Carbon,
Water and Energy of European Forests

formed into urban and suburban landscapes, and many tropical forests have been logged, burned, and converted to pasture. The role of humans in the carbon cycle is not new, as we have influenced it for thousands of years through agriculture, forestry, trade, and energy use in industry and transport. This is particularly true in Europe where millennia of civilizations have shaped the landscape and destroyed or created forests and agricultural regions. However, only over the past two or three centuries have the changes in global land use become sufficiently intensified, widespread, and far-reaching to match the great forces of the natural world.

At the same time, we are becoming aware of the fact that human societies are not unidirectional drivers of change: they are also impacted by changes in the carbon cycle and climate, and they respond to these impacts in ways that have a feedback potential on the carbon cycle itself.

Changes in land use alter the Earth's radiation balance by changing its albedo, the Bowen ratio, the leaf area index, and the physiological capacity to assimilate carbon and evaporate water (Pielke et al. 1998).

Changing a landscape from forest to agricultural crops, for instance, increases the surface's albedo and Bowen ratio, the ratio between the flux densities of sensible and latent heat exchange, and respiration. Forests have a lower physiological capacity to assimilate carbon and a lower ability to transpire water, as compared to crops. A change in the age structure of forests due to direct (deforestation) or indirect (climate-induced fire) disturbance alters it ability to acquire carbon and transpire water (Amiro et al. 1999; Schulze et al. 1999).

To illustrate the intrinsic interactions between the natural and human parts of the system, and the difficulty of separating them, several efforts have been carried out to identify the location and magnitude of carbon exchanges among atmosphere, land, and ocean. The location of current terrestrial sinks (uptake of CO_2 by land from the atmosphere) may be largely due to patterns of past land-use change, and their magnitudes a consequence of legacies of this change coupled with physiological processes punctuated by changing disturbance regimes. To understand these and other patterns in the global carbon cycle, many tools and many methods exist that are used by different research communities. For example, diverse entities such as satellite data, air sampling networks, and inverse numerical methods ("top-down" approaches) allow us to study the strength and location of the global-scale and continental-scale carbon sources and sinks. Surface monitoring and process studies ("bottom-up" approaches) provide estimates of land-atmosphere and ocean-atmosphere carbon fluxes at finer spatial scales, as well as examining the mechanisms that control fluxes at these regional and ecosystem scales (Fig. 13.1). A further suite of investigations is concerned with human-induced activities geared to manipulating components of the carbon cycle, including means of maximizing and tracking the net sequestration of carbon in the terrestrial biosphere, technologies for ocean sequestration, and the monitoring

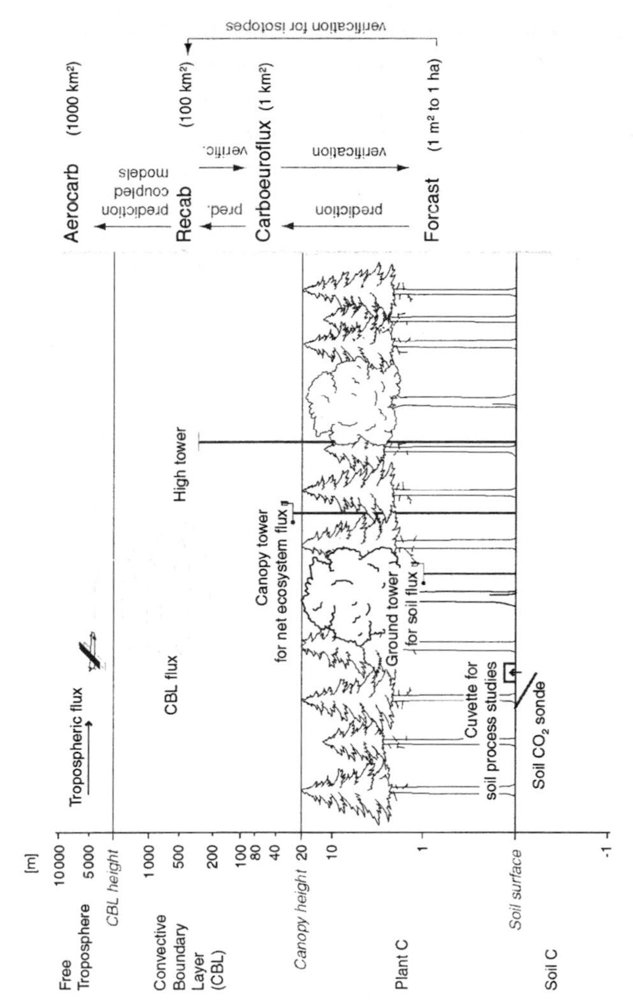

Fig. 13.1. Integration of observation techniques at several temporal and spatial scales: the concept of CarboEurope

and minimization of the massive emissions from fossil fuel combustion (Schulze et al. 2002). In the search for an integrated understanding of the carbon cycle and its current changes, each of these tools provides a piece of the puzzle. However, no single existing technique, approach, discipline, or regional outlook can perceive the whole picture.

The complexity of various scales of investigation is also intrinsically related to the different nature of carbon fluxes between the biosphere and atmosphere. Indeed, there are various definitions of what constitute productivity, e.g., the velocity of carbon dioxide reduction into organic material as the net of oxidation processes (respiration). Figure 13.2 presents a scheme that represents the various definitions of productivity according to the processes they incorporate.

Plants take up CO_2 from the air and convert it into carbohydrate by using photosynthetically active radiation (PAR). This process is called photosynthesis and translates into gross primary production (GPP). Globally, GPP is about 120 Gt C year^{-1}. Plants also respire for their maintenance, and the sum of photosynthesis and autotrophic plant respiration (R_a) is called net primary pro-

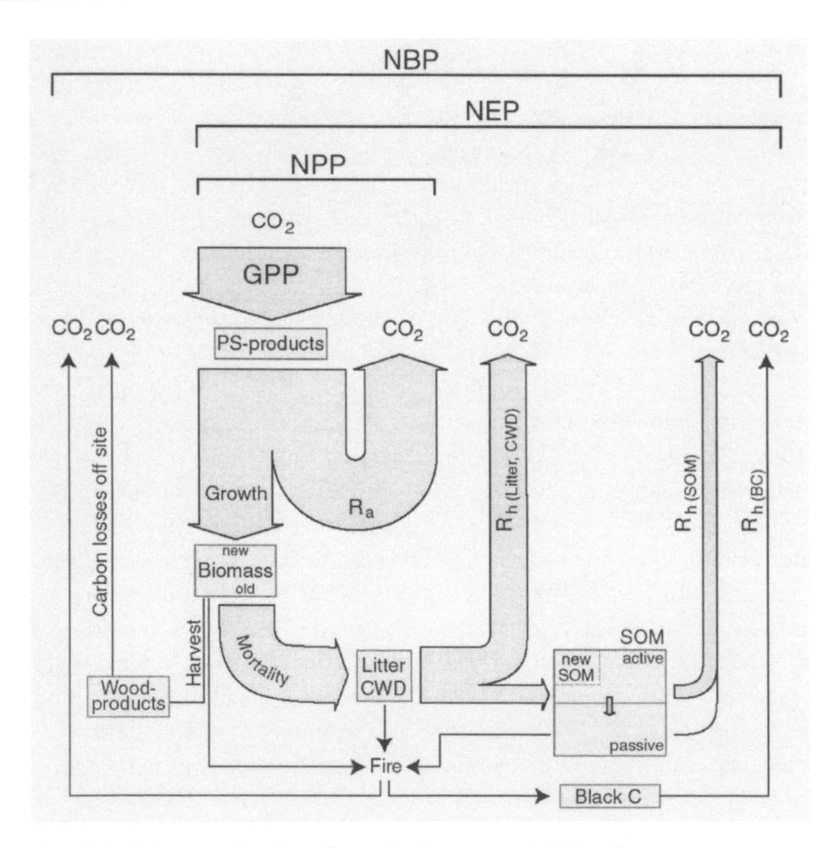

Fig. 13.2. Scheme of carbon fluxes in the terrestrial biosphere

duction (NPP), and this amounts globally to about 60 Gt C $year^{-1}$. Mortality of leaves and twigs produces dead organic material that is decomposed (SOM). The resulting term incorporating this decomposition, heterotrophic respiration (R_h), is net ecosystem production (NEP) and is considerable smaller than NPP, about 10 Gt C $year^{-1}$ globally. Net biome production (NBP) is the long-term carbon uptake, and takes into account disturbance cycles of harvest, forest fires (producing black carbon), and other disturbances such as pests. This is the long-term carbon storage and amounts to 0.7 Gt C $year^{-1}$ globally.

13.2 The Role of Canopy Flux Measurements in Global C-Cycle Research

In this book, we have made an attempt to contribute an understanding of part of the carbon-research puzzle, starting at the canopy level of ecosystems by means of an integrative technique – that is, by measuring biospheric fluxes (eddy covariance, see Chap. 2). The eddy covariance technique has the advantage of treating the stand scale as a single functional unit. This allows one to scale-up physiological processes usually investigated at the organ or individual scale. The assessment of canopy fluxes has proven to be important not only in ecosystem research, but also in atmospheric research, because it is an independent method that integrates the numerous and otherwise very complicated interactions between organisms in ecosystems. The method is based on the basic physics involved in measuring atmospheric turbulence. Thus, the eddy covariance technique bridges the gap between biological and atmospheric science, which is possibly the reason for its success in global C-cycle research. However, this success is not unjustified, as the various chapters in this book show.

Micrometeorologists have been making measurements of CO_2 and water vapor exchange between vegetation and the atmosphere since the late 1950s and early 1960s. The earliest measurements were conducted by Inoue (1958), Lemon (1960), Monteith and Sziecz (1960), and Denmead (1969) using the flux-gradient method over growing agricultural crops. Denmead (1964), Baumgartner (1969), and Jarvis et al. (1976) were among the first workers to apply micrometeorological methods to measure CO_2 exchange over forests. Since then new technological developments made the eddy covariance method more reliable and superior to the classical flux-gradient relationship (Verma et al. 1986). After initial studies conducted in the campaign mode during the peak of the growing season and for short intense periods (Valentini et al. 1991), long-term measurements became possible due to new gas analyzers, sonic anemometers, and other electronic devices that tolerate all climates.

The concept of a global network of long-term flux measurement sites was conceived at the 1995 La Thuile workshop (Baldocchi et al. 1996). At this meeting, the flux measurement community convened and discussed the possibilities, problems, and pitfalls associated with the idea of building a global network of monitoring sites. After the La Thuile meeting there was an acceleration in the establishment of flux tower sites, and regional flux networks on Earth started as the EUROFLUX project in 1996 (Aubinet et al. 2000; Valentini et al. 2000). In 1997 a similar network Ameriflux was established in North America.

This book comes at the peak of a worldwide process that currently encompasses a flux tower network (FLUXNET) of more than 160 sites distributed around the world (Baldocchi et al. 2001). In summarizing the results of the European network, the following novel results on process-level understanding are presented with special emphasis on European forest ecosystems. The EUROFLUX network demonstrated for the first time:

- It is possible to obtain a harmonized set of long-term measurements of NEE over a range of forest types within a continent.
- Temperate forests in Europe are considerable sinks of carbon, at a rate that exceeds estimates from the prior generation of models. The patterns of carbon and energy flux show a strong seasonality and interannual variability, which demonstrates the significant relevant role of climate variability on ecosystem biogeochemistry.
- The length of the growing season has a distinctive role in determining the net ecosystem exchange, particularly for broad-leaved deciduous forests; less evident is this factor for evergreen forests, as the definition of growing season becomes quite problematic in this case.
- Light use efficiency changes with the proportion of diffuse to direct light, and this has an important implication on future scenarios of atmospheric contamination by aerosols.
- Temperature response for canopy-scale CO_2 exchange shows a certain degree of adaptation, but temperature has a fundamental role in triggering the onset of net ecosystem C uptake (NEP), particularly for boreal forests. This effect is stronger than the temperature effect on photosynthesis (GPP) at the ecosystem level.
- GPP seems to be rather conservative across European forests with a similar management (in contrast to artificial plantation, where GPP is higher). An ecosystem's respiration seems to be the driving component in its carbon balance (NEP).
- Respiration, although showing a classical temperature response at single-site level, is not related to mean annual temperature, indicating that processes other than temperature are involved in determining the efflux of carbon.
- At least for the Mediterranean sites, the role of soil moisture on respiration processes is extremely important in driving the overall ecosystem carbon budget. Indeed, future predictions based on temperature response of respiration without accounting for water availability may likewise not be valid elsewhere.

Although the eddy covariance technique has proven to be quite successful in integrating ecosystem-level processes, it cannot be used alone to extrapolate point measurements to global-scale sink determinations. It needs to be integrated with several other techniques, and new attempts are under development (Papale and Valentini 2002) in which, by means of artificial neural networking, the capability is achieved of extrapolating single point measurements in space, and thus constructing regional and continental fluxes. Besides the scaling problem involved in going from points to regions, issues on the limits of application of the eddy covariance technique need to be addressed. Some of these issues (covered in Chap. 2, this Vol.) are: the uncertainties associated with nighttime fluxes, the lack of closure of energy balance experienced by many sites, and some discrepancies encountered between measurements of soil respiration by chambers and eddy flux data. Despite some of these issues, the eddy covariance technique is still the only method capable of investigating the ecosystem biospheric fluxes as an integration of several components at a scale constituting an interface between biological processes and the atmosphere.

As we write this chapter, over 60 site-years of carbon dioxide and water vapor flux data have been published in the literature. Several syntheses have been published covering most of the terrestrial biomes in the world (Schulze et al. 1999; Valentini et al. 2000; Baldocchi et al. 2001).

It is striking that most of the measurement sites are net sinks of CO_2 on an annual basis. Magnitudes of net carbon uptake by whole ecosystems range from near zero to up to 700 g C m^{-2} $year^{-1}$. The largest values appear to be associated with temperate forests (conifer and deciduous) and those in the tropics. Unexpected results are represented by the substantial rates of net carbon uptake by tropical forests sites, and by sites losing carbon dioxide such as boreal forests and tundra, which are experiencing disturbance of long-term carbon storage pools (Goulden et al. 1998; Lindroth et al. 1998; Oechel et al. 2000). The maximum carbon loss by these sites on an annual basis seems to be constrained and does not exceed 100 g C m^{-2}. Extrapolation of these data to a global scale, however, is not recommended without a consistent check on present observations with independent estimates of carbon budget.

One independent measure of ecosystem carbon balance can be provided by forest inventories (Kauppi et al. 1992; Gower et al. 1999). However, most forestry surveys have been intended to provide information on round wood for industry and not total NPP. In addition, these inventories do not address scientific questions relating to the ecosystem's carbon balance. Furthermore, inventory surveys generally provide information on decadal time scales only. They do not provide information on shorter-term physiological forces and mechanisms; they rarely measure belowground allocations of carbon. Only few experiments (Schulze 2000) have been carried out to directly measure the carbon budget of an ecosystem in order to determine the NEP by methods alternative to flux measurements. An example of such an effort is, however, the

Collelongo EUROFLUX site, where, around the tower, trees are equipped with dendrometer bands through which stem growth is monitored at weekly intervals (from summer 1996). During 1997, a number of sampling campaigns for leaves at different canopy depths were performed in order to analyze leaf development at a finer scale than that achieved by LAI measurement. The aims of these campaigns were to study the ecophysiological determinants of leaf and canopy level fluxes, and to study carbon allocation to leaves in close correlation with canopy fluxes. The results of these analyses are presented in Fig. 13.3 (where the traditional convention for NEE sign has been reversed to allow for a better readability of the graph). Scanning quickly, one can see that the NEE trend parallels the girth growth and that the carbon gain is effectively attributable to forest growth.

Looking at processes in the ecosystem, it is interesting to note that at the beginning of season all the carbon is allocated to leaf growth, which in turn increases the photosynthetic potential of the canopy. Later, carbon allocation to leaves decreases until maturation in late June or early July. Stem growth starts at the end of May, at which time the canopy is already absorbing 6–7 g C m^{-2} day^{-1} and reaches a peak in mid-July, when the forest shows its maximum NEE. The growth of dominant and codominant trees is timed differently, with the latter growing at a lower rate and reaching a maximum later. In the second half of August, both NEE and stem growth decline, probably

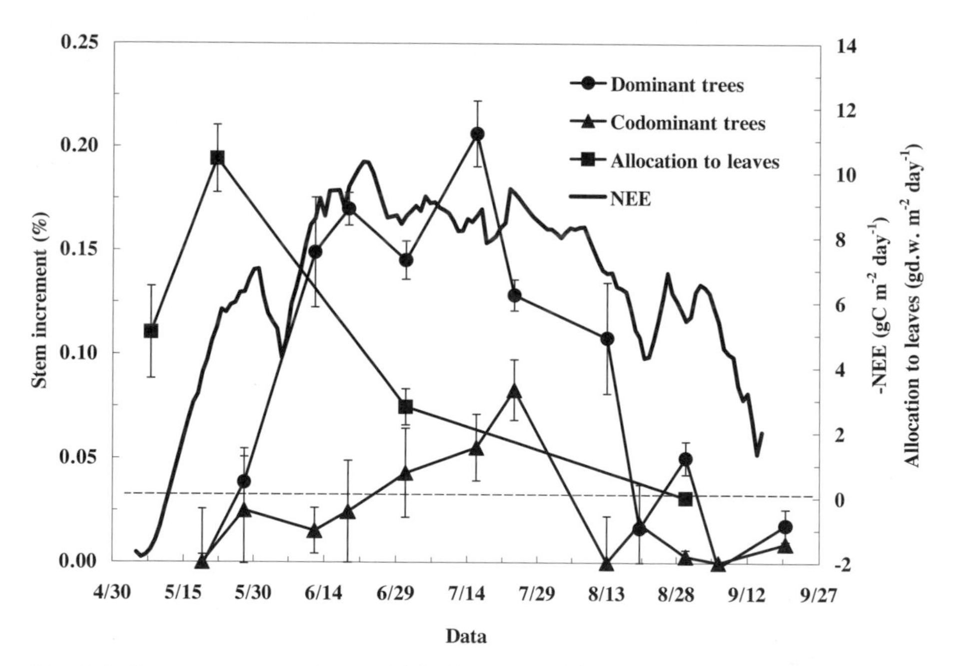

Fig. 13.3. Net ecosystem exchange (*tight line*: g C m^{-2} day^{-1}, smoothed), percentage stem increment of dominant (*circles*) and codominant (*triangles*) trees and allocation to leaves (*squares*: g dry wt. m^{-2} day^{-1})

because water stress exerts some effect on canopy photosynthesis. It is interesting to see that at the beginning of September, a new, lower peak of NEE is reflected in stem growth, particularly of the codominant trees.

The carbon budget of the Collelongo site as measured by eddy correlation (in 1993, 1996, and 1997) has been compared to the budget coming from biomass and soil carbon analysis (Table 13.1). The two values are in close agreement, signaling the overall comparability of the two approaches. It is important to note that only a "full" carbon inventory that includes the belowground components is able to match the eddy covariance data. Often only the aboveground components are measured, which obviously leads to underestimation of the total C flux.

The example indicates that whenever a detailed carbon budget is carried out, including above- and belowground components, the two independent estimates can converge to produce similar results.

Recent studies (Schulze 2000) show that both eddy covariance and carbon inventory data seem to allocate on average about 140 g C m^{-2} to the soil of European forests. The processes which account for such significant carbon accumulation in the soil are still under discussion; in any case, these findings, obtained by independent methods, open new questions and challenge the future research on ecosystem biogeochemistry.

The work described in this book constitutes a first step toward an integrated ecosystem analysis from functional perspectives.

For many years, ecophysiologists have tried to describe ecosystem processes by separating out the various ecosystem components at the level of organs or individuals, using methodologies such as leaves or branch cuvettes for gaseous exchanges or sap flow techniques for tree-level transpiration. This is the first time that an "integrated" approach using eddy covariance method-

Table 13.1. Carbon budgets by eddy correlation and analysis of biomass and soil carbon processes for the site Italy 1

Component	Carbon (g C m^{-2} $year^{-1}$)	Technique
Net Ecosystem Exchange (NEE)	567.4±134.9	Eddy correlation (93, 96–97)
1. Stem increment	136	Sampling and analysis
2. Branches and twigs increment	150	Sampling and analysis
3. Leaf litter production	141.2±17	Traps
4. Aboveground NPP (1+2+3)	427.2±17	
5. Coarse roots and stump increment	48	Sampling and analysis
6. Fine roots turnover (<2 mm)	184	Sequential coring
7. Belowground NPP (5+6)	232	
8. Total NPP (4+7)	659.2±17	
9. Litter and soil carbon mineralization	102±20	Bags, ^{14}C, incubation
10. NEP (8–9)	557.2±37	

ology has been able to provide process-level understanding at the ecosystem level , considering the integrated sum of biospheric fluxes as a single physiological unit. In this respect, the eddy covariance technique of measuring biospheric exchanges has proven to be very effective in providing an important interface between ecosystem and atmosphere studies. Thanks to integration at the ecosystem level, new interfaces have been opened with atmosphere scientists, the global scale modeling community, and the remote sensing end users.

With respect to the current Kyoto Protocol negotiations, new insights are also possible in the quantification of carbon sequestration by vegetation canopies at stand-scale level. The material in this book clearly shows how the eddy covariance technique can provide a tool that "verifies" other methods for estimating the carbon budget of the forest, and points out the importance of considering the full-carbon accounting when dealing with carbon management policies (Schulze et al. 2002).

The work contained in this book is the result of an extensive European effort in establishing a unique infrastructure of continuous observations (the tower flux network) that encompasses a range of disciplines from atmospheric physics to ecology. The story of successful results documented in this book is also that of a successful new cultural approach to ecological studies, in which the traditional barriers between biological and physical sciences have been broken down to form a new community of scientists with a common goal: the understanding of ecosystem functions and their role in the Earth systems. The work is, of course, not completed with the material presented in this book. At this moment, new flux towers are being established in a variety of ecosystems, including peat lands, grasslands, croplands, and in several other forests that have experienced natural disturbances and various forms of management. We hope that the experience shared in this book will serve as a basis for further improving our knowledge of ecosystem functions and for continuing the development of an interdisciplinary community in the search for even broader interfaces with other disciplines – including human sciences, which contribute valuable understanding of a significant component of the recent past, current, and future global changes.

References

Amiro BD, MacPherson JI, Desjardins RL (1999) BOREAS flight measurements of forest-fire effects on carbon dioxide and energy fluxes. Agric For Meteor 96:199–208

Aubinet M, Grelle A, Ibrom A, Rannik U, Moncrieff J, Foken T, Kowalski A, Martin P, Berbigier P, Bernhofer C, Clement R, Elbers J, Granier A, Grunwald T, Morgenstern K, Pilegaard K, Rebmann C, Snijders W, Valentini R, Vesala T (2000) Estimates of the annual net carbon and water exchange of European forests: the EUROFLUX methodology. Adv Ecol Res 30:113–175

Baldocchi DD, Valentini R, Running SR, Oechel W, Dahlman R (1996) Strategies for measuring and modeling CO_2 and water vapor fluxes over terrestrial ecosystems. Global Change Biol 2:159–168

Baldocchi DD, Falge E, Gu L, Olson R, Hollinger D, Running S, Anthoni P, Bernhofer C, Davis K, Fuentes J, Goldstein A, Katul G, Law B, Lee X, Mahli Y, Meyers T, Munger W, Oechel W, Paw UK, Pilegaard K, Schmid H, Valentini R, Verma S, Vesala T, Wilson K, Wofsy S (2001) FLUXNET: a new tool to study the temporal and spatial variability of ecosystem-scale carbon dioxide, water vapor and energy flux densities. Bull Am Meteor Soc 82:2415–2435

Baumgartner A (1969) Meteorological approach to the exchange of CO_2 between atmosphere and vegetation, particularly forests stands. Photosynthetica 3:127–149

Conway TJ, Tans PP, Waterman LS, Thoning KW, Kitzis DR, Masarie K, Zhang N (1994) Evidence for interannual variability of the carbon cycle from NOAA/CMDL global sampling network. J Geophys Res 99:22831–22855

Denmead OT (1964) Evaporation sources and apparent diffusivities in a forest canopy. J Appl Meteorol 3:393–389

Denmead OT (1969) Comparative micrometeorology of a wheat field and a forest of Pinus radiata. Agric Fore Meteor 6:357–371

Goulden ML, Wofsy SC, Harden JW, Trumbore SE, Crill PM, Gower ST, Fires T, Daube BC, Fan SM, Sutton DJ, Bazzaz A, Munger JW (1998) Sensitivity of boreal forest carbon balance to soil thaw. Science 279:214–217

Gower ST, Kucharik CJ, Norman JM (1999) Direct and indirect estimation of leaf area index, fpar and net primary production of terrestrial ecosystems. Remote Sensing Environ 70:29–51

Inoue I (1958) An aerodynamic measurement of photosynthesis over a paddy field. Proceedings of the 7th Japan National Congress of Applied Mechanics, pp 211–214

Jarvis PG, James GB, Landsberg JJ (1976) Coniferous forest. In: Monteith JL (ed) Vegetation and the atmosphere, vol 2. Academic Press, London, pp 171–240

Kauppi PE, Mielikainen K, Kuuseia K (1992) Biomass and carbon budget of European forests, 1971 to 1990. Science 256:70–74

Keeling CD, Whorf TP (1994) Atmospheric CO_2 records from sites in the SIO air sampling network. In: Trends'93: a compendium of data on global change. ORNL/CDIAC-65, Oak Ridge, TN, pp 16–26

Keeling CD, Whorf TP (2002) Atmospheric CO_2 records from sites in the SIO air sampling network. In: Trends: a compendium of data on global change. Carbon dioxide Information Analysis Center, Oak Ridge, TN, USA

Keeling CD, Whorf TP, Wahlen M, v d Plicht J (1995) Interannual extremes in the rate of rise of atmospheric carbon dioxide since 1980. Nature 375:666–670

Lemon ER (1960) Photosynthesis under field conditions. II. An aerodynamic method for determining the turbulent carbon dioxide exchange between the atmosphere and a corn field. Agric J 52:697–703

Lindroth A, Grelle A, Morén AS (1998) Long-term measurements of boreal forest carbon balance reveal large temperature sensitivity. Global Change Biol $:443–450

Monteith JL, Szeicz G (1960) The CO_2 flux over a field of sugar beets. Q J R Meteorol Soc 86:205–214

Myneni RB, Keeling CD, Tucker CJ, Asrar G, Nemani RR (1997) Increased plant growth in the northern high latitudes from 1981–1991. Nature 386:698–702

Oechel WC, Vourlitis GL, Hastings SJ, Zulueta RC, Hinzman L, Kane D (2000) Acclimation of ecosystem CO2 exchange in the Alaskan Arctic in response to decadal climate warming. Nature 406:978–981

Papale D, Valentini R (2002) Spatial and temporal assessment of biospheric carbon fluxes at continental scale by neural network optimization. In: Mencuccini M, Grace J,

Moncrieff J, Mc Naughton J (eds) Forest at land-atmosphere interface. A festschrift for Paul Jarvis. CAB Int, Wallingford, UK

Pielke RA, Avissar R, Raupach M, Dolman AJ, Zeng X, Denning AS (1998) Interactions between the atmosphere and terrestrial ecosystems: influence on weather and climate. Global Change Biol 4:61–476

Randerson JT, Thompson MV, Conway TJ, Fung IY, Field CB (1997) The contribution of terrestrial sources and sinks to trends in the seasonal cycle of atmospheric carbon dioxide. Global Biogeochem Cycles 11:535–560

Schulze ED (2000) Carbon and nitrogen cycling in European forest ecosystems. Ecological studies. Springer, Berlin Heidelberg New York, 499 pp

Schulze ED, Lloyd J, Kelliher FM, Wirth C, Rebmann C, Lukher B, Mund M, Milykova I, Schulze W, Ziegler W, Varlagin A, Valentini R, Dore S, Grigoriev S, Kolle O, Vygodskaya NN (1999) Productivity of forests in the Eurosiberian boreal region and their potential to act as a carbon sink. Global Change Biol 5:703–722

Schulze ED, Valentini R, Sanz MJ (2002) The long way from Kyoto to Marrakesh: implications of the Kyoto Protocol negotiations for global ecology. Global Change Biol 8:1–14

Valentini R, Scarascia Mugnozza G, De Angelis P, Bimbi R (1991) An experimental test of the eddy correlation technique over a Mediterranean macchia canopy. Plant Cell Environ 14:987–994

Valentini, R, Matteucci G, Dolman AJ, Schulze E-D, Rebmann C, Moors EJ, Granier A, Gross P, Jensen NO, Pilegaard K, Lindroth A, Grelle A, Bernhofer C, Grünwald T, Aubinet M, Ceulemans R, Kowalski AS, Vesala T, Rannik Ü, Berbigier P, Loustau D, Gudmundsson J, Thorgeirsson H, Ibrom A, Morgenstern K, Clement R, Moncrieff J, Montagnani L, Minerbi S, Jarvis PG (2000) Respiration as the main determinant of European forests carbon balance. Nature 404:861–865

Verma SB, Baldocchi DD, Anderson DE, Matt DR, Clement RE (1986) Eddy fluxes of CO_2, water vapor, and sensible heat over a deciduous forest. Bound Layer Meteor 36:71–91

Subject Index

A
Abies alba 62
acclimation 172
advection 10
aerodynamic resistance 104, 106, 111, 112, 207, 217
aerodynamic roughness 211, 213, 222
afforestation 1, 99, 100, 226
air density 17, 20
air temperature 11, 15, 40, 45, 59, 62, 66, 77, 78, 85, 95, 101, 108, 115, 118, 127, 129, 134, 136, 140, 142, 144, 145, 155, 160, 164, 203, 205, 229, 231, 240
albedo 102, 207–210, 222, 256
AMERIFLUX 2
annual precipitation 75, 112, 224, 242, 243
Arrhenius-type relationship 235
artificial neural networking 261
assimilation 61, 64, 66, 134, 135, 137, 138, 141, 142, 163, 165, 219
- rate 65
autotrophic 37, 38, 40, 48, 54, 219, 225, 258
- respiration 37, 38, 54
available energy 27, 107, 110, 117, 120, 215

B
basal ecosystem respiration 240
basal rate 240, 245, 246
beech 2, 7, 50, 52–59, 62, 64–66, 68–70, 75, 76, 99, 138, 147, 153, 155, 157, 158, 162, 163, 165, 166, 168, 174, 176, 181, 182, 185, 196, 202–206, 208, 209, 212, 215–217, 242, 249, 251–253
Betula pubescens 74
biomass temperature 109
biome-BGC 180, 181, 183, 186–189, 193, 201, 202, 205, 206
boreal forests 50, 99, 140, 226, 247, 260, 261
Bowen ratio 69, 81, 85, 86, 110–112, 138, 140, 215, 256
bud break 56, 61, 62

C
C/N ratios 243, 250
calibration 11, 12, 14, 17, 28, 47, 54, 108, 110, 179, 183
canopy conductance 58, 59, 68, 208, 220, 223
canopy evaporation 208, 213, 215
canopy heat flux 107, 108
canopy height 101, 104, 106, 111, 129, 212, 222
canopy resistance 111, 112
canopy-specific heat capacity 109
carbon flux 2, 6, 7, 37, 40, 54–56, 61, 67, 68, 96, 109, 112, 114, 116, 122, 130, 134–136, 148, 151, 153, 155, 157, 159, 161, 163, 165, 169, 171, 173, 175, 177, 185, 219, 225, 226, 229, 253, 256, 258, 265
carbonate dissolution 37
chamber system 41, 43, 45–48, 50, 51, 53, 54
chemical oxidation 37
climate 77, 1, 2, 5, 6, 15, 51–54, 56, 58, 68, 73, 77, 95, 96, 100, 101, 112, 119, 121, 122, 125–128, 147, 151–153, 155, 161, 165, 167–169, 174, 176, 183, 189, 198–200, 202–208, 214, 217, 219, 223, 226, 229–231, 233, 240, 241, 245, 246,

250, 253, 255, 256, 259–260, 265, 275
– variability 31, 96, 226, 260
CO_2 33, 54, 109, 116, 117, 157, 265
– efflux 54
– flux 85, 88–95
– – density 117
– mixing ratio 1
– storage 109
conductance 28, 58, 59, 68, 70, 144, 146, 155, 156, 159, 161, 175, 176, 208, 217–223
conservation equation 9, 10
coordinate rotation 19

D
decomposition 2, 10, 37, 39–41, 45, 51, 52, 64, 68, 69, 95, 133, 180, 183, 185, 186, 202–205, 226, 229, 240, 245, 247, 249, 253, 259
decoupling coefficient 107, 110, 111
deforestation 1, 205, 223, 226, 255, 256
dendrometer bands 262
diffuse radiation 15,
direct radiation 64, 210
displacement height 30, 104, 105, 122, 211–213
dissolved organic carbon 38
Douglas fir 52, 53, 99, 101, 119, 120, 153, 156, 162, 163, 165, 166, 171, 252
drought 58, 59, 65, 94, 121, 126, 141, 148, 151, 152, 174, 176, 185, 230, 231, 235, 236, 239, 240
dynamic chambers 43

E
Earth System 255, 264
ecosystem biogeochemistry 260, 263
ecosystem models 2, 176, 179, 180, 201, 226
ecosystem respiration 7, 30, 40, 41, 62, 63, 68, 69, 94, 95, 115, 116, 118, 140, 142–144, 147, 154, 159, 206, 229, 230, 242, 249, 251
eddy covariance 2, 7, 9–11, 17, 19, 24, 26, 28, 30, 32–35, 39, 41, 48, 52, 57, 61, 68, 69, 74, 81, 85, 95, 100, 103, 104, 106, 109, 110, 112, 126, 129, 140, 143, 144, 155, 159–161, 163, 165, 172, 174–176, 226, 231, 241, 252, 259, 261, 263, 264, 275
empirical models 233–235, 237, 238, 240, 241, 250
energy balance 7, 17, 26–28, 30, 35, 95, 107, 108, 110, 138, 207, 210, 215–217, 223, 261
energy budget 28, 81, 84
energy fluxes V, 5, 28, 110, 111, 208, 222, 264
EUROFLUX 1–7, 9, 11–17, 24, 27, 28, 30, 32, 33, 39, 41, 43, 45, 47–49, 55, 56, 62, 63, 67–69, 71–73, 75–80, 82–84, 86, 88–90, 92, 93, 95, 96, 99, 100, 122, 129, 147, 151–154, 174, 175, 180, 182, 184, 192, 196, 202–204
evaporative fraction 110, 111, 215–217, 222
evapotranspiration 57, 60, 68, 110, 112, 114, 120, 138, 146, 147, 190, 243

F
Fagus sylvatica L. 2, 50, 55, 69, 70, 192, 251
fetch 6, 12, 25, 29, 30, 34
filters 12, 13, 15, 21, 22, 24, 25, 129
flow rate 12, 13, 15, 129
flux divergence 19
FLUXNET 3, 28, 35, 76, 175, 203, 262, 267
footprint 8, 9, 29, 30, 33–35, 100, 110, 129, 154, 161, 163
fossil fuel emissions 225
frequency losses 13, 17, 28
friction velocity 28, 32, 104, 106, 107, 211

G
gap filling 17, 33, 79, 112, 114, 115, 120–122, 134, 148, 162, 168, 170, 172, 175
girdling 48, 51, 249, 251
global carbon budget 1
global circulation models 1
global radiation 15, 58, 59, 77, 80, 92–94, 190
gross primary production 96, 151, 179, 180, 196, 229, 258
growing season 52, 85, 217, 219, 221, 227, 229, 245, 255, 259, 260

H
heat storage 107, 109
heterotrophic respiration 37, 38, 68, 95, 183, 185, 186, 196, 225, 249

I
interannual variability 1, 95, 121, 255, 260, 265
IRGA 11–13, 24, 25, 32, 43, 44, 129

K
Karman's constant 211
Kyoto protocol VI, 1, 37, 51, 203, 231, 264, 266

L
LAI 4, 56, 58–60, 61, 65, 66, 68, 73, 75, 76, 94, 262
laminar flow 13
land-use change 176, 202, 229, 231, 256
latent heat flux 19, 81, 84, 86, 111–114, 117, 215, 216
latitude 2, 37, 70, 99, 103, 120, 125, 226–227, 229–231, 265
leaf area index 59, 61, 75, 96, 155, 181, 203, 212, 213, 222, 256, 265
light saturation 114, 115, 160
lignin 38, 183, 186
linear detrending 18, 21, 23, 32, 35
litter 39, 45, 49, 50, 54, 109, 129, 132, 133, 181–183, 185, 186, 201, 203, 205, 206, 240–242, 247, 250, 253, 263

M
maritime pine 68, 71, 72, 75, 77, 79, 81, 83, 85, 87, 91, 93, 95, 97
mean annual temperature 77, 168, 207, 220, 260
mediterranean 125–129, 131, 133–135, 138–140, 145, 147–149, 191, 221, 222, 226, 243, 245, 250, 260, 266, 273
metabolic heat 107
microbial respiration 48, 205, 233, 250, 253
moisture 27, 35, 38, 53, 64, 94, 115, 116, 130, 144, 147, 183, 205, 207, 221, 233, 235–241, 243, 248, 250–253, 260
molecular diffusion 10
momentum flux 17, 19, 211
Monin-Obukhov length 105
mycorrhizae 38

N
NEE 52, 61, 62, 64–68, 85, 94, 95, 114–116, 118, 120, 134, 151, 152, 154, 155, 160–163, 165–169, 171, 172
net biome productivity 5, 179
net radiation 15, 77, 80, 82, 84, 103, 107, 117, 129, 134, 138, 139, 207, 208, 210, 213–215, 221
night time fluxes 31, 40
nitrogen 49, 51–53, 73, 94, 121, 177, 179, 181–183, 203–206, 252, 253, 255, 266, 273
Norway spruce 76, 99, 100, 156, 158

O
oak 2, 75, 76, 99, 127, 147, 192, 199, 223, 231, 236, 242, 251, 273

P
Picea abies L. 4, 50, 74, 76, 99–101, 152, 153, 158, 164, 170, 171, 192, 199, 209, 250
Picea sitchensis Bong. 4, 99–101, 152–153, 164, 170, 171, 192
pine 7, 50–52, 54, 68, 70–73, 75–99, 156, 157, 192, 203–206, 208, 210, 212–223
Pinus pinaster Ait. 72
Pinus radiata D. Don 53, 72, 193, 250, 265
Pinus sylvestris L. 72, 176
plantations 71, 75, 96, 226, 227
Populus tremula L. 74, 192
PPFD 15, 76, 77, 80, 81, 83, 102, 114–118, 191
pressure fluctuations 13, 47
Pseudotsuga menziesii Franco 62, 99, 101, 152, 153, 164

Q
Q 10 143, 205, 229, 234–242, 244–246, 248
quality control 10, 17, 26, 35, 154
Quercus robur L. 76, 192, 199

R
radiation balance 208, 211, 227, 256
reforestation 226
remote sensing 2, 7, 121, 122, 180, 202, 203, 205, 219, 225, 226, 230, 231, 264, 265
Reynolds decomposition 10
root growth 240
root respiration 38, 39, 48, 49, 64, 248–250
roughness length 30, 104, 105, 211–213, 222
running mean 18, 21–23, 32

S

sap flow 56–58, 60, 68, 263
sapwood 57, 58
Scots pine 71, 72, 76, 157, 165, 210, 217, 218, 222, 249
sensible heat flux 19, 81, 84, 107, 111–113, 147, 215
site history 100, 121, 229
site productivity 245, 247, 250
Sitka spruce 68, 99, 100, 156, 165, 198, 226
software 10, 11, 15, 17, 20, 26, 32, 100, 129, 154
soil acidity 245
soil density 4, 108
soil frost 39
soil heat flux 15, 84, 107–109, 112, 129, 138, 139
soil moisture 38, 64, 94, 115, 130, 144, 147, 221, 222, 235–249
soil organic matter 21, 38, 40, 95, 127, 185, 226, 229, 245, 247
soil pH 239, 250
soil respiration 2, 7, 38–43, 45–49, 61, 63, 64, 69, 85, 94, 95, 116, 130, 137, 143, 159, 172, 174, 219, 226, 229, 230, 235, 245, 247, 261
soil specific heat 108
soil temperature 15, 38, 40, 45, 49, 63, 64, 75, 85, 94, 108, 109, 116, 129, 137, 143, 159, 161, 183, 233–241, 243–246, 248–250
soil water deficit 59, 60
solar radiation 46, 102, 116, 117, 138, 155, 207, 209
SOM 229, 245, 259
sonic anemometer 11, 13, 24, 32, 44, 129, 211, 259
spectral analysis 17
spruce 2, 7, 55, 68, 75, 76, 100–103, 116, 119–121, 156, 158, 165, 185, 198, 208, 212–214, 216–218, 221, 222, 226, 227, 229, 230
stability function 105
stable conditions 21, 24, 30, 31, 40, 47, 227
stable isotopes 49
stationarity 10, 22, 25, 26
stomata 81, 95, 134, 155, 156, 159, 161, 207, 208, 213, 215, 217, 219
surface resistance 207

T

temperature response 154, 158, 234, 235, 240, 241, 260
temperature sensitivity 229, 240, 244–246, 248, 250
temporal variability 49, 233, 240, 241, 246, 228, 250
terrestrial biota V, 225, 226
terrestrial sink 253, 256
thermal stratification 30
throughfall 15, 213–215
time average 10, 29
transfer function 20–25
transpiration 30, 56, 57, 59–61, 68, 81, 110, 112, 114, 120, 138, 146, 147, 190, 194, 207, 213, 215, 243, 249, 263

U

undergrowth 249
unstable conditions 21, 30

V

vapor density 109
vapor pressure deficit 56, 60, 68
von Karman's constant 211

W

water balance 56, 60, 68
water vapor 10, 11, 17, 20, 28, 32, 46, 58, 59, 81, 85, 95, 99, 100, 107, 112, 175, 213, 259
wind speed 11, 21, 22, 24, 31, 34, 46, 104, 106, 107, 111, 129, 134 155, 190, 211, 219

Ecological Studies

Volumes published since 1997

Volume 125
Ecology and Conservation of Great Plains Vertebrates (1997)
F.L. Knopf and F.B. Samson (Eds.)

Volume 126
The Central Amazon Floodplain: Ecology of a Pulsing System (1997)
W.J. Junk (Ed.)

Volume 127
Forest Decline and Ozone: A Comparison of Controlled Chamber and Field Experiments (1997)
H. Sandermann, A.R. Wellburn, and R.L. Heath (Eds.)

Volume 128
The Productivity and Sustainability of Southern Forest Ecosystems in a Changing Environment (1998)
R.A. Mickler and S. Fox (Eds.)

Volume 129
Pelagic Nutrient Cycles: Herbivores as Sources and Sinks (1997)
T. Andersen

Volume 130
Vertical Food Web Interactions: Evolutionary Patterns and Driving Forces (1997)
K. Dettner, G. Bauer, and W. Völkl (Eds.)

Volume 131
The Structuring Role of Submerged Macrophytes in Lakes (1998)
E. Jeppesen et al. (Eds.)

Volume 132
Vegetation of the Tropical Pacific Islands (1998)
D. Mueller-Dombois and F.R. Fosberg

Volume 133
Aquatic Humic Substances: Ecology and Biogeochemistry (1998)
D.O. Hessen and L.J. Tranvik (Eds.)

Volume 134
Oxidant Air Pollution Impacts in the Montane Forests of Southern California (1999)
P.R. Miller and J.R. McBride (Eds.)

Volume 135
Predation in Vertebrate Communities: The Białowieża Primeval Forest as a Case Study (1998)
B. Jędrzejewska and W. Jędrzejewski

Volume 136
Landscape Disturbance and Biodiversity in Mediterranean-Type Ecosystems (1998)
P.W. Rundel, G. Montenegro, and F.M. Jaksic (Eds.)

Volume 137
Ecology of Mediterranean Evergreen Oak Forests (1999)
F. Rodà et al. (Eds.)

Volume 138
Fire, Climate Change and Carbon Cycling in the North American Boreal Forest (2000)
E.S. Kasischke and B. Stocks (Eds.)

Volume 139
Responses of Northern U.S. Forests to Environmental Change (2000)
R. Mickler, R.A. Birdsey, and J. Hom (Eds.)

Volume 140
Rainforest Ecosystems of East Kalimantan: El Niño, Drought, Fire and Human Impacts (2000)
E. Guhardja et al. (Eds.)

Volume 141
Activity Patterns in Small Mammals: An Ecological Approach (2000)
S. Halle and N.C. Stenseth (Eds.)

Volume 142
Carbon and Nitrogen Cycling in European Forest Ecosystems (2000)
E.-D. Schulze (Ed.)

Volume 143
Global Climate Change and Human Impacts on Forest Ecosystems: Postglacial Development, Present Situation and Future Trends in Central Europe (2001)
J. Puhe and B. Ulrich

Volume 144
Coastal Marine Ecosystems of Latin America (2001)
U. Seeliger and B. Kjerfve (Eds.)

Volume 145
Ecology and Evolution of the Freshwater Mussels Unionoida (2001)
G. Bauer and K. Wächtler (Eds.)

Volume 146
Inselbergs: Biotic Diversity of Isolated Rock Outcrops in Tropical and Temperate Regions (2000)
S. Porembski and W. Barthlott (Eds.)

Volume 147
Ecosystem Approaches to Landscape Management in Central Europe (2001)
J.D. Tenhunen, R. Lenz, and R. Hantschel (Eds.)

Volume 148
A Systems Analysis of the Baltic Sea (2001)
F.V. Wulff, L.A. Rahm, and P. Larsson (Eds.)

Volume 149
Banded Vegetation Patterning in Arid and Semiarid Environments (2001)
D. Tongway and J. Seghieri (Eds.)

Volume 150
Biological Soil Crusts: Structure, Function, and Management (2001)
J. Belnap and O.L. Lange (Eds.)

Volume 151
Ecological Comparisons of Sedimentary Shores (2001)
K. Reise (Ed.)

Volume 152
Global Biodiversity in a Changing Environment: Scenarios for the 21st Century (2001)
F.S. Chapin, O. Sala, and E. Huber-Sannwald (Eds.)

Volume 153
UV Radiation and Arctic Ecosystems (2002)
D.O. Hessen (Ed.)

Volume 154
Geoecology of Antarctic Ice-Free Coastal Landscapes (2002)
L. Beyer and M. Bölter (Eds.)

Volume 155
Conserving Biological Diversity in East African Forests: A Study of the Eastern Arc Mountains (2002)
W.D. Newmark

Volume 156
Urban Air Pollution and Forests: Resources at Risk in the Mexico City Air Basin (2002)
M.E. Fenn, L. I. de Bauer, and T. Hernández-Tejeda (Eds.)

Volume 157
Mycorrhizal Ecology (2002)
M.G.A. van der Heijden and I.R. Sanders (Eds.)

Volume 158
Diversity and Interaction in a Temperate Forest Community: Ogawa Forest Reserve of Japan (2002)
T. Nakashizuka and Y. Matsumoto (Eds.)

Volume 159
Big-Leaf Mahogany: Genetic Resources, Ecology and Management (2003)
A. E. Lugo, J. C. Figueroa Colón, and M. Alayón (Eds.)

Volume 160
Fire and Climatic Change in Temperate Ecosystems of the Western Americas (2003)
T. T. Veblen et al. (Eds.)

Volume 161
Competition and Coexistence (2002)
U. Sommer and B. Worm (Eds.)

Volume 162
How Landscapes Change: Human Disturbance and Ecosystem Fragmentation in the Americas (2003)
G.A. Bradshaw and P.A. Marquet (Eds.)

Volume 163
Fluxes of Carbon, Water and Energy of European Forests (2003)
R. Valentini (Ed.)

Volume 164
Herbivory of Leaf-Cutting Ants: A Case Study on *Atta colombica* in the Tropical Rainforest of Panama (2003)
R. Wirth, H. Herz, R.J. Ryel, W. Beyschlag, B. Hölldobler